GeoPlanet: Earth and Planetary Sciences

More information about this series at http://www.springer.com/series/8821

Ryszard Staroszczyk

Ice Mechanics for Geophysical and Civil Engineering Applications

Springer

Ryszard Staroszczyk
Institute of Hydro-Engineering
Polish Academy of Sciences
Gdańsk, Poland

The GeoPlanet: Earth and Planetary Sciences Book Series is in part a continuation of Monographic Volumes of Publications of the Institute of Geophysics, Polish Academy of Sciences, the journal published since 1962 (http://pub.igf.edu.pl/index.php).

ISSN 2190-5193 ISSN 2190-5207 (electronic)
GeoPlanet: Earth and Planetary Sciences
ISBN 978-3-030-03037-7 ISBN 978-3-030-03038-4 (eBook)
https://doi.org/10.1007/978-3-030-03038-4

Library of Congress Control Number: 2018961226

This Springer imprint is published by the registered company Springer Nature Switzerland AG
The registered company address is: Gewerbestrasse 11, 6330 Cham, Switzerland

Series Editors

Managing Editor

Anna Dziembowska
Institute of Geophysics, Polish Academy of Sciences

Advisory Board

Kon-Kee Liu
Institute of Hydrological
and Oceanic Sciences
National Central University Jhongli
Jhongli, Taiwan

Teresa Madeyska
Research Centre in Warsaw
Institute of Geological Sciences
Warszawa, Poland

Stanisław Massel
Institute of Oceanology
Polish Academy of Sciences
Sopot, Poland

Antonio Meloni
Instituto Nazionale di Geofisica
Rome, Italy

Evangelos Papathanassiou
Hellenic Centre for Marine Research
Anavissos, Greece

Kaja Pietsch
AGH University of Science and
Technology
Kraków, Poland

Dušan Plašienka
Prírodovedecká fakulta, UK
Univerzita Komenského
Bratislava, Slovakia

Barbara Popielawska
Space Research Centre
Polish Academy of Sciences
Warszawa, Poland

Tilman Spohn
Deutsches Zentrum für Luftund
Raumfahrt in der Helmholtz
Gemeinschaft
Institut für Planetenforschung
Berlin, Germany

Krzysztof Stasiewicz
Swedish Institute of Space Physics
Uppsala, Sweden

Ewa Szuszkiewicz
Department of Astronomy
and Astrophysics
University of Szczecin
Szczecin, Poland

Roman Teisseyre
Department of Theoretical Geophysics
Institute of Geophysics
Polish Academy of Sciences
Warszawa, Poland

Jacek Tronczynski
Laboratory of Biogeochemistry
of Organic Contaminants
IFREMER DCN_BE
Nantes, France

Steve Wallis
School of the Built Environment
Heriot-Watt University
Riccarton, Edinburgh
Scotland, UK

Wacław M. Zuberek
Department of Applied Geology
University of Silesia
Sosnowiec, Poland

Piotr Życki
Nicolaus Copernicus Astronomical
Centre
Polish Academy of Sciences
Warszawa, Poland

Preface

This book is intended for geophysicists interested in the theoretical analysis and numerical modelling of processes occurring in sea ice and grounded polar ice sheets on geophysical scales, as well as for civil engineers involved in the design of engineering structures subjected to the action of floating ice. The main purpose of the book is to provide readers with knowledge of the concepts and tools of ice mechanics and to present examples of its application in glaciology, climate research and civil engineering in cold regions; some of the results are also of relevance to materials science. Accordingly, the book gives an account of the most important physical properties of sea and polar ice treated as a polycrystalline material and reviews relevant results of field observations and experimental measurements. The major part of the book presents theoretical descriptions of the material behaviour of ice observed in different stress, deformation and deformation-rate regimes, on spatial scales ranging from that of a single ice crystal, through those of typical civil engineering applications, up to those of thousands of kilometres, characteristic of large polar ice sheets in Antarctica and Greenland. In addition, the book offers a range of numerical formulations based on either discrete (finite-element, finite-difference and smoothed particle hydrodynamics) methods or asymptotic expansion methods, which can be used by geophysicists, theoretical glaciologists and civil engineers for solving problems of their interest. The numerical formulations presented here have been employed to simulate the behaviour of ice in a number of problems of importance to glaciology and engineering, and the results of these simulations are discussed throughout the book.

The readers are assumed to possess a standard knowledge of theoretical and structural mechanics and to be familiar with the formalism of continuum mechanics. Some knowledge of materials science could also be useful, though it is believed that the concepts introduced and then gradually developed in the course of the text are presented in a way that is adequate for understanding the content of this book.

Gdańsk, Poland — Ryszard Staroszczyk

Preface

Contents

Symbols

a	Crack half-length
$a(T), a_T$	Temperature-dependent scaling factors
A	Area; ice concentration in Chap. 5
A_f	Critical ice concentration
b, b_0	Ice plate width
$\boldsymbol{B}$	Left Cauchy–Green deformation tensor
$c(T)$	Temperature-dependent scaling factor
$\boldsymbol{c}$	Crystal c-axis unit vector
C	Specific heat capacity of ice
C_a, C_w	Dimensionless wind and water drag coefficients
d	Mean ice grain size
d_c	Critical grain size
D	One-dimensional strain-rate
D_0, D^*	Strain-rate units
D_{eq}	Equivalent strain-rate
D_p	Plate flexural rigidity
$\boldsymbol{D}$	Strain-rate tensor
e	Parameter in viscous plastic model for sea ice
E	Young's modulus
$\boldsymbol{e}^{(r)}$ $(r = 1, 2, 3)$	Unit vectors of principal stretch axes
E_a, E_s	Enhancement factors for compression and shear
$\boldsymbol{E}$	Euler–Almansi finite strain tensor
$f, \hat{f}, \bar{f}, \breve{f}, \tilde{f}$	Material response functions in Chap. 7
$f(A)$	Ice–Ice contact length function in Chap. 5
$f(x), f(r)$	Ice sheet bed elevation in Chap. 8
f_c	Coriolis parameter
$F(X), F(R)$	Normalized ice sheet bed elevation
$\boldsymbol{F}$	Material deformation gradient tensor
g	Gravitational acceleration
$g(\cdot)$, $\hat{g}$, $\bar{g}$, $\breve{g}$	Material response functions in Chap. 7

G	Shear modulus
$G(\cdot), \hat{G}, \bar{G}, \breve{G}, \tilde{G}$	Material response functions in Chap. 7
h	Floating ice thickness
$h(x), h(r)$	Ice sheet free surface elevation in Chap. 8
h^*	Ice sheet length scale
$H(\cdot)$	Heaviside unit function
$H(X), H(R)$	Normalized ice sheet free surface elevation
H_D	Ice divide elevation
$I, I_2, \tilde{I}, \bar{I}$	Strain-rate tensor invariants
I_k $(k = 1, \ldots, 24)$	Invariants of second-order tensors
$\bar{I}_c, \bar{I}_{cl}, \bar{I}_{cu}$	Critical values of the strain-rate invariant $\bar{I}$
$\boldsymbol{I}$	Unit tensor
J, J_2	Deviatoric stress tensor invariants
K	Sea ice strength parameter; strain invariant
K_{IC}	Fracture toughness of ice
$\boldsymbol{K}$	Stiffness matrix
l	Wing crack length
L	Plate span in Chap. 4; latent heat of ice melting in Chap. 5
L, L_c, L_0	Buckle half-wavelengths in Chap. 4
$\boldsymbol{L}$	Spatial velocity gradient tensor
m	Mass
M, M_x, M_y	Bending moments in a sea ice plate
M_{xy}	Twisting moment in a sea ice plate
$\boldsymbol{M}$	Structure tensor
n	Creep power law exponent
n_c	Stress correction factor for porous sea ice
n_s	Sea ice salinity
$\boldsymbol{n}$	Unit normal vector
N, N_x, N_y	In-plane axial forces in a sea ice plate
N_{xy}	In-plane shear force in a sea ice plate
$\boldsymbol{N}$	Depth-integrated horizontal stress tensor
$\boldsymbol{O}$	Zero tensor
q	Transverse load intensity; ice accumulation rate
Q, Q_x, Q_y	Transverse shear forces in a sea ice plate
Q_a	Creep activation energy
p	Pressure
P	Ice plate buckling force; normalized pressure
P_1, P_2	Compressive and tensile strength of sea ice
r	Radial coordinate
$r(\bar{I})$	Fabric strength scaling factor
R	Universal gas constant; normalized radial coordinate
R_0	Ice floe radius
R_M	Ice sheet margin radius
$\boldsymbol{R}$	Rotation tensor

$\boldsymbol{s}$	Unit tangential vector
S_1, S_2	State variables in primary creep models
S_{eq}	Equivalent deviatoric stress
$\boldsymbol{S}$	Deviatoric Cauchy stress tensor
t	Time
t^*	Time scale
t_n, t_s	Traction vector components
T	Absolute temperature
T_c	Centigrade temperature
$\bar{T}$	Dimensionless temperature
T_m	Ice melting point
u	Horizontal ice velocity
$\boldsymbol{u}_a, \boldsymbol{u}_w$	Wind and water current velocity vectors
U, U_0	Material coefficients in secondary creep laws
$U; U_b$	Dimensionless horizontal velocity; basal value
$\boldsymbol{U}$	Right stretch tensor
υ^*	Ice sheet velocity unit
$\boldsymbol{\upsilon}$	Velocity vector
V, V_0	Material coefficients in secondary creep laws
V_n, V_s	Dimensionless velocity vector components
$\boldsymbol{V}$	Left stretch tensor
w	Sea ice plate deflection; vertical ice velocity
$W; W_b$	Dimensionless vertical velocity; basal value
$\boldsymbol{W}$	Spin tensor
x, y, z	Spatial Cartesian coordinates
$x_i \ (i = 1, 2, 3)$	Spatial Cartesian coordinates
$\boldsymbol{x}$	Position vector in spatial coordinates
X	Dimensionless horizontal coordinate
$X_i \ (i = 1, 2, 3)$	Material Cartesian coordinates
$\boldsymbol{X}$	Position vector in material coordinates
z_0	Position of a neutral plane in a plate
Z	Dimensionless vertical coordinate
$\alpha(A)$	Sea ice ridging function
$\beta(X), \beta(R)$	Scaled ice sheet bed slope $F'(X)$, $F'(R)$
γ	Shear strain-rate invariant
$\Gamma(X), \Gamma(R)$	Scaled ice sheet free surface slope $H'(X)$, $H'(R)$
δ	Half-span of the critical range of invariant $\bar{I}$
δ_{ij}	Kronecker symbol
Δ, Δ_c	Strain-rate invariants
$\Delta(R)$	Normalized ice sheet thickness
ε	One-dimensional infinitesimal strain
$\varepsilon^e, \varepsilon^d, \varepsilon^v$	Elastic, delayed elastic and viscous creep strains
ε_f	Critical failure strain
ϵ	Ice sheet aspect ratio

$\epsilon_{ij}\ (i,j = 1,2,3)$	Strain tensor components
ϵ_{ijk}	Permutation symbol
η	Dilatation rate in Chap. 5; fluidity in Chaps. 6 and 7
η_0	Isotropic ice fluidity
$\eta_{ij}\ (i,j = 1,2,3)$	Instantaneous fluidities
ζ	Bulk viscosity of ice; strain invariant in Sect. 7.5
ζ_m	Bulk viscosity of ice
ζ_{eq}	Normalized equivalent deviatoric stress
θ	Angular coordinate; zenith angle in Chap. 6
κ	Shear strain
κ_x, κ_y	Curvatures of a plate deflection surface
κ_{xy}	Twist of a plate deflection surface
λ	Lamé elastic constant; friction coefficient
Λ	Normalized basal friction coefficient
$\lambda_i\ (i = 1,2,3)$	Principal stretches
μ, μ_m	Shear viscosities of ice
μ_0	Isotropic ice viscosity
$\mu_{ij}\ (i,j = 1,2,3)$	Instantaneous viscosities
μ^*	Viscosity unit
ν	Poisson's ratio
ξ_{eq}	Normalized equivalent strain-rate
$\xi_r\ (r = 1,2,3)$	Strain invariants
ϱ	Ice density
ϱ_a, ϱ_w	Air and water densities
σ	One-dimensional stress
σ_c	Ice clearing stress
σ_f	Ice fracture strength
σ_f^*	Normalized ice fracture strength
σ^*	Effective/reduced stress in primary creep laws
σ_1, σ_2	Principal stresses
$\boldsymbol{\sigma}$	Cauchy stress tensor
Σ	Normalized deviatoric stress tensor
τ^*	Ice sheet stress unit
τ_x, τ_y	Tangential traction components
$\boldsymbol{\tau}_a, \boldsymbol{\tau}_w$	Wind and water drag surface tractions
ϕ	Latitude angle
$\phi(I)$	Viscosity scaling function
ϕ_b	Brine content in sea ice
$\phi_i\ (i = 1,2,3)$	Sea ice response functions
φ	Longitude (azimuth) angle
$\boldsymbol{\Phi}$	Interpolation (shape) function
$\psi(J)$	Viscosity scaling function
$\boldsymbol{\Omega}$	Earth's rotation angular velocity vector

Chapter 1
Introduction

Sea ice, subject to seasonal variations, covers an area of approximately 20–25 million km^2 of Arctic and Antarctic waters, which is about the size of the whole continent of North America and accounts for around 6–7% of the total area of the World Ocean. Polar ice caps in Greenland and Antarctica, on the other hand, occupy together an area of nearly 16 million km^2, which is about 11% of the total land area on Earth and accounts for around 90% of fresh water on our planet.

People became acquainted with sea ice a long time ago, with the first, though not very reliable, reports of its sightings coming from ancient Greeks. For centuries, sailors, fishermen and whalers were exposed to risks associated with navigation in polar and subpolar regions, and engineers had to learn how to construct bridges across ice-covered waters and how to protect harbours from sea ice. Finding the shipping routes connecting the Atlantic and Pacific Oceans through the Arctic Ocean (the Northeast and Northwest Passages) was a great challenge in the 18th and 19th centuries. Then, in the second half of the 20th century, the exploitation of the oil and natural gas fields off the coasts of Alaska began. Sea ice in the Arctic was also of special interest to the navies of the nuclear powers during the cold war. In the meantime, with intensifying economic activities in polar regions, engineers had to deal with such problems as the construction of ice roads, aircraft runways and material storage places on ice covers.

The first serious attempts to describe the large-scale motion and deformation of sea ice started in the 1930s, though in these early efforts a sea ice pack was treated as a collection of rigid bodies freely drifting on the ocean surface, and no interactions between ice floes were considered. This situation changed only in the 1950s and 1960s, when extensive exploration of natural resources in polar regions started, and an interest in the studying of the weather and climate systems of the planet emerged. Numerous experiments and field observations were then carried out, and many scientists working in related disciplines, such as continuum mechanics, fracture mechanics and materials science, became attracted to this new research field. As a result, rigorously formulated theories describing the behaviour of sea ice were developed, and one can assume that around the year 1960 a new discipline of ice mechanics was born.

R. Staroszczyk, *Ice Mechanics for Geophysical and Civil Engineering Applications*, GeoPlanet: Earth and Planetary Sciences, https://doi.org/10.1007/978-3-030-03038-4_1

Regarding the land-based polar ice masses in Greenland and Antarctica, an interest in their, first, economic and then scientific exploration developed, for obvious reasons, much later than in the case of sea ice. The ice cap in Greenland was first seen by Vikings around the year 1000, and the ice cap in Antarctica was discovered as late as in 1819. For a long time afterwards little was known and understood about the large-scale mechanical behaviour of polar glacier ice. Only after the Second World War the first theories attempting to describe the features and behaviour of ice were developed, mostly by British scientists, with the famous Glen's flow law, describing the creep of ice, being formulated in 1955. However, for nearly the thirty following years, progress in this research area was modest. Only in the late 1970s and early 1980s did the research on polar ice gain a significant impetus, largely due to interest shown by climatologists. First theoretical models based on the rigourous methods of continuum mechanics were developed to describe the constitutive behaviour of polar ice on long geophysical scales. Even though the science of glaciology had already been well established by then, it was probably around the year 1980 that the discipline of theoretical glaciology, in the sense it is understood today, took its present form. The efforts to formulate theoretical descriptions of ice motion and deformation, with the material treated first as isotropic and then anisotropic, were continued for another twenty or so years, before interest in this research started to wane around 2005. It seems that at present the main thrust in the field of polar ice mechanics is directed towards the development of large-scale computational models for polar ice caps, rather than towards the description of the fundamental properties of ice, which is the material that belongs to the group of the most anisotropic natural materials on our planet.

This book is an attempt to present, in a single volume, the theoretical tools of ice mechanics and glaciology that can be useful in the analysis of the mechanical behaviour of ice in a wide array of spatial and time scales, ranging from the scales typical of civil engineering applications, up to those which are characteristic of geophysical and climate phenomena occurring in large parts of the planet, such as continental ice sheets and the surrounding ocean waters. Hence, the scope of the book covers the concepts, methods and theoretical results that might be of interest to civil engineers, glaciologists and geophysicists.

The book is organized as follows. Chapter 2 describes the processes of formation of natural ice masses on Earth and characteristics of the most common types of ice encountered on our planet. First, various mechanisms involved in the formation of sea ice from water are discussed, with a focus on the influence of environmental conditions prevailing during the development of ice on its structure and mechanical properties. Next, entirely different mechanisms that are involved in the transformation of snow into glacier ice in Antarctica and Greenland are described, together with an outline of the processes which continuously change ice properties as ice particles descend through a polar ice sheet from its surface to depth over very long geophysical time scales. Some aspects of the macroscopic anisotropy of ice developing in polar glaciers are signalled in this chapter; albeit very briefly, since the topics of the formation and evolution of the anisotropic properties of polar ice are discussed in great detail in Chaps. 6–8. For completeness, the chapter is concluded with a short presentation of the most important features of polar ice shelves and icebergs; these types of natural ice, however, will not be considered in this book.

In Chap. 3, the properties of ice and its mechanical behaviour are discussed. First, the basic facts concerning various forms of ice are presented, and relevant physical parameters are given. Then, the crystalline microstructure of ice is described, with an emphasis on the anisotropic properties of a single ice crystal and their effect on various types of its microscopic deformation. This is followed by the presentation of the macroscopic properties of polycrystalline ice and its behaviour in various stress and deformation regimes. Thus, the elastic, viscoelastic, creep and brittle behaviour of the material is discussed, and constitutive equations describing all these types of the response of ice to stress are given. At this stage polycrystalline ice is treated as an isotropic body. The constitutive models for the ice treated as a macroscopically anisotropic material with an evolving microstructure are developed in Chaps. 6 and 7.

Chapters 4 and 5 deal with problems involving sea ice. The first of these chapters is concerned with the behaviour of ice on civil engineering length scales. Several problems of the interaction between a coherent sea ice cover and an engineering structure are analysed. First, the problem of elastic response of ice during its short-time interaction (measured in seconds) with a rigid vertical structure is analysed, with the aim to evaluate maximum horizontal forces that can be exerted by ice on the structure. These forces are assumed to be those which cause an elastic buckling failure of a floating ice plate under compressive and bending loadings. Next, ice-structure interaction events lasting for hours and days are investigated, in which the deformations of ice are dominated by its creep. Thus, the mechanism of creep buckling of a floating ice plate is analysed, the values of time at which the flexural failure of ice occurs are determined, and the magnitudes of forces acting on the structure are calculated. Further, the interaction problems involving cylindrical structures are considered, and the effects of different sea ice rheologies on the predicted values of forces exerted by ice on structure walls are examined. In the concluding part of this chapter, a dynamic impact of floating ice on the structure is analysed, during which the ice behaves in a brittle manner. For an adopted set of parameters defining the limit failure stresses in ice, the history of loads sustained by the structure during a typical impact event is determined. Furthermore, probability distributions for maximum impact forces as functions of the floe size, its thickness and initial horizontal velocity are also determined.

In Chap. 5 the behaviour of sea ice on geophysical length scales is considered. Thus, the motion and deformation of a large ice pack driven by wind drag and ocean current stresses is investigated. The behaviour of the pack, consisting of a multitude of ice floes interacting with each other or separated by water, is analysed by treating the ice cover as a two-dimensional continuum having horizontal dimensions of the order of tens to hundreds kilometres, with local properties defined by the ice thickness and the ice area concentration. The equations governing the macroscopic behaviour of the ice pack are solved in the material coordinates by applying two discrete methods: a finite-element method and a smoothed particle hydrodynamics method. The results of numerical simulations, carried out for several constitutive models describing the large-scale rheology of sea ice, illustrate the evolution of the pack under the action of wind, including variation in the ice thickness and ice concentration, with a particular focus on changes in the position of open sea-ice pack boundaries.

The remaining part of the book is devoted entirely to the analysis of creep phenomena in grounded polar ice treated as an anisotropic polycrystalline aggregate. Thus, two families of constitutive theories are presented, based on two fundamentally different approaches: micro-mechanical and phenomenological. The constitutive laws derived by applying either approach are subsequently used to simulate flows of large polar ice sheets. In Chap. 6, micro-mechanical constitutive models describing the evolving anisotropy of polycrystalline ice are presented. Based on some assumptions regarding the anisotropic properties of an individual ice crystal and slip systems active during its deformation, frame-indifferent constitutive laws for the creep response of the crystal are formulated. By applying homogenization methods, the microscopic laws are then used to derive the macroscopic constitutive relations for polycrystalline ice. These relations are employed to simulate the creep behaviour of ice in simple flows in order to correlate parameters in the macroscopic flow laws with the observed anisotropic behaviour of polar ice, and also to evaluate directional viscosities of the material depending on its current deformation. The concluding part of this chapter is concerned with the important mechanism of dynamic (migration) recrystallization of polycrystalline ice. This mechanism, usually occurring in regions close to the bottom of polar ice sheets, leads to the weakening, and often to total destruction, of the anisotropic microstructure that develops at earlier stages of the ice descent through the depth of an ice sheet. This process considerably modifies the macroscopic viscosities of ice, and thus affects the overall flow of an ice sheet. Three alternative dynamic recrystallization models are formulated, in which the weakening of the anisotropic microstructure of ice is modelled by introducing a scalar strength factor depending continuously on a single, temperature-dependent invariant (of stress, strain-rate or strain). These three alternative models are used in the simulations for simple flow configurations to investigate the effect of the recrystallization process on the evolution of macroscopic viscosities of ice.

The following Chap. 7 deals with a phenomenological description of polycrystalline ice, in which the macroscopic creep response of ice is determined solely in terms of the macroscopic stress, strain-rate, and deformation. However, the microscopic mechanism of re-orientation of crystal slip systems during the deformation of ice is accounted for in order to model the evolution of the internal structure of the material. General forms of frame-indifferent constitutive flow laws, which express either the stress in terms of the strain-rate, or the strain-rate in terms of the stress, are derived on the assumption that the type of anisotropy which develops in polar ice sheets is close to orthotropy. Again, the parameters in the derived constitutive models are determined by the correlation of model predictions with available experimental data. Then, the phenomenological approach is applied to model the mechanism of the dynamic recrystallization of polar ice. All phenomenological models developed in this chapter are applied to simulate the evolution of the macroscopic viscous properties of polycrystalline ice with increasing shear and axial strains.

In the last Chap. 8, the micro-mechanical and phenomenological constitutive models presented in Chaps. 6 and 7 are used in numerical simulations of plane and radially-symmetric ice sheet flows. First, a simpler case of a gravity-driven flow of an isothermal ice sheet with a prescribed free-surface elevation is analysed. Assuming a horizontal bedrock and no-slip basal conditions, free-surface ice accumulation rates required to maintain the prescribed geometry are calculated. By employing a finite-element method, numerical simulations have been carried out for different sets of constitutive model parameters to examine their effect on the flow field. In particular, the influence of ice anisotropy on stress and velocity depth-profiles across the ice sheet is demonstrated. In addition, the effect of the dynamic recrystallization mechanism on ice sheet behaviour is also investigated.

In the second part of Chap. 8, a more realistic, and more complex, steady-state, radially-symmetric flow problem of a polythermal ice sheet is analysed, in which accumulation (precipitation) rates are prescribed, and an a priori unknown free surface elevation is to be determined as part of the solution. Assuming that ice viscosities depend on temperature and strain-rate in a non-linear manner, a computational model is constructed by applying the method of asymptotic expansions in a small parameter defining the ratio of an ice sheet's thickness to its lateral span. The results of numerical simulations of isotropic and anisotropic ice flow on a flat bedrock are compared to show the effects of ice anisotropy on both the free surface geometry and the velocity field in an ice sheet. The results obtained for the flow of ice over an undulating bedrock are also presented to show the influence of the bed topography features on the overall flow of an ice sheet. In addition, the effect of varying basal melt rates and temperature distributions on an ice sheet profile is investigated. The chapter is complemented with the presentation of results illustrating the effects of the dynamic recrystallization process on the free-surface profiles and flow velocities in a polar ice sheet.

This work summarizes the results obtained by the author in the past two decades, and most of these results have been published in a number of peer-reviewed journal papers (see the bibliography lists in successive chapters of the book). The results of research on sea ice, presented in Chaps. 4 and 5, were obtained mainly in 1995–1997, 2001–2006 and 2016–2017, and those relating to polar ice, presented in Chaps. 6–8, were obtained mostly in the periods of 1997–2005 and 2010–2013. During those years, the author was supported not only by his home Institute of Hydro-Engineering of the Polish Academy of Sciences in Gdańsk, Poland, but also by a number of research projects. These projects included *Modelling of sea ice dynamics* granted by the British NERC (1995–1996), *Fabric development and rheology of anisotropic ice for ice sheet modelling* granted by the European Union (1997–1999), *Evaluation of forces on engineering structures exerted by floating ice* granted by the Polish Committee for Scientific Research (KBN) (2001–2002), *Evolving anisotropy in ice sheet flows* granted by the British EPSRC (2001–2003), and *Development of a model for dynamic recrystallization of ice and its application to large polar ice sheet flows* granted by the Polish Ministry of Science and Higher Education (2010–2011). While working on the projects funded by the British research councils and the EU, the author was employed by the University of East Anglia in Norwich. The support provided at various stages by the European Science Foundation in

Strasbourg, France, and the British Antarctic Survey in Cambridge, Great Britain, is also acknowledged.

My special thanks go to Leslie W. Morland, now Emeritus Professor at the School of Mathematics of the University of East Anglia in Norwich, who brought me to the field of ice mechanics in the mid-1990s. Without his friendly guidance and help during my very productive six years in Norwich and thereafter, this book would have never been conceived and written.

Chapter 2
Formation and Types of Natural Ice Masses

Floating ice (sea, lake and river ice) and glacier ice (polar ice caps in Antarctica and Greenland and mountain glaciers) are the two most common kinds of natural ice on the surface of Earth, with sea ice and polar ice sheets and shelves occupying vast areas of ocean and land in high-latitude regions in both hemispheres of the planet. Despite obvious similarities, though, these two forms of ice are quite distinct materials in terms of their origin, internal structure, mechanical properties and typical life-span. Sea ice, as in general any floating ice, forms by the process of solidification of liquid water at its freezing temperature. The process starts on the free surface of a water body and then progresses downwards, creating a floating sheet of ice that is made up of elongated crystals, with typical diameters of 5–10 mm and lengths of 10 and more centimetres, provided that the environmental conditions are sufficiently still. Usually, a sea ice cover needs a period of weeks to months to fully develop. The resulting ice, if it survives the first summer season, can last for several years before disintegrating.

Land-based polar ice, on the other hand, develops from snow that falls on the upper surface of a glacier. Due to various thermodynamic processes, snow transforms first to firn (a form of high-porosity ice), and then to bubble-free glacier ice. It usually takes hundreds to thousands of years to form polar ice from snow. The resulting ice is composed of fairly regular grains a few millimetres across. Once ice has formed, its particles begin to descend slowly from near-surface layers of the glacier to its depth. During this descent, which is driven by gravity, ice undergoes several processes that change considerably its internal structure, leading to the development of a strong anisotropy of the material (polar ice is one of the most anisotropic natural materials on Earth). Time scales required for ice particles to descend from the surface of an Antarctic glacier to its base are of the order of 100,000 years. Thus, these scales are by four orders of magnitude longer than those characteristic of sea ice.

This chapter provides an account of the major mechanisms involved in the formation of both sea and polar ice, together with a description of the most characteristic types of ice and features that can develop in it. Only the aspects which seem the

R. Staroszczyk, *Ice Mechanics for Geophysical and Civil Engineering Applications*,
GeoPlanet: Earth and Planetary Sciences, https://doi.org/10.1007/978-3-030-03038-4_2

most relevant to the subject matter of this book are addressed, so this chapter is by no means an exhaustive overview of the physical properties of sea and glacier ice. Readers interested in more detailed treatments can find them elsewhere in the literature; for instance, in the books on general ice physics by Schulson and Duval (2009) and Hobbs (2010), on sea ice by Weeks (2010), and on glacier ice by Paterson (1994) and Greve and Blatter (2009). A very thorough review of the properties of polar ice, with the focus on its microstructure (addressed in the next Chap. 3 of this book) can be found in two papers by Faria et al. (2014a, b), both providing extensive lists of references.

2.1 Sea Ice

Sea ice is formed from freezing salt water, with the freezing temperature decreasing with increasing salinity of water. The so-called standard ocean water, which has a salinity of 35 ppt (parts per thousand) by weight, freezes at −1.91 °C; for comparison, water with a salinity of 30 ppt freezes at −1.63 °C. Due to the significant concentration of salt in water, the process of sea ice formation is quite distinct from that occurring in the case of fresh (lake or river) water. In fresh water, which has an anomalous maximum density at +4 °C, the process of freezing is relatively stable, since as the surface layer of water cools down (and becomes lighter than the underlying denser water), there is no convective mixing of the liquid. Under such conditions, water freezes easily and forms a stably growing layer of ice with a density of 917 $kg\,m^{-3}$. In contrast, as salt water cools, its density increases and it sinks convectively. The sinking water is replaced by deeper, warmer water, so that a surface layer 10–20 m thick is usually involved in the water mixing process (lower layers of ocean, deeper than 20 m, are generally denser due to their higher salt content and usually do not mix with the upper layers). This means that the entire surface column of ocean water has to reach a uniform temperature of the freezing point to initiate the process of ice formation. Generally, it takes weeks to months before a layer of floating ice forms on the sea surface.

Depending on local weather and environmental conditions, there are two possible ways for sea ice cover to form (Weeks 2010). The first, less common situation arises when the conditions are calm, and appreciable winds and ocean waves are absent. Under such conditions, the process of sea ice formation is initiated by the nucleation of fine crystals, gradually transforming into small *platelets* of ice floating on the sea surface. This initial form of ice is called *frazil*, and, as the freezing progresses, frazil develops into a continuous crust of ice on the water surface. When the ice thickness increases to a few centimetres, it becomes *black ice*, because of its dark colour. This term is due to the transparency of the ice cover, through which the underlying ocean water can be seen. With time, the ice platelets flex and slip relative to each other to form brittle and shiny *ice rind*. This type of ice often forms in coastal regions where the water salinity is low; for instance, because of the proximity of a river mouth. All the above initial types of sea ice are referred to as *new ice*. As the ice rind thickens to about 10 cm, it becomes *young ice*, with a subdivision into *grey ice* (10–15 cm thick) and *grey-white ice* (15–30 cm thick). Young ice has a greenish-blue appearance, and is often moist on the top surface. With further growth of the ice cover, the *first-year ice* (FY ice) forms. It is usually assumed that its thickness ranges from 30 to as much as 200 cm. Sometimes, first-year ice is subdivided into the categories of thin (30–70 cm), medium (70–120 cm), and thick (>120 cm) FY ice.

The second, and more common, process in which sea ice is formed occurs in the presence of strong winds, and hence high ocean waves. Under such conditions, typical of the open ocean, the frazil platelets which have initially formed on the surface of freezing water do not merge into a coherent layer of black ice and subsequently ice rind, but rather they form a soupy layer of loose and disconnected frazil crystals, termed *grease ice*. Another type of the initial form of the sea ice, called *slush*, develops during heavy snowfalls, when snow crystals are deposited directly on the sea surface. Yet another, very rare, type of ice, called *shuga* can develop as a result od white lumps of ice, a few centimetres in size, rising from the seabed to the water surface. As the freezing of sea water continues, lumps of grease ice, slush or shuga gradually coalesce, to form a very characteristic type of sea ice known as *pancake ice*. The pancakes are circular pieces of ice about 30 cm to 3 m in diameter, and about 10 cm thick. Pancake ice is quite a frequent feature during the initial phase of an ice cover development in the Southern Ocean around Antarctica. Subsequently, especially when the weather becomes calmer, pancake ice transforms into a coherent layer of young ice and then first-year ice, whose structure is similar to that of ice formed in more calmer weather conditions.

The reason for choosing the ice terminology break (young ice vs. first-year ice) at a thickness of 30 cm is the observation (Sanderson 1988; Weeks 2010) that the properties of sea ice change significantly just around this limit thickness. It turns out that, largely because of the violent history of its development on the water surface, young ice is made up of a conglomerate of randomly oriented crystals, typically forming grains of a size 1–3 mm. Because of the randomness of the crystal orientations, this ice is macroscopically isotropic in terms of its mechanical behaviour. Such ice is referred to as *granular ice*, also known as random polycrystalline or *T1 ice*.

Once a solid layer of young ice has developed on the sea surface, a stable growth of ice at the bottom of an ice sheet takes place. As the conditions immediately under the floating ice sheet are much calmer than the rough conditions during the earlier stages of ice formation, new crystals of ice grow in a more regular and ordered pattern, which to a large extent is governed by local thermodynamic conditions. The crystal growth process is relatively slow, and typically occurs at a daily rate of about 1 cm. This growth rate decreases as the ice thickens, since the ice cover serves as a thermal barrier separating the sea water at the freezing temperature from the much colder air above the ice. The crystal growth process gives rise to the formation of large, regular, and vertically elongated crystals of diameters ranging from 3 to 100 mm. These elongated crystals are composed of horizontal platelets which are approximately 1 mm thick, and their mechanical properties are the same in all directions in the horizontal plane. Accordingly, these platelets are laterally isotropic. Since, in the vertical directions, the properties of ice can, generally, vary, ice which develops under the upper

30 cm-thick layer of new ice is macroscopically transversely isotropic. Such ice is known as *columnar ice* or *S2 ice*. A similar type of transversely isotropic columnar ice, known as *S1 ice*, forms on the surface of fresh water under calm conditions, typical of lakes and water reservoirs. This type of ice also consists of large vertically elongated crystals, but, compared to S2 ice, it differs in the internal oriented alignment of characteristic axes of elementary ice crystals. Freshwater ice usually develops relatively quickly, in a matter of days rather than weeks or months as in the case of sea ice, and is generally thinner than sea ice. Therefore, freshwater ice is frequently composed of only S1 ice, and its macroscopic mechanical properties are the same across the whole depth profile. This is in contrast to a more complex structure of first-year sea ice, which reflects the history of all processes to which the ice was subjected during its formation.

Under certain environmental conditions, a somewhat more complex structure of first-year ice than described above can develop. This may occur under stable conditions in which uniform directional currents flow in water on which ice is formed (such conditions can be encountered near sea shores or in rivers). As ice grows under such flow conditions, its internal structure reflects the presence of the prevailing sea current direction, and that direction becomes one of preferred orientations in newly formed anisotropic ice (Weeks and Gow 1978; Stander and Michel 1989). Such ice is referred to as *oriented columnar ice* or *S3 ice*. Ice of this kind is macroscopically orthotropic, with the strength of anisotropy increasing with depth through the ice sheet.

In the light of what has been said above, a typical depth profile of first-year ice forming under Arctic conditions can be schematically illustrated as shown in Fig. 2.1. The top layer, about 30 cm thick, consists of isotropic granular T1 ice. Below it there is usually a layer of transversely isotropic columnar S2 ice, which can extend to a depth of about 2–2.5 m. Depending on the local sea current conditions, a layer of orthotropic columnar S3 ice can form, though it must be admitted that its occurrence is relatively rare. Such a regular pattern of layers as depicted in Fig. 2.1 is characteristic of the so-called *land-fast ice* (or *fast ice*); that is, ice which is 'fastened' to a coastline. Further offshore, due to wave and wind action, the internal structure of

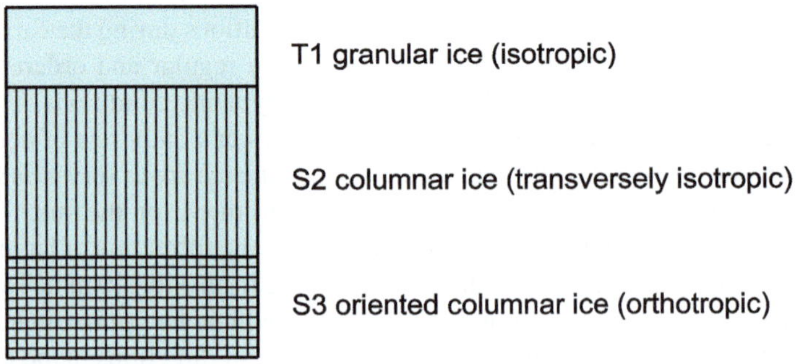

Fig. 2.1 Typical depth profile of first-year sea ice

ice becomes less ordered. Inclusions of both granular and columnar ice may occur at any depth and be incorporated into the ice cover during the freezing process, resulting in a complicated structure of the ice cover. To make the picture even more confusing, some amount of ice can be formed on top of already existing coherent sea ice. This kind of ice is due to sea water that floods the ice surface, and is called *infiltration ice*. It is composed of randomly seeded crystals and therefore resembles granular ice.

During the growth process of sea ice, some amount of salt water is being trapped in ice and forms spherical pockets in its structure. Also some small amount of solid salt can be deposited between ice crystals (typically, there is no salt inside ice crystals). Thus, sea ice exhibits some degree of salinity, which depends on the rate of sea water freezing. In the granular ice layer on top of an ice sheet, typical salinities are of the order of 8–12 ppt (Sanderson 1988). Lower down in the ice, the freezing process is slower, resulting in salinities in the range of 5–8 ppt.

As a coherent layer of young/first-year ice continues to grow, it is continuously subjected to the action of ocean swell and water waves, which often results in floating ice sheets being broken up into *ice floes*. Individual floes can have diameters ranging from tens of metres to several kilometres, depending on prevailing meteorological conditions. Their planar shape is usually irregular, and their thickness, which is initially that of the original ice sheet they are formed from, can increase as a result of floe over-riding and rafting. Large fields of sea ice, composed of a mosaic of floes separated by open water and smaller ice forms, can extend for hundreds and thousands of kilometres, forming an *ice pack*. The region of an ice pack which is close to the boundary of the open (ice-free) ocean is called the *marginal ice zone* (MIZ). Due to an increased action of ocean waves in this region, floe sizes are typically smaller than those further inside the ice pack, ranging from 10 m to several kilometres (Dempsey 2000).

All the above-said concerns first-year ice, that is, ice before its first summer season. As summer temperatures increase, the process of ice deterioration begins. The thickness of ice decreases due to melting, its surface becomes smoother, and the coherent winter ice cover disintegrates into floes before the next winter starts. Ice which has survived at least two summer seasons is termed *multi-year ice* (MY ice) or *old ice*, and much of the polar ice pack in the Arctic is about 5–10 years old. Sometimes a separate category of *second-year ice* is distinguished. This ice forms a characteristic two-layer system, in which the top layer, consisting of ice that has survived one summer melt season, is underlain by a layer of first-year ice. The salinity of this ice is in the range 1–4 ppt. With increasing age of ice, and with successive cycles of its melting and refreezing, the structure of multi-year ice becomes increasingly complex and variable. If, however, multi-year ice grows under sufficiently stable conditions, then, similarly to second-year ice, the oldest ice is that on the top, while at the bottom there is the youngest first-year ice. Such ice often shows a recognizable annual layer structure, with annual layers being usually about 30–50 cm thick (Sanderson 1988). In general, multi-year ice is less porous than first-year ice, and therefore stronger. The salinity of multi-year ice, due to the continuing expulsion of salt during the growth process, is typically very small, in the range 0.5–4 ppt. Therefore, in many aspects, the properties of old sea ice resemble those of freshwater ice and glacier ice rather than those of first-year sea ice (Timco and Weeks 2010).

There are some visible differences between sea ice in the Arctic and Antarctica. In the enclosed Arctic Ocean, there is a considerable convergence of the ice pack, resulting in a practically continuous process of dynamic interactions between individual floes. For this reason, the ice in the Arctic is largely composed of relatively old multi-year ice, with its average thickness reaching an equilibrium mean value of about 2–6 m. There are, however, some locations in the bays and fjords of islands off the northern coasts of Canada at which ice is over 10 m thick. Smaller values of annually-averaged ice thickness, equal to 2–3 m, are observed off the coasts of Siberia. A different situation occurs off the coast of Antarctica, where, because of the absence of natural obstacles, there is much less interactions between ice floes, and the ice pack is easily blown by strong winds into warmer regions. Therefore, the ice cover in the Southern Ocean is composed mainly of relatively smooth and unbroken first-year ice, and is typically only about 1 m thick.

The structure of sea ice in a pack is continuously modified not only by the interaction of floes and repeating cycles of melting and refreezing in multi-year ice, but also by the presence of pockets of *brine* (a solution of salt and water). As a result of some complex thermodynamic processes, these inclusions of brine migrate through ice and expand, evolving gradually into vertical, roughly cylindrical channels of approximately 0.1–1 cm in diameter (Nakawo and Sinha 1981; Sanderson 1988). This significantly increases the porosity of ice and weakens its mechanical properties. However, it can be assumed that for the most common cases of loading acting in the horizontal plane, the effect of the brine inclusions on the macroscopic structure of sea ice is small, so the types of ice anisotropy that have developed during the ice cover formation remain essentially unchanged.

Apart from trapped brine, the porosity of sea ice is also due to the entrapment of gas, mainly air. Typical gas concentrations in the upper layer of an ice cover (above sea level) vary from 1 to 5%, while in the lower, more consolidated layers of ice, the gas content is about 0.5%. Because of the trapped brine and gas, sea ice density differs from that of pure, bubble-free freshwater ice ($917\,\mathrm{kg\,m^{-3}}$). However, since brine is denser than solid ice, the effects of gas and brine on sea ice density cancel each other, so a typical density of sea ice is still close to that of pure ice and is in the range $915–920\,\mathrm{kg\,m^{-3}}$. Only in the upper, more porous layer of ice, the density may be lower, of the order of $890–920\,\mathrm{kg\,m^{-3}}$ (Nakawo 1983).

2.2 Polar Ice Sheets

Snow that has fallen on a grounded polar ice sheet surface undergoes a long and complex process of transformation to ice (Paterson 1994). In the first phase of the process, fresh snowflakes break into tiny snow crystals. These crystals, through the mechanism of *sintering* (Schulson and Duval 2009), begin to cluster to form spherical particles. As these particles move downwards, they are subjected to increasing pressures leading to the densification of the medium by elimination of air spaces between the grains. Through certain thermodynamical processes, such as molecular diffusion and sublimation, larger crystals start to grow at the cost of smaller ones, and merge. As a result, the material gradually transforms into *firn* with a density exceeding 400 kg m^{-3}, compared to the initial density of 50–70 kg m^{-3} for newly deposited snow. With the pressure growing during the downward passage of firn, the density of the material steadily increases and the air spaces between the grains close, forming bubbles. When the density reaches a value of about 830 kg m^{-3}, the material eventually transforms into bubbly ice. Under typical Antarctic conditions, the firn–ice transition occurs at depths of 50–100 m below the free surface of the ice sheet, and the age of ice there ranges from 100 to 200 years. In the coldest and high-altitude regions of this continent, though—for example near the Vostok polar station, where the snow precipitation is extremely low—the age of ice at the firn–ice transition depth (of about 100 m) can actually reach 4000 years (Paterson 1994). As ice descends further into the ice sheet, the ice density increases slowly with depth because of the gradual compaction of trapped air bubbles, which eventually transform into hydrate crystals. When this occurs, at a typical depth of around 500 m, glacier ice attains the pure ice density of 917 kg m^{-3}.

A typical vertical cross-section of a large grounded polar ice sheet is shown in Fig. 2.2. The maximum thickness of ice in Antarctica is about 4.8 km (in East Antarctic Ice Sheet), with an average thickness of 2.1 km; for Greenland the corresponding values are about 3 and 1.7 km. The horizontal spans of polar ice caps are of the order of 10^3 km. At its top, the ice sheet is bounded by the atmosphere, and its bottom either rests or slides on the underlying bedrock. Due to varying temperatures and precipitation rates, the mass of the ice sheet changes, as does the free surface position. At its free surface, the ice sheet can be subject to either *accumulation*, when mass is added to the glacier, mainly by snowfalls, or *ablation*, when the glacier losses mass, mainly through ice melting, water evaporation or *iceberg calving*. The line separating the accumulation and ablation regions on the glacier free surface, along which the net mass flux is zero, is termed the *equilibrium line*, or the *snow line*. Mean accumulation

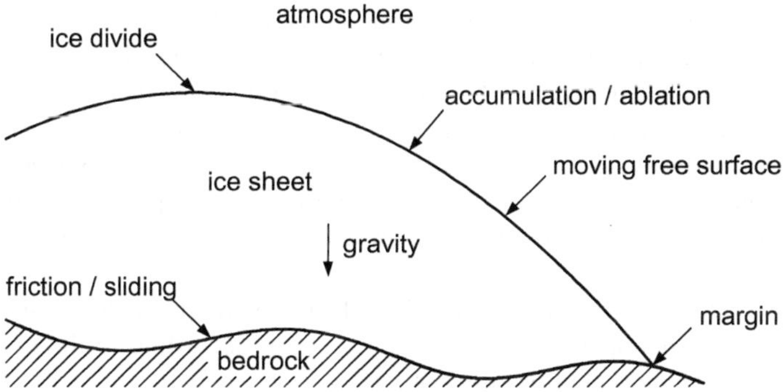

Fig. 2.2 Vertical plane cross-section of a grounded polar ice sheet

rates for the Antarctic ice cap, expressed as water equivalent, amount to about 140 mm per year (Paterson 1994), with only 20 mm per year at high altitudes in the central part of the ice cap. The corresponding rates for Greenland are about 300 mm per year as the mean value, and 100 mm as the minimum value in the north-central area of the ice cap. At its bottom, the ice sheet usually loses mass through basal melting, though a mass gain is also possible through basal refreezing. The overall behaviour of the entire sheet, with a typical temperature difference of about 30 °C between the glacier free surface and its base, is significantly affected by thermal processes, such as heat fluxes across the boundaries (for example, a geothermal flux from the lithosphere at the glacier base), or heat production due to the friction between the ice and the bedrock.

The motion of the polar ice sheet is driven by gravity. Typical horizontal velocities of large polar glaciers are of the order of 100 m per year. The location, where the horizontal velocity component is zero, is called an *ice divide*, as it separates regions in which ice moves in opposite directions. As the particles of ice descend from the free surface of the ice sheet to its depth (a process that can take as long as 400,000 years), ice is subject to different stress regimes. Near the divide, ice undergoes mainly vertical compression, while further away from the divide it is mainly sheared, and the magnitude of shearing increases with the depth of ice and the distance from the divide. As ice passes through varying stress and deformation fields, its macroscopic properties evolve in response to current conditions by a mechanism referred to as *induced anisotropy*. As a result, the initially isotropic ice at the free surface becomes increasingly anisotropic as it moves downwards through the sheet. The evidence of such increasing anisotropy of the material are ice cores retrieved from boreholes drilled in Antarctica and Greenland (Gow and Williamson 1976; Russel-Head and Budd 1979; Herron and Langway 1982; Lliboutry and Duval 1985; Lipenkov et al. 1989; Thorsteinsson et al. 1997; Faria et al. 2014a), showing a considerable alignment of ice crystals along some preferential directions.

The formation of polar ice and the subsequent evolution of its anisotropic properties is caused by processes occurring at the ice crystal level. As these micro-processes will be discussed in detail in Chaps. 3 and 6, only the most general features of glacier ice are briefly outlined at this point. A result of the mechanisms operating on the crystal scale is the evolution of the internal structure of ice; in glaciology, the ice microstructure is often called *ice fabric*. Typical ice fabrics developing in large polar ice sheets are sketched in Fig. 2.3. The fabrics are presented by means of the so-called equal-area *Schmid diagrams*, in which the dots represent the positions of individual crystal c-axes (axes of crystal rotational symmetry) on the unit hemisphere projected onto the plane of the plot (refer to Sect. 3.2 and Fig. 3.3 on p. 24 for explanations; here we only note that the more the dots are clustered, the stronger is the ice fabric anisotropy).

Lliboutry and Duval (1985) and Alley (1992) distinguish three characteristic regions along the depth of a typical polar ice sheet, according to the mechanisms dominating the development of fabric in ice.

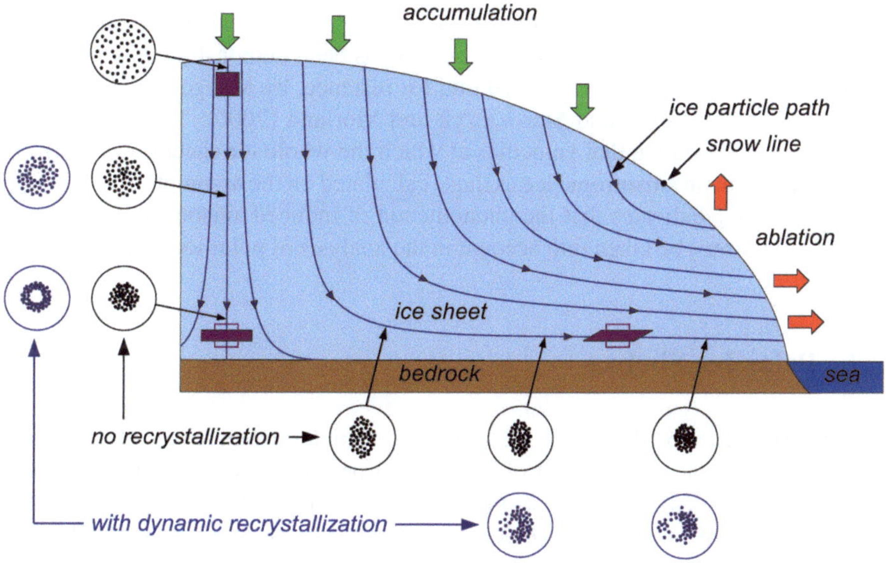

Fig. 2.3 Polar ice sheet cross-section and typical ice fabrics

(1) In the upper part of the ice sheet, extending downwards from about 100 m under the free surface to about one-third of its thickness, the anisotropic structure of ice starts to form. As shear stresses (which are responsible for fabric evolution) are relatively small in this region, the anisotropy of ice is moderate. The average size of crystals increases approximately linearly with the age of ice, from about 1 to about 5 mm, and no new grains are formed at this stage of fabric development.

(2) In the middle part of the ice sheet, shear stress magnitudes increase significantly, leading to the development of strong anisotropic fabrics, especially in the deepest regions of the glacier. A characteristic feature is the mechanism of *polygonization*, by which ice crystals undergo splitting, so that new crystals, with spatial orientations very similar to those of old crystals, are produced. The existing old crystals continue to grow, but since the total number of grains increases, both processes balance each other, so the average size of grains changes very little with depth in this part of the ice sheet (Thorsteinsson et al. 1997).

(3) In the region directly over the bedrock, shear stresses, and hence shear deformations in ice, attain very large magnitudes. This, combined with high (close to the ice melting) temperatures initiates a process known as *migration* (or *dynamic*) *recrystallization*, through which new crystals, oriented favourably for further deformation, are created at the expense of old grains that disappear. Usually abrupt changes in the average crystal size occur in this region, so the crystals generally become much larger than those in region (2), with their typical diameters exceeding 10 mm (Gow et al. 1997; De La Chapelle et al. 1998).

The fact that polar ice is macroscopically anisotropic has a significant effect on the overall behaviour of polar ice caps, which has been demonstrated by numerical simulations of ice sheet flows carried out, for instance, by Mangeney et al. (1996), Mangeney et al. (1997), and Staroszczyk and Morland (2000). The results of the simulations have shown that velocities at which the whole ice sheet flows are about twice as high for an anisotropic ice as those calculated on the assumption of isotropic ice. This clearly indicates that the phenomenon of induced anisotropy is important and therefore must be taken into account in the analysis of polar ice sheet behaviour.

2.3 Polar Ice Shelves

An interesting feature of polar ice are *ice shelves*, which surround much of Antarctica. These are large and thick sheets of ice floating on the sea surface and attached to grounded glaciers, as illustrated in Fig. 2.4. An ice shelf is fed by ice coming from the glacier and by snow falling on its top surface. It loses mass through melting on its bottom surface and iceberg calving occurring at its seaward margin (at the ice shelf tongue). The motion of the shelf is driven by gravity forces and forces exerted by the grounded ice sheet. Except a relatively narrow *transition zone*, extending a few ice thicknesses along the *grounding line* (the place where ice starts to float), the movement of ice in the floating shelf is largely horizontal, with constant velocity profiles across the shelf thickness. Because of the uniformity of the velocity field, the deformation pattern within the ice shelf is that of longitudinal stretching, in stark contrast to the simple shear deformation pattern governing the flow of the feeding ice sheet.

The structure and properties of the ice shelf resemble those of its parent continental ice sheet. It has the same pattern of characteristic layers, whose densities increase with depth. The top layer consists of firn of density as low as about 400 kg m^{-3}, so that air and water can pass freely through spaces between ice grains. Typically, a true ice begins at a depth of approximately 50 m, where its density reaches about

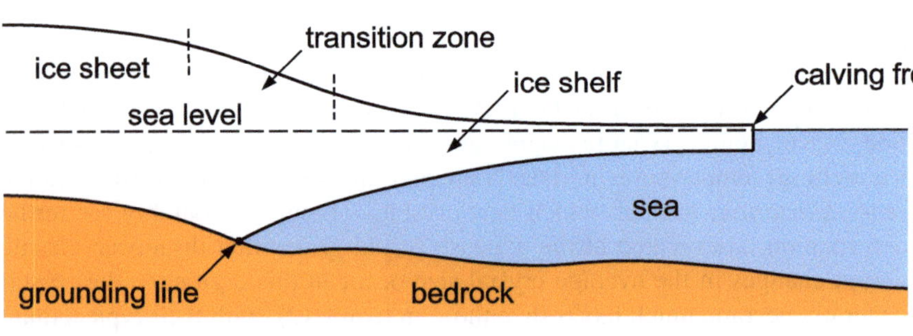

Fig. 2.4 Vertical plane cross-section of an ice shelf

800 kg m^{-3}, and at depths below 200 m pure bubble-free ice, of a density exceeding 900 kg m^{-3}, is found.

The largest of the Antarctic ice shelves is the Ross Ice Shelf. It has an area of about 0.5 million km^2, so it is approximately the size of France and extends up to some 800 km off the coast of the continent. The thickness of this shelf varies from about 1000 m near the grounding line to about 250 m at the margin, and it moves out into the Ross Sea at an average speed of 1.5–3 m per day (about 500–1000 m per year). The seaward edge of the Ross Ice Shelf is formed of ice cliffs which are up to 50 m high. It is the place where the world's largest ever recorded iceberg, named Iceberg B-15, was calved in March 2000. This iceberg was nearly 300 km long and 37 km wide, so its surface area of about 11,000 km^2 was larger than that of the island of Jamaica. The iceberg, whose movement was accurately traced by satellites, drifted in the Southern Ocean for over six years, colliding with other icebergs and gradually breaking into smaller pieces. One of these pieces, still rising some 30 m from the sea surface, was seen in November 2006 as far as near the coast of New Zealand. More information on icebergs is provided in the next section.

2.4 Icebergs

Icebergs are large masses of floating ice which have calved from the seaward edges of ice shelves or land-based glaciers. Icebergs can be found in the oceans around Antarctica, in the seas of the Arctic, and in fjords, bays and lakes fed by glaciers (for instance, by the glaciers of Svalbard, Ellesmere and Baffin Islands). The main sources of icebergs are the Antarctic ice caps in the Southern Hemisphere and the Greenland ice cap (especially its west coast) in the Northern Hemisphere. In Antarctica, most icebergs break off from floating polar ice shelves, which occupy about one-third of the Antarctic coastline.

There are two main types of icebergs (Sanderson 1988): *tabular icebergs* and *blocky icebergs*. Because of its characteristic shape and size, Iceberg B-15, with a flat top surface and steep sides, belongs to the class of tabular icebergs. Since such icebergs originate from polar ice shelves, they are common in waters surrounding Antarctica. Iceberg B-15 was an extreme example of a tabular iceberg. They are usually smaller in size and have the form of slabs of ice, with typical diameters ranging from 0.5 to several kilometres. In general, their length to thickness ratio is about 1 : 10. Blocky icebergs, on the other hand, are formed by ice breaking off from grounded polar glaciers. Such icebergs, in contrast to tabular icebergs, are irregular in shape and typically smaller. Their characteristic size ranges from 50 to 500 m, with a typical length to thickness ratio between 1 : 1 and 1 : 2.

The upper part of a newly calved iceberg is usually much warmer than its parent ice shelf or grounded glacier, which is caused by its intensive melting when the iceberg drifts into warmer waters and higher air temperatures in regions off the coasts of Antarctica or Greenland. The meltwater created at the surface percolates downwards through the uppermost part of the iceberg, thus transfers heat, and gradually warms the deeper regions of ice. As a result of this process, the whole above-water part of the iceberg is relatively quickly brought to temperatures which are close to the melting point. Due to this warm temperature and high porosity of ice, a relatively quick disintegration of the upper part of the berg occurs. Contrary to that, the underwater core of the berg, with typical centigrade temperatures around 15°–20° below zero, remains cold, and therefore mechanically strong, for a long time (up to several years).

The above-described thermodynamic processes can significantly change the initial shape and structure of an iceberg. This gives rise to two types of transformed icebergs: *dome icebergs* and *dry-dock icebergs*. A dome iceberg results from a berg which has become unstable and has therefore inverted in the water, exposing its rounded and smoothed surfaces which were initially below the sea surface. A dry-dock iceberg is a result of heavy erosion of the upper part of a berg by sun and wind, whereby its top surface becomes very irregular and often exhibits sharp spikes and pinnacles of very picturesque shapes. At final stages of the ice disintegration process, small forms known as *bergy bits* and *growlers* can develop. The former are about 10 m in diameter, while the latter are about 2–3 m in size.

The motion of an iceberg is generally governed by water currents within the top 100–200 m of the sea, with a significant contribution from winds (the latter can increase the speed of an iceberg by 0.1 to 0.3 ms^{-1} above the speed resulting from the sole action of currents). In addition, the motion of icebergs is considerably influenced by the Coriolis effect. In general, the trajectories of icebergs are very complex and hardly predictable, so that coherent iceberg drift patterns are rarely observed (Sanderson 1988).

References

Alley RB (1992) Flow-law hypotheses for ice-sheet modelling. J Glaciol 38(129):245–256

De La Chapelle S, Castelnau O, Lipenkov V, Duval P (1998) Dynamic recrystallization and texture development in ice as revealed by the study of deep ice cores in Antarctica and Greenland. J Geophys Res 103(B3):5091–5105. https://doi.org/10.1029/97JB02621

Dempsey JP (2000) Research trends in ice mechanics. Int J Solids Struct 37(1–2):131–153. https://doi.org/10.1016/S0013-7944(01)00032-7

Faria SH, Weikusat I, Azuma N (2014a) The microstructure of polar ice. Part I: highlights from ice core research. J Struct Geol 61:2–20. https://doi.org/10.1016/j.jsg.2013.09.010

Faria SH, Weikusat I, Azuma N (2014b) The microstructure of polar ice. Part II: state of the art. J Struct Geol 61:21–49. https://doi.org/10.1016/j.jsg.2013.11.003

Gow AJ, Meese DA, Alley RB, Fitzpatrick JJ, Anandakrishnan S, Woods GA, Elder BC (1997) Physical and structural properties of the Greenland Ice Sheet Project 2 ice core: a review. J Geophys Res 102(C12):26559–26575. https://doi.org/10.1029/97JC00165

Gow AJ, Williamson TC (1976) Rheological implications of the internal structure and crystal fabrics of the West Antarctic ice sheet as revealed by deep core drilling at Byrd Station. Geol Soc Am Bull 87(12):1665–1677

Greve R, Blatter H (2009) Dynamics of ice sheets and glaciers. Springer, Berlin, Heidelberg

Herron SL, Langway CC (1982) A comparison of ice fabrics and textures at Camp Century, Greenland and Byrd Station, Antarctica. Ann Glaciol 3:118–124

Hobbs PV (2010) Ice physics. Oxford University Press, Oxford
Lipenkov VY, Barkov NI, Duval P, Pimienta P (1989) Crystalline texture of the 2083 m ice core at Vostok Station, Antarctica. J Glaciol 35(121):392–398
Lliboutry L, Duval P (1985) Various isotropic and anisotropic ices found in glaciers and polar ice caps and their corresponding rheologies. Ann Gheophys 3(2):207–224
Mangeney A, Califano F, Castelnau O (1996) Isothermal flow of an anisotropic ice sheet in the vicinity of an ice divide. J Geophys Res 101(B12):28189–28204
Mangeney A, Califano F, Hutter K (1997) A numerical study of anisotropic, low Reynolds number, free surface flow for ice sheet modeling. J Geophys Res 102(B10):22749–22764
Nakawo M (1983) Measurements on ice porosity of sea ice. Ann Glaciol 4:204–208
Nakawo M, Sinha NK (1981) Growth rate and salinity profile of first-year ice in the high Arctic. J Glaciol 27(96):315–330
Paterson WSB (1994) The physics of glaciers, 3rd edn. Butterworth-Heinemann, Oxford
Russel-Head DS, Budd WF (1979) Ice-sheet flow properties derived from bore-hole shear measurements combined with ice-core studies. J Glaciol 24(90):117–130
Sanderson TJO (1988) Ice mechanics. Risks to offshore structures. Graham and Trotman, London
Schulson EM, Duval P (2009) Creep and fracture of ice. Cambridge University Press, Cambridge
Stander E, Michel B (1989) The effect of fluid flow on the development of preferred orientations in sea ice: Laboratory experiments. Cold Reg Sci Technol 17(2):153–161
Staroszczyk R, Morland LW (2000) Plane ice-sheet flow with evolving orthotropic fabric. Ann Glaciol 30:93–101
Thorsteinsson T, Kipfstuhl J, Miller H (1997) Textures and fabrics in the GRIP ice core. J Geophys Res 102(C12):26583–26599. https://doi.org/10.1029/97JC00161
Timco GW, Weeks WF (2010) A review of the engineering properties of sea ice. Cold Reg Sci Technol 60(2):107–129. https://doi.org/10.1016/j.coldregions.2009.10.003
Weeks WF (2010) On sea ice. University of Alaska Press, Fairbanks
Weeks WF, Gow AJ (1978) Crystal alignments in the fast ice of Arctic Alaska. J Geophys Res 85(C2):1137–1146

Chapter 3
Properties and Mechanical Behaviour of Ice

In this chapter the properties of ice which are important in the context of this book are described. First, the basic data concerning various forms of ice are presented, and relevant physical parameters are given. Then, in Sect. 3.2, the crystalline microstructure of ice is discussed, with the focus on the anisotropy of a single crystal and the consequences it has for the microscopic deformations experienced by the crystals. This is followed by the discussion of the macroscopic mechanical properties of polycrystalline ice and its ductile, Sect. 3.3, and brittle, Sect. 3.4, responses to loading. In the concluding Sect. 3.5 thermal properties of ice are briefly discussed.

3.1 Basic Properties of Ice

Ice is water in a solid state. It exists in either crystalline or amorphous form. Virtually all ice present in natural conditions on Earth is of the crystalline form. Amorphous ices can be made only at low temperatures in laboratory conditions, by employing a number of methods. For instance, by condensing water vapour below $-160\,°\mathrm{C}$, or by applying high pressure to ordinary ice at $-196\,°\mathrm{C}$. In this way three amorphous forms of ice can be produced: *low-density ice* (940 kg m^{-3} at $-196\,°\mathrm{C}$ at 1 atmosphere), *high-density ice* (1170 kg m^{-3} under same conditions) (Schulson and Duval 2009), and *very-high-density ice* (1260 kg m^{-3} at $-196\,°\mathrm{C}$ at pressures between 1 and 2 GPa). Essentially, natural amorphous ice can occur only in extra-terrestrial conditions (for example on some satellites of Jupiter or in comets), though some traces of it can be detected in the coldest regions of Earth's atmosphere.

There are 16 crystalline forms of ice known in physics so far. They exist at various temperatures and pressures, as illustrated in a phase diagram of water presented in Fig. 3.1. It is known that the largest possible pressures to which ice is subjected on Earth are those occurring at the bottom of the Antarctic ice cap (at present of a

R. Staroszczyk, *Ice Mechanics for Geophysical and Civil Engineering Applications*,
GeoPlanet: Earth and Planetary Sciences, https://doi.org/10.1007/978-3-030-03038-4_3

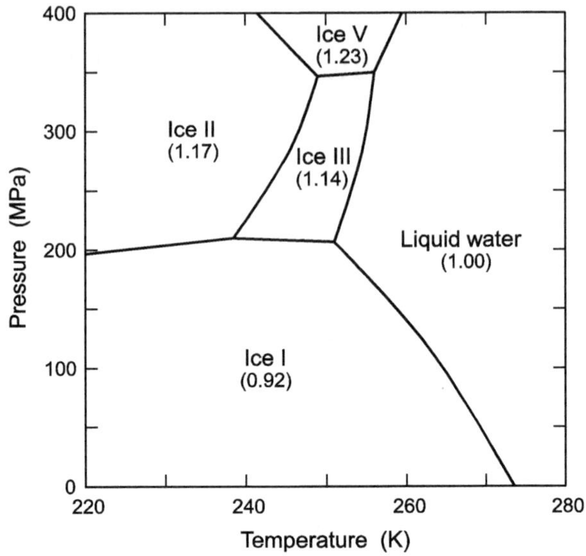

Fig. 3.1 Phase diagram of water. The numbers in parentheses denote densities expressed in $\mathrm{g\,cm^{-3}}$ (Schulson and Duval 2009)

maximum thickness of about 4800 m), and the magnitude of these pressures does not exceed 45 MPa. Hence, it follows from the phase diagram that the only stable form of ice that can exist on our planet is the one called ice I. The other forms can be produced only in a laboratory, or observed on the surfaces of extra-terrestrial bodies (Hobbs 2010; Schulson and Duval 2009). Depending on how the oxygen atoms are arranged in ice crystals, two related variants of ice I exist: hexagonal ice, denoted by Ih, and cubic ice, denoted by Ic. The hexagonal ice, the microstructure of which is described in detail in the following Sect. 3.2, is the one which is formed when liquid water is cooled down to 0 °C (273.15 K) at Earth's normal atmospheric pressure (101.325 kPa). The density of such ice at 0 °C is 916.7 $\mathrm{kg\,m^{-3}}$. The cubic crystalline variant of ice I is a rare form in which the oxygen atoms are arranged in a diamond structure. It can be made in a laboratory at temperatures between −143 and −53 °C. Like the amorphous ice, some traces of ice Ic can be present in the upper layers of Earth's atmosphere. Since this book is concerned entirely with the mechanics of the sea and land-based ice, whenever the term 'ice' is used throughout the rest of this work, it will always signify ice Ih.

Owing to its regular and ordered crystalline structure, natural ice can be considered a mineral, or a rock. As it exists in natural conditions on Earth at very high homologous temperatures, usually above 0.9 (which corresponds to about 246 K, or −27 °C), therefore it behaves in a way that is very similar to the behaviour of the majority of metals and rocks close to their melting point. When ice is subjected to stress, it displays a wide range of mechanical responses, depending on the stress magnitude and strain, as well as on strain-rate and temperature. Schematically, the typical history of stress in polycrystalline ice during its sustained loading can be sketched as in Fig. 3.2 (adapted from Sanderson 1988). At low stress levels, ice shows an instantaneous elastic response and is a creeping, ductile material, the behaviour of which

Fig. 3.2 Typical history of stress in polycrystalline ice under sustained loading

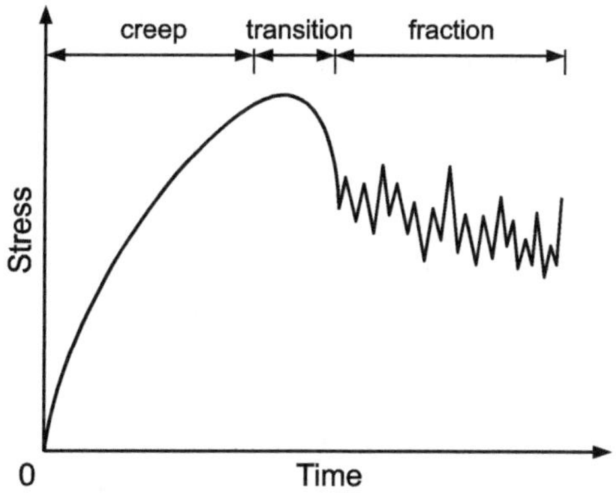

depends on the rate of deformation and the deformation history. At high stresses and strain-rates, in turn, ice is an extremely brittle material (more so than glass) which may fail by brittle fracture as a result of the formation and propagation of cracks. The distinction between ductile and brittle behaviour of ice is one of the central aspects of ice mechanics, with the *ductile-to-brittle transition* phase playing a key role, since usually during this phase the stresses in ice attain maximum magnitudes. Therefore, this transition stage is of special interest in engineering problems, in which an interaction between sea ice and a man-made structure occurs, and forces exerted by ice on the structure need to be determined. In land-based polar ice flow problems, on the other hand, due to long time scales involved, viscous creep is a dominant mode of ice deformation. Hence, the brittle effects in ice are then of little importance on the scale of large polar ice sheets, as these effects are essentially confined to relatively small regions near the glacier margins, where the calving of icebergs takes place.

Due to the complexity of the material response depending on so many factors, and the fact that the internal structure of the material can evolve under stress and strain, it is impossible to describe the behaviour of ice by applying simple classical constitutive relations known from other disciplines of mechanics; therefore, special approaches are required. As concerns the ductile response of ice, and in particular its creep, this can already be regarded as a relatively well-understood topic, owing to numerous laboratory experiments and field observations carried out in the past decades. Also, satisfactory theoretical models have already been developed to explain the observed behaviour of ice. This can be said at least for the case of isotropic ice, since for strongly anisotropic ice there are still no well-established constitutive laws that would be generally accepted by the community of theoretical glaciologists. As regards the fracture behaviour of ice, on the other hand, the picture is still far from clear and complete. Firstly, because experiments on the brittle behaviour of ice are very difficult to conduct and are less repeatable than those performed to examine the ductile behaviour of ice, and in addition the mechanisms controlling the process of brittle failure are insufficiently understood. Secondly, the theory necessary to describe the fracture mechanism is complicated and requires the knowledge of advanced methods of mechanics.

This chapter gives an account of fundamental physical properties of ice, with the focus on its mechanical properties, and briefly presents some basic constitutive relations that can be used as a first approximation to describe the behaviour of ice in different regimes. We start from the description of the microstructure of the material, concentrating on the most important properties of individual ice crystals and on micro-mechanisms underlying various types of macroscopic deformations. This is followed by a description of the ductile behaviour of polycrystalline ice, and some basic constitutive equations are presented for the elastic and creep response of the material. Further, the brittle behaviour of ice is considered, by first discussing the ductile-to-brittle transition in ice, and then by addressing the issues of fracture toughness of ice, followed by the description of the brittle behaviour of ice under tension and compression. The chapter is concluded with some remarks on thermal properties of ice, since thermodynamic effects are important in many problems involving both sea and grounded polar ice.

3.2 Microstructure of Ice

The macroscopic mechanical behaviour of ice is essentially determined by the microstructure and properties of individual crystals, and the way they are oriented relative to one another in a polycrystalline aggregate. As already mentioned in the previous section, an ice crystal possesses hexagonal symmetry, and therefore it belongs to the same family of hexagonal crystals which includes, among others, such metals as cadmium, magnesium, titanium and zinc, and also many minerals, ceramics, etc.

Since ice forms from liquid water, its crystalline structure reflects the geometrical features of the water molecule, by repeating a tetrahedral coordination of oxygen atoms, which are bonded through hydrogen to four adjacent oxygen atoms. Schematically, the hexagonal ice crystal lattice is illustrated in Fig. 3.3. In the crystal lattice, oxygen atoms are situated at each of the vertices of the hexagonal rings, and the edges of these rings are formed by hydrogen bonds. The spacing between oxygen

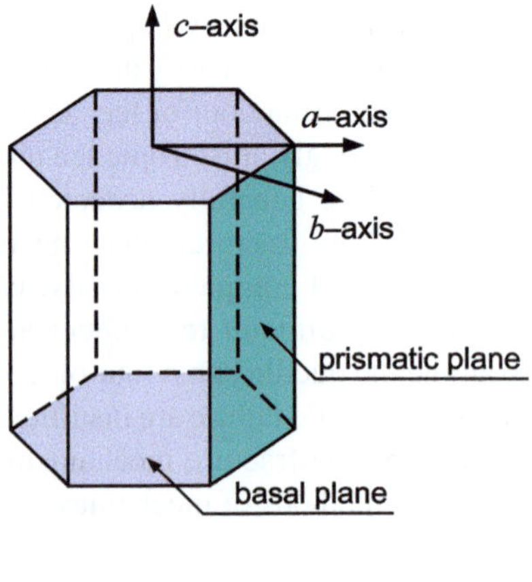

Fig. 3.3 Hexagonal crystal of ice

atoms along each bond is about 0.276 nm, and is the same between any two bonded oxygen atoms in the lattice. The planes parallel to the crystal hexagonal bases are called basal faces, or more often *basal planes*, the lateral sides of the crystal are called *prismatic planes*, and the direction which is perpendicular to the basal planes is referred to as the crystal *c-axis* (or *optic axis*). The orientation in space of the latter axis is important, as it defines various properties of ice, in the first place those relating to the anisotropy of ice. There are also two other characteristic axes that can be distinguished in the ice Ih crystal, *a* and *b*, but these play little role in the mechanics of ice.

A closer insight into the microstructure of ice reveals, though, that hexagonal rings are not perfectly planar, since the six oxygen atoms forming a ring are arranged not in one plane, but in two parallel planes, so that there are three oxygen atoms in each of the planes, with alternate atoms in the 'upper' and 'lower' planes (Paterson 1994), see Fig. 3.4 for illustration. The distance between any pair of such planes, in which oxygen atoms forming a hexagonal ring lie, is equal to 0.0923 nm, which is much less than the distance of 0.276 nm separating adjacent pairs of planes. For comparison, the distance between any two parallel prismatic faces in a crystal is 0.4523 nm. The latter quantity is often used as the length of the *Burgers vector* for ice (the Burgers vector describes the direction and magnitude of the crystal lattice distortion due to the presence of a dislocation in the lattice). When viewed along the *c*-axis direction, the consecutive parallel planes in which oxygen atoms lie are arranged in an alternate pattern, so that they are mirror images of each other. Any of these parallel planes can be regarded as a crystal basal plane, though one can assume for simplicity that the basal plane is that lying midway between the two planes containing oxygen atoms belonging to hexagonal rings.

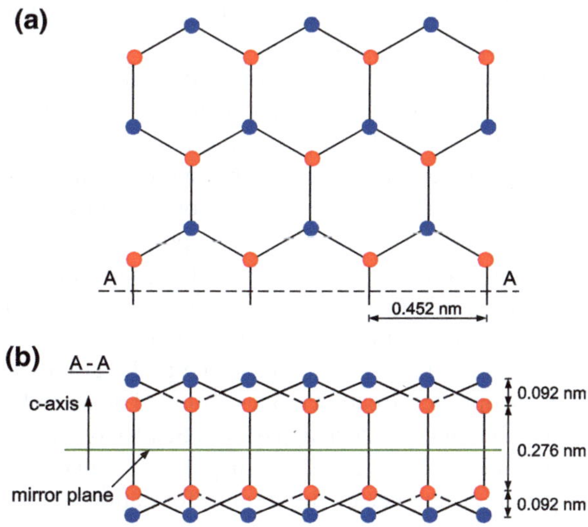

Fig. 3.4 The structure of hexagonal ice crystals: **a** the projection on the plane parallel to the crystal basal planes, **b** the projection on the plane parallel to the *c*-axes and the dashed line A–A in (**a**). The black/white circles denote oxygen atoms in the upper and lower planes, respectively, and the numbers denote corresponding atoms in (**a**) and (**b**). After Paterson (1994)

Since oxygen atoms have much stronger bonds within the layers of hexagonal rings than across the layers (there are three bonds within a layer but only one across to the next layer), creep deformation by basal slip (resulting from the movent of dislocations in the crystal lattice) requires a stress of much smaller magnitude than that needed for non-basal deformations. Therefore, the basal plane can be also called the *easy glide plane*. The existence of such preferential glide planes in an elementary crystal of ice Ih is the reason why the latter exhibits very strong anisotropy in terms of its creep shearing resistance in different slip planes. This property has very important implications for the macroscopic behaviour of a polycrystalline aggregate, since any alignment of individual ice crystals which is different from random results in the macroscopic anisotropy of the material as a whole.

Ice crystals are usually of extreme purity, irrespective of the composition of water from which they are formed, since only few chemical substances are able to fit into the ice crystal lattice. Therefore, even in the case of sea ice which is formed from a strong saline solution of sea water, the ice itself contains a negligible amount of salt incorporated into its crystalline structure, which means that the mechanical properties of such ice are not significantly different from those of pure ice. Individual elementary ice crystals form larger crystallites, or monocrystals, whose characteristic dimensions are of the order of millimetres in the case of sea ice, and a few centimetres in the case of polar ice. In natural ice masses, crystallites are usually clustered into larger grains. The grains are of random shape and are approximately of similar sizes, and may consist of up to several crystallites, though in polar ice sheets grains are often made up of only one or two of them (some authors do not make distinction between the crystallites and the grains, and treat them as equivalent entities). Grains, in turn, form larger structures called polycrystals which, in general, contain many grains varying in shape, size and orientation. When the orientations of individual crystal *c*-axes in a polycrystalline aggregate are distributed at random then the polycrystal can be considered macroscopically isotropic, otherwise it displays some form of macroscopic anisotropy.

As individual ice grains in a polycrystal are stressed due to applied loading, they deform themselves and interact with each other in a way that depends considerably on how the crystal basal planes are oriented to the applied stress field. The crystal *c*-axes, in turn, gradually rotate in response to the current stress configuration, giving rise to the evolution of the preferred orientations of *c*-axes, or *fabric*. Such a process of the evolution of the macroscopic properties of polycrystalline ice, known as the *induced anisotropy*, is of a paramount significance for the polar ice creep behaviour. Since this mechanism is very slow and occurs over time scales of thousands of years, it plays no role in the sea ice applications.

The micro-processes taking place at the grain level lie behind various types of macroscopic deformation modes observed in the bulk ice. A simplified picture of this is sketched in Fig. 3.5, adapted from Sanderson (1988), and illustrates the behaviour of a single grain as a sample of polycrystalline ice is subjected to uniaxial compression. A number of different deformation mechanisms occurring on the grain micro-scale can be distinguished. Initially, immediately after the sample has been stressed, a grain deforms in a purely elastic and reversible manner due to either

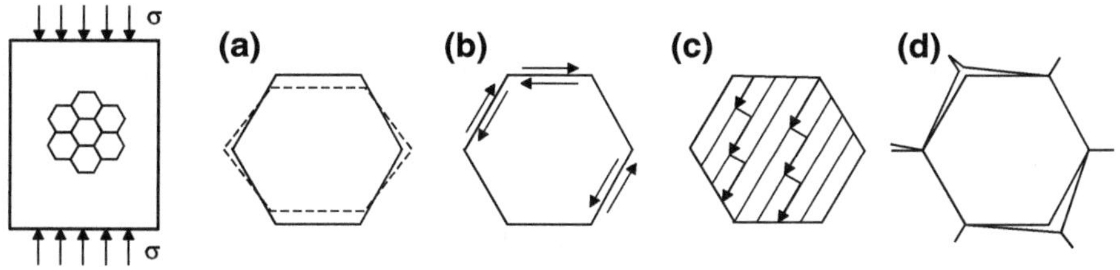

Fig. 3.5 Schematic illustration of processes occurring at the grain level during loading of an ice sample: **a** purely elastic deformation; **b** delayed elastic deformation; **c** viscous deformation; **d** brittle deformation due to crack formation and propagation. Adapted from Sanderson (1988)

lengthening or shortening of atomic bonds (Fig. 3.5a), giving rise to the bulk instantaneous elastic strain. Simultaneously, as a result of shear stresses generated between grains (Fig. 3.5b), sliding takes place on the grain boundaries (Sinha 1979). This sliding is accommodated by the elastic deformation of the grain, with no permanent deformation inside the grain, since all the atomic rearrangement processes take place only at the boundaries. This means that some elastic energy is stored in the crystal during its deformation, and this energy can be recovered if the applied compressive stress is relaxed. In order to do so, some work on the grain boundaries has to be done to reverse all the sliding that has taken place, and this requires time. Hence, there is some delay in the material response to changing stresses, and the macroscopic deformation associated with this micro-mechanism, which is fully reversible (Sinha 1983), is known as the *delayed elastic strain.*

Apart from the elastic, reversible deformations in ice, which are relatively small, also permanent changes usually occur in and between the grains, resulting in much larger irreversible macroscopic deformations. These deformations occur at all stress levels, and take place in characteristic discrete bands parallel to the crystal basal planes, in a mechanism that resembles a pack of cards sliding on each other. The irreversible deformation of ice, usually referred to as creep, is due to the presence of point defects in the crystal lattice, where a bonding between the oxygen atoms is formed by two hydrogen atoms, or there is no hydrogen atom at all. These irregularities in the crystal lattice, called *dislocations*, enable the deformation of the material through the mechanism of switching the hydrogen atoms from one bond-site to another, in a process known as the *dislocation glide* (Goodman et al. 1981) . This is illustrated in Fig. 3.5c, in which the dislocation movements are represented by arrows. As the defects propagate (migrate) through the grain, they can gradually accumulate, or pile up, at the grain boundaries, and thus resist further creep, making the polycrystalline ice harder to deform. To soften again, the material tends to recrystallize, a process in which either the crystal basal planes change their orientation to facilitate further deformation, or the existing grains are divided, or entirely new, favourably oriented grains are formed at the expense of old grains that vanish. This mechanism, which can considerably modify the macroscopic creep properties of ice, is termed *migration recrystallization* and is observed in ice cores retrieved from bottom layers of ice sheets in Antarctica and Greenland (Gow et al. 1997; Thorsteinsson et al. 1997).

Finally, when the stress applied to the ice sample acts for a sufficiently long time, the dislocations which piled up at the grain boundaries move towards the junction of three grains (Fig. 3.5d). This initiates the process of nucleation of cracks, which dramatically accelerates macroscopic strain-rates and may eventually lead to a *brittle fracture* of the material.

Although the single ice crystal is hexagonally symmetric, its mechanical properties, with a high degree of accuracy (Kamb 1961), can be considered to be those of a *transversely isotropic medium*, with the crystal c-axis being the axis of material symmetry, and the crystal basal plane being the plane of isotropy. Thus, the elastic properties of the crystal can be described by five independent constants (Green and Zerna 1992). Experimental measurements of these constants, carried out by Gammon et al. (1983), have shown that the degree of elastic anisotropy of the single crystal of ice is relatively small. At $-16\,°\mathrm{C}$, the values of Young's moduli along and normal to the c-axis are equal to 11.85 GPa and 9.69 GPa, respectively; that is, they differ by about 22%. The shear moduli in the planes containing the c-axis (the prismatic planes) and normal to it (the basal plane) are 3.01 GPa and 3.42 GPa, respectively, showing a difference of about 14%. Within the range of 0 to $-50\,°\mathrm{C}$, the single crystal elastic constants display a relatively weak dependence on temperature, with about a 1.4% increase in their respective values for every $10\,°\mathrm{C}$ decrease in temperature.

As regards the creep behaviour, the anisotropy of ice Ih is much stronger than it is in the case of the elastic response. This has been demonstrated by Duval et al. (1983), who compiled and compared the results of creep tests on ice monocrystals then available in the literature. Fig. 3.6, adapted from their paper, summarizes the creep data for basal and non-basal glide in crystals at a temperature of $-10\,°\mathrm{C}$. The data for basal glide have been obtained in simple shear tests, while those for non-basal glide come from axial compression tests conducted in the direction normal to the crystal c-axis; therefore all the data have been converted to equivalent stresses and strain-rates. For comparison, the creep data for an isotropic polycrystal (that is, the one with a random orientation distribution of constituent crystals) are also displayed. It can be seen from the diagram that, at a prescribed strain-rate, the stresses required for non-basal deformation are of up to two orders of magnitude greater than those needed for basal slip. It is also evident from the plots that the creep response of ice is strongly non-linear.

Also in contrast to the elastic response, the creep of ice is a strongly temperature-dependent phenomenon. At low temperatures, below, say, $-20\,°\mathrm{C}$, the creep temperature dependence is commonly described in glaciology by means of an Arrhenius-type law (Glen 1955; Paterson 1994; Hutter 1983), relating the creep strain-rates to the absolute temperature in an exponential manner. At higher temperatures, though, especially very close to the ice melting point, experimental evidence (Mellor and Testa 1969b) indicates that the Arrhenius law is inadequate, and therefore other relations have been derived by correlation with empirical data; for instance, the relations proposed by Smith and Morland (1981).

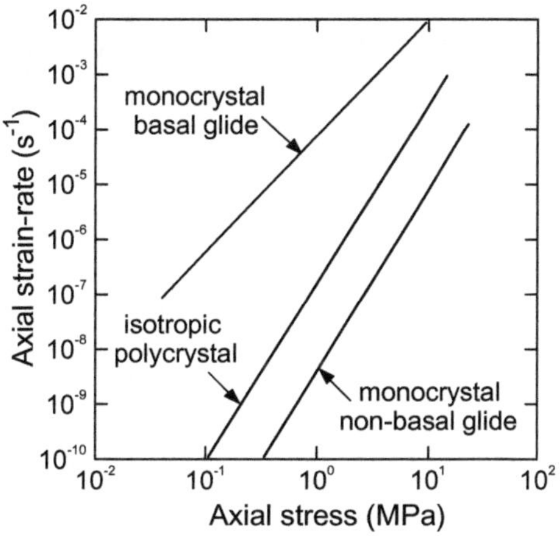

Fig. 3.6 Creep data for basal and non-basal creep of ice monocrystals, and for creep of an isotropic polycrystal, at −10 °C. Adapted with permission from Duval et al. (1983). Copyright 1983 by the American Chemical Society

3.3 Ductile Behaviour of Polycrystalline Ice

In principle, the macroscopic ductile (elastic, viscoelastic and creep) properties of polycrystalline ice can be tested by applying the same, or very similar, techniques that have already been developed in other branches of experimental solid mechanics. Hence, laboratory tests usually consist in subjecting ice specimens to prescribed stress, strain and/or strain-rate regimes in simple geometric configurations, such as uniaxial compression and tension or simple shearing, etc. However, what is specific of ice and distinguishes it from other materials is (1) the necessity of conducting experiments at negative Celsius temperatures, and (2), in order to reproduce small strain-rates occurring in natural conditions, the necessity of performing measurements over long periods of time, ranging from hours and days in the case of sea ice, up to several years in the case of polar ice. Such long measurements times are required to develop anisotropic fabrics in creeping ice from initially isotropic ice samples.

An important conclusion following from many experiments is that, basically, the elastic properties of natural sea ice are very similar to those of pure ice, provided that an appropriate normalization of experimental results is performed in order to account for the possible porosity of ice (due to the presence of brine), as well as for other factors that cause the weakening of the structure of sea ice (Timco and Weeks 2010). Such similarities significantly simplify the analysis of the empirical data, since the most important rheological parameters pertaining to sea ice can be inferred from tests conducted on freshwater ice that can be easily prepared in laboratory conditions. Creep tests on ice are most frequently carried out by placing an ice specimen under constant stress or constant load, and measuring its deformation as a function of time, possibly with additional measurements of ice porosity and ice flaws (in the case of sea ice), and crystal size, ice anisotropy parameters, etc., (in the case of polar ice). In the case of the creep properties of sea ice, in contrast to elastic properties, there is no simple correspondence between the parameters for real sea ice and the parameters obtained in laboratory experiments, as it is difficult to reproduce large-scale natural

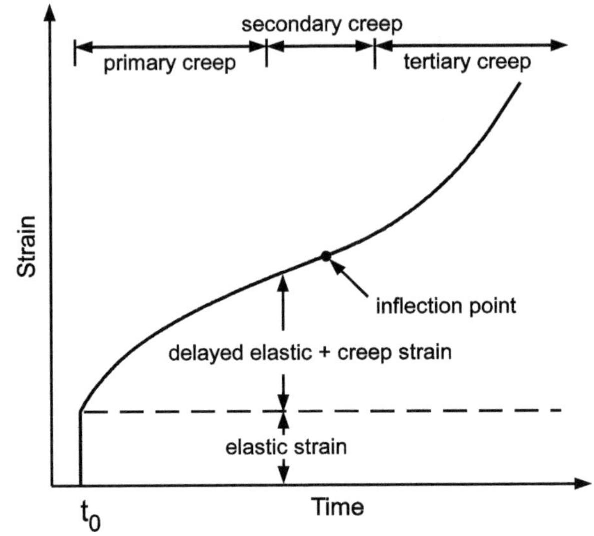

Fig. 3.7 History of strain in polycrystalline ice under constant stress suddenly applied at time t_0

conditions in a laboratory. On the other hand, the parameters measured during tests on small specimens of ice retrieved from boreholes agree well with those of natural polar ice, as long as the stress relaxation in the ice samples is accounted for.

A typical deformation history during a constant-stress test on polycrystalline ice is illustrated in Fig. 3.7, which plots strain as a function of time t. The general features of the strain curve shown in this figure (recalling the discussion of micro-processes occurring in ice on the grain level presented in Sect. 3.2, and also referring to Fig. 3.5) are as follows. When the loading is applied to a sample at time t_0, then it undergoes an instantaneous elastic strain, fully recoverable and increasing approximately linearly with stress. Typically, this strain is of a magnitude 10^{-4} at a stress of 1 MPa. After the initial elastic impulse, a transient-time delayed elastic strain develops, which is also recoverable upon the stress removal. This first stage of the strain increase in ice under the applied loading is termed *primary creep* or *transient creep*, and it decelerates with time. Simultaneously, an irreversible deformation in the material has already started, as the latter takes place in polycrystalline ice at any stress level. With still increasing strain, in which the irreversible creep component gradually overtakes the elastic counterpart, the material enters the stage in which the total strain varies about linearly with time; that is, the strain-rate is approximately constant. At this point (corresponding to the inflection point of the curve shown in Fig. 3.7), the strain-rate attains its minimum value. This phase is called *secondary creep*, or *steady-state creep*. Secondary creep is present throughout the whole loading history of ice, but it becomes apparent only when the delayed elastic strain-rate approaches zero, thus allowing the creep strain to dominate the deformation. Once the stage of apparently stable secondary creep has been completed, which typically occurs at strains equal to about 0.1 (10%), the behaviour of polycrystalline ice becomes more complex. Two distinct situations can arise, primarily depending on stress levels in ice. At smaller

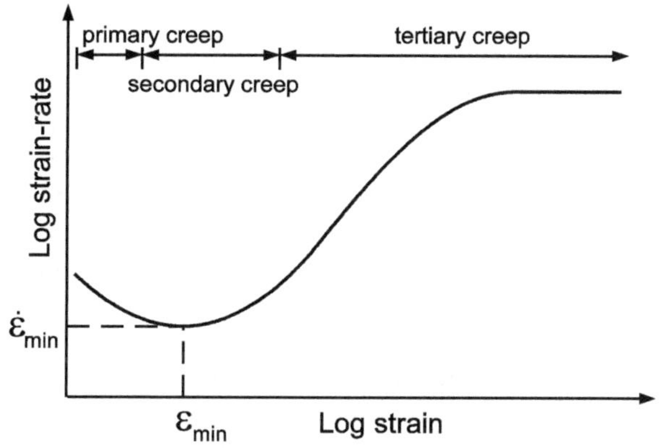

Fig. 3.8 Strain-rate dependence on strain during creep of polycrystalline ice under constant stress and temperature

stress magnitudes, ice continues to deform in a ductile manner, entering the stage of *tertiary creep*. Typically, strain-rates in the first phase of tertiary creep accelerate, before reaching another steady state in which very large deformations can develop in ice without causing any damage to the material. The acceleration of the creep process is due to the adjustment of the internal structure of ice to the current stress and strain configurations by the recrystallization mechanisms, as described in Sect. 2.2. All these mechanisms contribute to the development of anisotropic features in ice. Another situation arises at larger stress magnitudes, at which microcracks form at grain boundaries and subsequently coalesce and propagate through the material. As a result, the deformation of ice becomes much more unpredictable, and eventually the material fractures.

Corresponding to the plot of the strain history shown in Fig. 3.7 is the *creep curve* displayed in Fig. 3.8, illustrating the dependence of strain-rate on strain under the constant-stress conditions (Mellor 1980; Mellor and Cole 1982; Budd and Jacka 1989). It has been demonstrated by Mellor and Cole (1982) and Jacka (1984) that the general form of the creep curve shown in the figure holds over wide ranges of stress, strain-rate and temperature.

The three above-described stages of creep are shown in Fig. 3.8. Primary (decelerating) creep turns out to be a grain-size dependent process (Sinha 1983; Ashby and Duval 1985): the smaller the average grain diameter, the larger the strain-rate generated by a given stress. In contrast, secondary (steady-state) creep does not depend on the grain size (Duval and Lorius 1980). A characteristic feature is that the strain at which a minimum strain-rate $\dot{\varepsilon}_{min}$ (indicated in the figure) develops in polycrystalline ice during secondary creep varies very little with the applied stress. This strain, denoted by ε_{min} in the plot, is of order 10^{-2}. For instance, Jacka (1984), after conducting a series of creep tests on ice at temperatures ranging from -32 to $-5\,°C$ and stress magnitudes varying from 0.05 to 1.2 MPa, has concluded that the (octahedral) strain ε_{min} is equal to about 0.6×10^{-2}. By the time the minimum strain-rate $\dot{\varepsilon}_{min}$ has been reached during secondary creep, the internal structure of ice essentially does not change; therefore, isotropic/anisotropic properties of ice can be regarded as constant. This situation can change during tertiary creep, when the mechanisms of recrystallization can take place, giving rise to the development of strong anisotropic fabrics in ice. It is seen in Fig. 3.8 that strain-rates which can be generated in polycrystalline ice during the steady stage of tertiary creep (the horizontal straight line segment of the creep curve) can be much greater than the minimum strain-rates occurring during secondary creep. This is proved by the results of experiments conducted by Duval (1981) and Budd and Jacka (1989), showing that the ratio of the maximum to minimum strain-rates during creep in polycrystalline ice can be as large as about 10. The latter ratio of the maximum to minimum strain rates is termed in glaciology *enhancement factor*.

It is well known that most of natural materials, such as metals and rocks, creep due to both plastic and viscous irreversible strains developing in the material subjected to loading. In this context, it is interesting to analyse the types of deformations that occur in polycrystalline ice in different stress and temperature regimes. Goodman et al. (1981) investigated this problem experimentally, and the results which they obtained are presented in Fig. 3.9 adapted from Duval et al. (1983). The figure shows dominant deformation mechanisms in ice depending on the homologous temperature T/T_m and the normalized shear stress σ/G, where T and T_m denote, respectively, temperature and melting point, both expressed in (K), σ is shear stress, and G is the shear modulus. It is seen in the diagram that for typical temperatures ($T/T_m \gtrsim 0.7$) and stresses ($\sigma/G \sim 10^{-4} - 10^{-2}$) encountered in natural conditions on Earth, polycrystalline ice creeps by viscous deformation, without exhibiting any apparent plastic effects developing in the material at earlier stages of its loading. For this reason, it seems fully justified to treat ice as a viscous material; that is, the material which creeps, or flows, at any stress level. Such an approach has been a common practice in theoretical glaciology since the formulation of a viscous flow law for isotropic ice by Glen (1955).

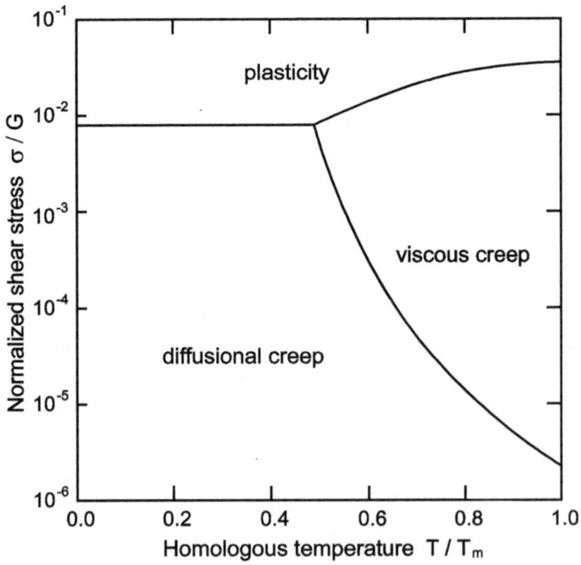

Fig. 3.9 Deformation mechanism diagram for polycrystalline ice of a 1 mm grain size. Adapted with permission from Duval et al. (1983), Fig. 1. Copyright 1983 by the American Chemical Society

In what follows, the elastic, viscous creep, and brittle behaviour of polycrystalline ice is discussed in turn. The characteristic properties of ice for the above three types of the behaviour are presented, and basic constitutive equations that describe the material response to stress are given. At this point it is assumed that ice is either isotropic or anisotropic, but with a non-evolving microstructure in the latter case. The constitutive modelling of polycrystalline ice with *evolving anisotropy*, which at present is one of the most important problems in mechanics of polar ice, will be extensively treated further in this book.

3.3.1 Elastic Behaviour

In general, two different methods can be applied to measure the elastic properties of ice. In the first, more traditional method, a sample of ice is compressed, extended or sheared in simple configurations, or a beam of ice is bent, and relevant displacements are measured in order to determine elastic constants. The problem with this method is, however, that the load has to be released very quickly (in a matter of seconds) after its application. Otherwise, irreversible creep deformations that develop in the material, even in such a short period of time, can considerably obscure the results. Therefore, a better approach is to evaluate the required elastic parameters by measuring the speed of sound waves propagating through the material (Gammon et al. 1983; Schulson and Duval 2009). As the displacements induced then in ice are very small, irreversible effects are small as well. The parameters determined by applying such a method are called dynamic constants, and their numerical values are widely regarded as the most accurate.

The purely elastic behaviour of polycrystalline ice is commonly described by Hooke's classical linear law

$$\sigma_{ij} = C_{ijkl}\,\epsilon_{kl} \quad (i, j, k, l = 1, 2, 3), \tag{3.1}$$

where σ_{ij} and ϵ_{ij} are, respectively, the Cauchy stress and infinitesimal strain tensor components, C_{ijkl} are the components of the forth-order elastic moduli tensor, and repeated indices imply summation. In the most general case of elastic anisotropy, the material is characterized by 21 independent elastic constants, the number of which reduces to nine for orthotropic materials, to five for transversely isotropic materials, and to two for isotropic materials (Green and Zerna 1992).

The values of some elastic parameters for a single crystal of ice Ih are already given in Sect. 3.2 on p. 28; the full lists can be found, for example, in the books by Hobbs (2010) and Schulson and Duval (2009). The analogous parameters for a polycrystal, of any kind of its anisotropy, can be derived from the elastic parameters of the crystal by applying one of several homogenization techniques developed in continuum mechanics (Hill 1952, 1965). Some of these averaging techniques will be described in detail further in this book when dealing with the viscous creep properties of polycrystalline polar glacier ice. Here the details of these techniques are omitted and only the quantitative results are presented.

One of the homogenization methods is based on the assumption of stress uniformity throughout a polycrystalline aggregate. Sinha (1989) employed this method, and after making use of the experimental data obtained by Dantl (1969) he evaluated the values of the macroscopic elastic constants and their temperature dependence of the most common types of natural ice. Sinha's estimates of the Young and shear moduli for granular T1 ice (which is isotropic) and columnar S2 ice (which is transversely isotropic) are as follows:

(a) granular T1 ice

$$\begin{aligned} E(T) &= 8.93 + 1.2 \times 10^{-2}(T_m - T) \quad [\text{ GPa }], \\ G(T) &= 3.41 + 4.5 \times 10^{-3}(T_m - T) \quad [\text{ GPa }]; \end{aligned} \tag{3.2}$$

(b) columnar S2 ice

$$\begin{aligned} E_V(T) &= 9.61 + 1.1 \times 10^{-2}(T_m - T) \quad [\text{ GPa }], \\ E_H(T) &= 9.39 + 1.3 \times 10^{-2}(T_m - T) \quad [\text{ GPa }], \\ G_H(T) &= 3.37 + 4.7 \times 10^{-3}(T_m - T) \quad [\text{ GPa }]; \end{aligned} \tag{3.3}$$

where E and G are, respectively, the Young and shear moduli for isotropic ice, E_V and E_H are the Young moduli in the directions parallel and normal to the length of the column, respectively, G_H is the shear modulus in the plane of isotropy of columnar ice, T is the absolute temperature, and $T_m = 273.15$ K is the ice melting temperature. The above relations describe the obvious increase in the strength of ice with decreasing temperature. It can be noticed that the shear moduli are much less sensitive to temperature than the Young moduli.

Another, and a more general approach than that by Sinha (1989), was applied by Nanthikesan and Shyam Sunder (1994). They used two extreme averaging approximations, namely the stress and the strain homogeneity conditions in the aggregate, to estimate the lower and upper bound limits on the elastic constants to be determined. Using the experimental results obtained by Gammon et al. (1983), Nanthikesan and Shyam Sunder have found that the two bounds are very close to each other: for isotropic polycrystalline ice they lie within 2.5% to each other, while for columnar S2 ice the bounds for corresponding constants differ by no more than 4.2%. Since the currently available measurement techniques have a comparable percentage resolution (which is about 1% in the technique applied by Gammon et al.), it seems that for practical purposes both limit bounds, or an average of these, can be used to provide reliable approximations to the exact values of elastic constants of polycrystalline ice. Obviously, the results by Sinha (1989) and Nanthikesan and Shyam Sunder (1994) are close to each other; that is, the differences are within the above-mentioned percentage limits.

In the case of an isotropic material, the Hooke law is commonly expressed by the equation

$$\sigma_{ij} = \lambda \epsilon_{kk} \delta_{ij} + 2G\epsilon_{ij} \quad (i, j, k = 1, 2, 3), \tag{3.4}$$

in which λ denotes the first Lamé constant (G is the second one), and δ_{ij} is the Kronecker symbol. An alternative form of the isotropic Hooke's law, traditionally used by engineers, is

$$\sigma_{ij} = \frac{E}{1+\nu}\left[\epsilon_{ij} + \frac{\nu}{1-2\nu}\,\delta_{ij}\epsilon_{kk}\right], \tag{3.5}$$

where ν stands for Poisson's ratio. In terms of the Young and shear moduli, the Lamé coefficient λ and the Poisson ratio are given by

$$\lambda = \frac{G(E-2G)}{3G-E}, \quad \nu = \frac{E}{2G} - 1. \tag{3.6}$$

Typical values of the Poisson ratio for isotropic polycrystalline ice are slightly above 0.30. For instance, the values of E and G given in (3.2) yield $\nu = 0.31$ for ice at a temperature of $-10\,°\mathrm{C}$. The Poisson ratio is little sensitive to ice temperature. Sinha (1989) proposed the following approximation of the dependence of ν on temperature:

$$\nu(T) = 0.308 + 7 \times 10^{-5}(T_m - T). \tag{3.7}$$

As indicated by Hutter (1983), in typical sea ice problems the above temperature dependence can be ignored, and the Poisson ratio can be treated as a temperature-independent quantity.

The values of elastic constants given above apply to pure (bubble-free) freshwater ice. The physical properties of sea ice differ considerably from those of freshwater ice, mainly due to the brine and salt inclusions in the ice sheet structure, making such an ice weaker (Schwarz and Weeks 1977). As there is a great variety of sea ice structures encountered in natural conditions, it is difficult, and practically impossible, to come out with the values of physical parameters that would describe all possible cases. Instead, if such a need arises, the required parameters are measured in situ. In general, the elastic modulus of sea ice can be determined by measuring the velocity of sound waves propagating in the ice sheet, or by measuring flexural waves in ice. As the delayed elasticity effects in sea ice are much more pronounced than in pure ice (Timco and Weeks 2010), the values of elastic parameters measured for sea ice in *in situ* conditions are not correct. For this reason, the term *effective modulus*, or *strain modulus*, is used. As might be expected, the effective modulus is always smaller than the elastic (Young's) modulus. Weeks (2010) reported the values of the effective modulus ranging from 1.7 to 5.7 GPa when determined by flexural waves, and ranging from 1.7 to 9.1 GPa when measured by body-wave velocities. This shows how much the elastic parameters can differ between various types of sea ice.

The above considerable differences in the values of the effective modulus are, to a large extent, due to the differences in the values of ice porosity. The pores within the sea ice cover are filled with brine liquid. The latter exists in a complex thermodynamic equilibrium with the surrounding material, and the volume of the brine changes in time depending on the salinity of ice (the content of the salt inclusions trapped in the ice) and temperature. According to Nakawo and Sinha (1981), the salinity of first-year sea ice decreases at a typical rate of 0.5 ppt per month, so the ice becomes less saline with its age. Given the current ice salinity, the gross volume of brine inclusions in the bulk of ice can be evaluated by the empirical formula (Sanderson 1988):

$$\phi_b = n_s(0.00053 - 0.0492/T_c), \tag{3.8}$$

holding for temperatures ranging from −23 to −0.5 °C. In (3.8), ϕ_b denotes a fraction of brine liquid per unit volume of ice, n_s is the ice salinity expressed in ppt, and T_c is temperature given in centigrade degrees. For example, the ice of salinity 10 ppt (a typical initial salinity in the upper layer of first-year ice) has a brine content of about 0.25 (that is, 25%) at a temperature of −2 °C, and about 0.05 (5%) at −10 °C.

The value of the elastic modulus decreases very rapidly with increasing brine content ϕ_b. For smaller values of ϕ_b, not exceeding ~ 0.15, it was established that the effective modulus E decreases linearly with ϕ_b; for larger values of ϕ_b the process of the degradation of the stiffness of ice becomes slower, so that $E \to 0$ as $\phi_b \to 0$. This reduction in the effective modulus can be quantified by the following approximate formulae derived by Hutter (1975, 1983) as a fit to empirical results:

$$\frac{E}{E_0} = \begin{cases} 1 - 5\phi_b, & \text{for } 0 \le \phi_b < 0.15; \\ 47 \cdot (0.15 - \phi_b)^3 - 46 \cdot (0.15 - \phi_b)^2 + & \\ +0.5 \cdot (0.15 - \phi_b) + 0.25, & \text{for } 0.15 \le \phi_b < 0.4; \\ 0.06 \cdot (1 - \phi_b), & \text{for } 0.4 \le \phi_b \le 1; \end{cases} \tag{3.9}$$

where E_0 is the Young modulus of pure ice at a given temperature. It follows from (3.9) that, for example, a 10% brine content in sea ice leads to a 50% reduction in the value of its elastic modulus.

3.3.2 *Creep Behaviour*

At relatively low stress levels, below 0.1 MPa, when loading is applied in a quasi-static manner and the time scales involved are short enough to prevent excessive irreversible creep, polycrystalline ice can be regarded to a good approximation as a linearly viscoelastic material. Such a simplification seems to describe reasonably well the behaviour of floating lake and sea ice sheets, and has been employed in a number of engineering applications (Sjölind 1985; Sanderson 1988; Morland 1996).

In principle, two general methods (Flügge 1967; Findley et al. 1976) can be followed to describe the viscoelastic behaviour of ice. The first, simpler method, consists in representing the material by a set of elastic springs and viscous dashpots that are connected in parallel and/or in series. For such an idealized mechanical system, constitutive differential equations relating stress, strain, and their time rates are formulated and subsequently solved in order to obtain a creep or relaxation

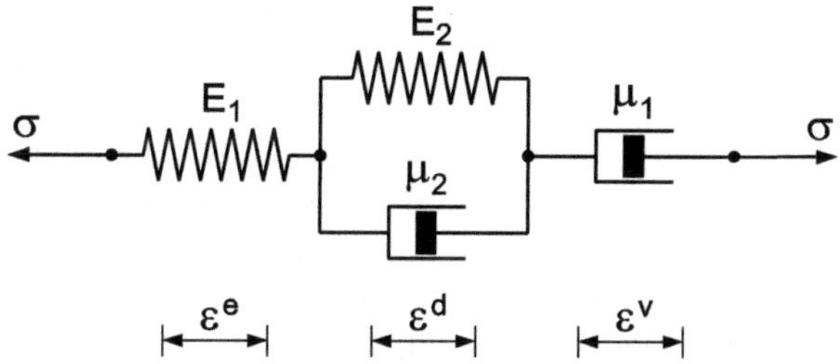

Fig. 3.10 Burgers viscoelastic model for polycrystalline ice rheology

function for the material considered. An example of such a model which has found application in ice mechanics, is the *Burgers model*, also known as the four-parameter fluid model, presented in Fig. 3.10. The model consists of a series combination of two simpler rheological units, representing the Kelvin-Voigt solid and Maxwell fluid. Despite its apparent limitations, Burgers' model is capable of capturing quite satisfactorily the three basic modes of the ice behaviour, namely, the purely elastic, delayed-elastic and secondary creep responses.

The general differential equation which relates one-dimensional stress σ and strain ε, together with their rates, in the Burgers model is given by

$$\frac{\mu_2}{E_1}\ddot{\sigma} + \left(1 + \frac{E_2}{E_1} + \frac{\mu_2}{\mu_1}\right)\dot{\sigma} + \frac{E_2}{\mu_1}\sigma = \mu_1\ddot{\varepsilon} + E_2\dot{\varepsilon}, \tag{3.10}$$

where E_1 and E_2 are the elastic spring constants, μ_1 and μ_2 are viscous dashpot parameters, and the superposed dots denote time derivatives. When a stress of constant magnitude is suddenly applied to a system at time $t = 0$ and maintained thereafter, then a time-dependent strain $\varepsilon(t)$ develops in viscoelastic material. This strain is obtained by solving (3.10) and is expressed in the form:

$$\varepsilon(t) = \frac{\sigma}{E_1} + \frac{\sigma}{\mu_2}\left[1 - \exp\left(-\frac{E_2}{\mu_2}t\right)\right] + \frac{\sigma}{\mu_1}t. \tag{3.11}$$

It is easy to see that the above strain consists of three components, that is,

$$\varepsilon = \varepsilon^e + \varepsilon^d(t) + \varepsilon^v(t), \tag{3.12}$$

with ε^e, ε^d and ε^v representing, in turn, the instantaneous elastic response, the delayed elastic strain (primary creep), and the secondary creep accumulated strain. By correlating the above solution with experimental data in order to determine the material parameters E_1 and E_2 (corresponding to the ice elastic moduli), and μ_1 and μ_2 (corresponding to the ice viscosities), one can attempt to model the real viscoelastic behaviour of polycrystalline ice. It appears that the material behaviour predicted by the four-element fluid model is unable to reproduce all the features of creep curves presented in Figs. 3.7 and 3.8. Therefore, more elaborate spring-and-dashpot rheological models have been proposed (Hutter 1983) and some, though not fully satisfactory, improvements have been achieved.

The method in which the viscoelastic behaviour of a material is approximated by the behaviour of a system of springs and dashpots, has, however, two serious disadvantages. First of all, the models of this type are essentially one-dimensional, and their extension to three dimensions by employing simple methods (for instance, by replacing one-dimensional stresses and strains entering the model equations by relevant tensor variables) is, in general, very difficult and not always possible. The second disadvantage of the spring-and-dashpot models is due to the fact that the relaxation functions that originate from their application have very specific forms, restricting their use to a relatively narrow class of materials. For these reasons, another approach is usually preferred, in which constitutive relations which connect stress and strain measures are expressed by means of hereditary integrals. Such methods usually result in the necessity of solving integral equations of the Volterra type, which proves to be mathematically more difficult than it is in the case of the first, more traditional approach. Despite this additional effort, however, the second method offers greater generality and flexibility when it comes to determining the creep and relaxation functions describing the viscoelastic behaviour of a material. General aspects of this method are discussed in detail in the book by Hutter (1983), and examples of its application can be found in the papers by Spring and Morland (1983) and Morland (1996).

At large stress levels and for slow processes occurring over long time scales, such as the flow of polar glaciers, the behaviour of polycrystalline ice behaviour is by far too complex to be analysed in the framework of the linear theory of viscoelasticity, as has been evidenced by the results of numerous creep tests conducted, among others, by Glen (1955), Mellor and Testa (1969a), Goodman et al. (1981), Mellor and Cole (1982), and Budd and Jacka (1989). These results clearly show that polycrystalline ice is a rate-dependent material, for which the minimum strain-rate which is attained during secondary creep (see Fig. 3.8) is a function of applied stress, thus implying that the constitutive response is non-linear.

Commonly, it is assumed in theoretical glaciology that the non-linear behaviour of polycrystalline ice during its secondary creep is governed by a *power law*, by analogy to many other materials, including metals. Accordingly, for the simple one-dimensional case, the creep of ice is described by the law expressed in the form

$$D = (\sigma/V)^n, \tag{3.13}$$

where D and σ are axial strain-rate and stress, respectively, V is a constant, and n is the power law exponent. The parameter V, representing the creep resistance of the material, is a temperature-dependent parameter. Usually, the latter is assumed to obey an Arrhenius-type relation given by

$$V = V_0 \exp(Q_a/nRT), \tag{3.14}$$

where V_0 is a temperature-independent constant, Q_a is the activation energy for creep, which at $T = 263$ K is equal to 6.7×10^4 J mol^{-1} (Sinha 1983), R = 8.314 J K^{-1} mol^{-1} is the universal gas constant and T is absolute temperature.

Relation (3.13) is known in engineering as the *Norton-Hoff creep* power law. It was introduced to the field of theoretical glaciology by Glen (1955) who, for uniaxial deformation, formulated it in a slightly different way, namely

$$D = U\sigma^n. \tag{3.15}$$

In view of Eqs. (3.13) and (3.14), the coefficient U varies with temperature in accordance with the relation

$$U = U_0 \exp(-Q_a/RT), \tag{3.16}$$

where U_0 is a temperature-independent constant. A three-dimensional generalization of Glen's flow law (3.15) is usually expressed in tensorial form as

$$\boldsymbol{D} = C\tau_e^{n-1}\boldsymbol{S}, \tag{3.17}$$

where $\boldsymbol{D}$ is the strain-rate tensor, $\boldsymbol{S}$ is the deviatoric Cauchy stress tensor, and C is a temperature-dependent material parameter. The coefficient τ_e is a stress tensor invariant that describes the observed dependence of the material creep resistance (ice viscosity) on a stress state in ice. Often, the second deviatoric stress invariant is adopted to serve this purpose.

In his original formulation of the power law (3.15), Glen (1955) suggested a value of $n \sim 3.5$ as best agreeing with (then) available experimental data. Other authors obtained values of n varying from 2 to 4 for stresses exceeding about 1 MPa. Nowadays, a generally accepted value of the creep power law exponent is $n = 3$ (Sinha 1978a; Goodman et al. 1981; Weertman 1983; Duval et al. 1983; Budd and Jacka 1989; Treverrow et al. 2012), valid for stresses ranging from about 0.2 MPa to about 1.5 MPa. At lower stresses, however, which are more relevant to polar glaciers in which typical deviatoric stress magnitudes are usually smaller than 0.1 MPa, both laboratory and field measurements (Mellor and Testa 1969a; Doake and Wolff 1985; Lliboutry and Duval 1985; Alley 1992) give indications that the exponent n can have a value below 2, and at yet smaller stress magnitudes n possibly approaches a value close to 1. The latter creep regime corresponds to that of *Harper-Dorn creep* observed in many metals (Lliboutry and Duval 1985). Such nearly Newtonian viscous flow, with $n \sim 1$, may be a dominant material behaviour in the upper parts of the polar ice sheets, though a definite conclusion, whether thc most appropriate value of the power law index at low stresses is 3, 1, or some value in between, still awaits more experimental support than is available (Baral et al. 2001). A possible solution is to derive a viscous creep flow law of a more complex form, in which strain-rates are expressed in terms of polynomial functions of stress magnitudes/invariants. The parameters of a flow equation could be determined by correlation with the observed creep behaviour of ice, though one must be aware of significant inconsistencies existing between various experimental data sets. Examples of the application of such a method can be found in the papers by Lliboutry (1969), Colbeck and Evans (1973), Smith and Morland (1981) and Morland (1993).

The parameter C in Eq. (3.17) is often assumed to depend on temperature in the very same way as U depends on it by relation (3.16), that is, exponentially in terms of temperature T. However, Eqs. (3.14) and (3.16) are known to hold only in the case of cold ice, being at temperatures well below the ice melting point $T_m = 273.15$ K (that is, for temperatures, say, lower than $T \sim 260$ K). At higher temperatures, which are of considerable importance to natural ice masses, the experimental evidence (Mellor and Testa 1969b; Mellor 1980) indicates that the Arrhenius-type law is inappropriate; therefore, a different type of relationship is required. An example of such a relationship, derived by correlation with experimental data of Mellor and Testa, was proposed by Smith and Morland (1981). In its modified form (Morland 1993, 2001) it is expressed as

$$a(\bar{T}) = 0.68 \exp(12\bar{T}) + 0.32 \exp(3\bar{T}), \tag{3.18}$$

where the normalized temperature $\bar{T}$ is defined by

$$\bar{T} = (T - T_m)/\Delta_T, \quad \Delta_T = 20 \text{ K}. \tag{3.19}$$

The above function, with the properties $a(T_m) = 1$ and $a(T) < 1$ for $T < T_m$, is valid for temperatures up to 60 K below the melting point T_m. The rate factor $a(T)$ scales, at a given stress, the time-dependent ice viscosity $\mu(T)$ by the relation

$$\mu(T) = \mu(T_m)/a(T), \tag{3.20}$$

where $\mu(T_m)$ denotes the near-melting point ice viscosity. Referring to Eq. (3.17), the ice viscosity is inversely proportional to the factor $C\tau_e^{n-1}$.

The process of non-linear primary (transient) creep is a more complex deformation mechanism than secondary creep, and for this reason is more difficult to be investigated experimentally and described analytically. In principle, two distinct approaches have been pursued to construct constitutive equations for primary creep. The first approach is purely phenomenological, and consists in relating only macroscopically observable variables, such as stress, strain, strain-rate, temperature and time, without considering any quantities that describe the internal state and structure of the material. This method has been applied, among others, by Sinha (1978a, b, 1979, 1983) and Gold and Sinha (1980). An example is a one-dimensional constitutive relation describing the evolution of strain in polycrystalline ice in terms of stress, temperature and grain size, formulated by Sinha (1979, 1983). This relation is expressed as

$$\varepsilon = c_1 \left(\frac{d_0}{d}\right)\left(\frac{\sigma}{E}\right) \left\{1 - \exp\left[-(a_T t)^{1/n}\right]\right\}, \tag{3.21}$$

where ε is a uniaxial strain, σ denotes stress which is applied to previously undeformed ice at time $t = 0$ and then held constant, $c_1 = 9 \times 10^{-3}$ is an empirically determined constant, d is a mean ice grain diameter, d_0 is a unit of the grain size, and E is the Young modulus. The coefficient a_T entering (3.21) accounts for the temperature-dependence of the primary creep process in a manner described by the Arrhenius law (3.16), with U and U_0 being replaced, respectively, by a_T and a_{T_0}, where the latter is a temperature-independent constant of the value $2.5 \times 10^{-4}\ \mathrm{s}^{-1}$. Subsequently, Zhan et al. (1994) have extended the law (3.21) by formulating a three-dimensional constitutive model for columnar ice, expressed in the following rate form:

$$D_{ij} = \frac{a_T}{n}\left[\left(\frac{c_1 d_1}{d}\right)^n \left(\frac{\bar{\sigma}}{E\bar{\epsilon}}\right)^{n-1} \frac{1}{2E}\frac{\partial \bar{\sigma}^2}{\partial \sigma_{ij}} - \epsilon_{ij}\right] \quad (i, j = 1, 2, 3), \tag{3.22}$$

in which D_{ij} and ϵ_{ij} $(i, j = 1, 2, 3)$ are, respectively, strain-rate and delayed elastic strain tensor components, and $\bar{\sigma}$ and $\bar{\epsilon}$ are some functions of stress and strain tensor components, respectively, and of four model parameters describing the anisotropy of ice in primary creep.

Another, more general method to formulate constitutive laws consists in considering, besides the macroscopically observable quantities, also variables which describe the internal state of the material. In the context of modelling non-linear creep of ice, this method has been applied by Le Gac and Duval (1980), Ashby and Duval (1985) and Shyam Sunder and Wu (1989a, b, 1990b). The constitutive equations derived are all based on experimental results obtained by Mellor and Cole (1983), indicating that the strain-rate at the onset of primary creep, when measured at different stress levels, is approximately equal to a constant multiple of the corresponding minimum strain-rate occurring during secondary creep. This has prompted an idea that the deformation-rate in non-linear primary creep, by analogy to secondary creep, is also governed by a power law similar in form to (3.13). Thus, the strain-rate in ice, D, is given by

$$D = (\sigma^* / V^*)^n, \tag{3.23}$$

in which V^*, the temperature dependence of which is described by the Arrhenius-type relation (3.14), is the creep resistance at the onset of primary creep, before the material starts to harden (see the creep curve in Fig. 3.8). The meaning of the stress σ^* in the power law (3.23), however, differs from that in the corresponding equation (3.13) for secondary creep. While in the latter case the stress σ does not change in time (provided that loading is constant in time), in the case of primary creep the stress σ^* evolves (also under constant external loading). In this way the material hardening mechanism is incorporated in the constitutive model. Two kinds of hardening can be accounted for in the model: kinematic and isotropic. In order to describe these two types of hardening, two internal state variables are introduced. One is the elastic and fully recoverable *back stress* (or *rest stress*), S_1, the evolution of which describes kinematic hardening of the material, whereas the other state variable, S_2, describes the evolution of isotropic hardening.

Two distinct ways of expressing the stress σ^* have been pursued. In the first method, due to Le Gac and Duval (1980), σ^* is defined by

$$\sigma^* = |\sigma - S_1| - S_2, \tag{3.24}$$

where σ^* is termed the *effective stress*, and S_2 the *isotropic-hardening stress*. The evolution of internal stresses S_1 and S_2 is described by the relations

$$\dot{S}_1 = a_1 D - b_1 S_1^n, \quad \dot{S}_2 = a_2 |D| - b_2 S_2^n, \tag{3.25}$$

where a_1 and a_2 are strain-hardening functions, and b_1 and b_2 are recovery coefficients, all to be determined by fitting to experimental data.

The model proposed by Le Gac and Duval (1980) has subsequently been modified by Ashby and Duval (1985). Their approach is based on the assumption that two different deformation mechanisms operate as polycrystalline ice aggregate creeps. One of them is a *soft system* and corresponds to the slip on crystal basal planes, and the other is a *hard system*, associated with the slip on non-basal planes. The authors have developed a spring-and-dashpot model consisting of two Maxwell rheological units (representing the soft and the hard systems, respectively), connected in parallel. The model allows a clear physical interpretation of the kinematic hardening process; unfortunately, the isotropic hardening has not been considered.

In the second method, developed by Shyam Sunder and Wu (1989a), the stress σ^* in the creep power law (3.23) is defined in an alternative form as

$$\sigma^* = (\sigma - S_1)/S_2, \tag{3.26}$$

where σ^* is called the *reduced stress*, and the scalar-valued dimensionless parameter S_2, describing isotropic hardening, is called the *drag stress*. The evolution of the internal state variables is given by

$$\dot{S}_1 = c_1 E D, \quad \dot{S}_2 = c_2 E D, \tag{3.27}$$

where the parameters c_1 and c_2 are temperature-independent material parameters. Altogether, the model includes six material parameters: n, c_1, c_2, E, V_0^* and an initial value of the drag stress S_2. The uniaxial constitutive model by Shyam Sunder and Wu (1989a) has been generalized to three dimensions by the same authors in their paper (1989b), in which polycrystalline ice is treated as an anisotropic material displaying orthotropic symmetries. The state equations have been derived from a Helmholtz free energy potential, and the evolution relations for the transient strain have been derived from a set of independent potential functions accounting for the dissipation processes occurring in the material. Besides the six material constants in the one-dimensional model, the extended version requires five more parameters needed to describe the orthotropic structure of ice during its creep deformation.

The above constitutive theories have been generalized in another paper by Shyam Sunder and Wu (1990b), in which a number of criteria have been formulated that should be used to assess primary creep models for polycrystalline ice. These criteria address such important questions as: (1) the correspondence between the constant stress and constant strain-rate creep responses, conjectured by Mellor and Cole (1982), (2) the kinematic consistency between strain, strain-rate and time, and (3) the consistency of the evolution of the internal variables with underlying microstructural mechanisms.

Essentially, the results presented thus far in this section are relevant to the creep processes in bubble-free freshwater ice, typical of land-based ice, and therefore any direct extension of these results to sea ice must be carried out with caution. This is, first of all, because of a great variety of possible types of internal structures and features of sea ice, as already discussed in this work, which makes generalization and systematization of results difficult. Here, we return once more to the effect of brine inclusions in floating ice. These inclusions not only affect the elastic properties of ice, see Eq. (3.9) on p. 36, but also have an influence on its creep behaviour. This is due to the phenomenon of local stress concentration occurring near the boundaries of brine pockets and channels in ice. The result is that the actual stresses sustained by the solid sections of an ice cover are much larger than the gross stresses applied to the bulk of ice. Due to these higher stresses, the creep rates in such weakened ice can substantially exceed those in pure ice. One can devise various methods to account for increased creep rates in porous sea ice. Certainly, one of the simplest possible approaches, well suited to engineering applications, is that proposed by Weeks and Assur (1967). In this approach, a stress correction factor, $n_c \geq 1$, is introduced, by means of which bulk stresses in ice are scaled to estimate the magnitudes of local stresses. On the basis of some geometrical consideration, Weeks and Assur (1967) have showed that in the case of horizontally loaded columnar ice with vertically arranged brine channels, the correction factor is given by

$$n_c = \frac{1}{1 - \sqrt{\phi_b / \phi_0}}, \tag{3.28}$$

whereas for granular ice containing spherical brine pockets this factor is expressed by

$$n_c = \frac{1}{1 - \phi_b / \phi_0}. \tag{3.29}$$

In the latter two formulae, ϕ_b is the brine volume that can be evaluated, for given ice salinity and temperature, from Eq. (3.8), and ϕ_0 is a normalizing parameter that can be determined by experiment. By comparing the results of uniaxial compression tests conducted on pure and sea ice, it has been established that the best approximations are $\phi_0 = 0.16$ for columnar ice, and $\phi_0 = 0.10$ for granular ice (Sanderson 1988).

Tertiary creep, which is a much more complex physical process than primary and secondary creep mechanisms discussed in this section, is not addressed at this point. There are a number of reasons for not doing this. Firstly, tertiary creep, which is the main mode of ice deformation in polar glaciers, occurs on time scales that are by a few orders of magnitude longer than the time scales on which primary and secondary creep strains typically develop. Secondly, several recrystallization mechanisms operate during this type of creep, giving rise to significant changes in the microstructure of the material. And thirdly, due to the evolution of its internal structure, polycrystalline ice becomes increasingly anisotropic during tertiary creep. These mechanisms are difficult to be investigated experimentally and treated formally, and also are very different from those acting during primary and secondary creeps. Therefore, their detailed discussion is deferred until Chaps. 6 and 7 which are entirely devoted to the description of the phenomenon of creep induced anisotropy in polar ice.

3.4 Brittle Behaviour of Polycrystalline Ice

Contrary to the ductile behaviour of polycrystalline ice, which is macroscopically scale-independent so that the same laws apply whatever the size of an ice sample, the fracture mechanism exhibits a pronounced scale effect. The stress under which a particular sample of ice fails in a brittle manner depends on its geometrical dimensions, as well as on the size, shape and the distribution of flaws (cracks) in the material (Sanderson 1988). The scale-dependence of the brittle behaviour of ice can be illustrated schematically by the plot in Fig. 3.11, summarizing the results obtained both in the laboratory and during field observations of sea ice. This plot, adapted from Sanderson (1988), is known as a *pressure–area curve*, and shows an average compressive stress in ice at its failure as a function of the area on which the stress acts. It is seen in the figure, which covers a very wide range of scales, that the typical failure stresses measured in small-scale laboratory tests are in the range of 10–20 MPa, while on the large scales in the field these stresses can be much lower than 1 MPa.

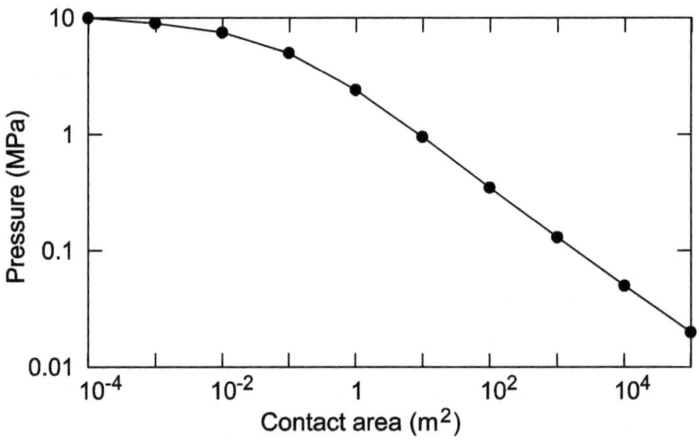

Fig. 3.11 Pressure–area curve for sea ice, showing the dependence of failure stress on the area on which the stress is applied. Based on data from Sanderson (1988), Fig. 5.36

There are a number of processes underlying the above scale effect, so clearly displayed by the pressure-area curve in Fig. 3.11. It appears that, apart from the phenomena of imperfect contact between different parts of an ice body (for instance, between floes in a sea ice pack) causing non-simultaneous fracture of the material, the major factor are the statistical features of the brittle fracture mechanism. Since this process is essentially associated with the nucleation and propagation of cracks, and is controlled by the largest of them, in particular under tensile stresses, the probability that larger samples of ice contain larger flaws (and therefore are weaker) is greater than in the case of smaller samples.

There are a variety of reasons for which polycrystalline ice begins to display brittle behaviour and crack formation instead of continuum creep. Generally, ice starts to fail when either stress, strain, or strain-rate exceeds certain critical magnitude. Typical critical levels for the stress are about 1 MPa in tension and about 5 MPa in compression, for the (elastic) strain the limit value is about 0.01 (1%), and for the strain-rate it is about 10^{-4} to 10^{-3} s^{-1} (Barnes et al. 1971; Hawkes and Mellor 1972; Sanderson 1988). Two main mechanisms that control the brittle behaviour of ice are: crack nucleation (formation) and crack propagation. These are two quite distinct processes, and it occurs that the fracture of ice may be either *nucleation-controlled* or *propagation-controlled.* The nucleation-controlled fracture is characteristic of coarse-grained ice and arises in situations in which an applied stress induces the formation of cracks that are large enough to propagate immediately after they have appeared. In the case of fine-grained ice a typical situation is different. The development of first microcracks usually leads to increased ductility of the material, but does not lead to its fracture, so ice still behaves in a continuum manner. It is only after the stress has been further increased to make the microcracks merge and propagate that the ice starts to fail, and the process becomes propagation-controlled. Which of these two mechanisms actually controls the brittle fracture of ice depends largely on the size of cracks developed in the material, and this in turn depends on the size of ice grains.

The processes of ice failure due to crack nucleation and propagation develop in very different ways under tensile and compressive stresses. Under tension, the failure mechanism is governed by a single crack alone, the largest in a specimen, and typically the process occurs in a very unstable manner. Under compression, the presence of cracks in the material does not necessarily lead to its fracture. Only after new cracks have develop, or existing oncs have coalesced and started to expand, the material starts to deteriorate and ultimately fails. For these reasons, the brittle compressive fracture is, in general, a more complicated process, and thus more difficult to investigate and describe formally, than the fracture due to tensile stresses. In what follows these two, tensile and compressive, fracture mechanisms are described. The emphasis is on sea ice engineering applications, therefore only the most basic mechanisms and equations are discussed. A more thorough and rigourous treatment requires the knowledge of advanced tools of fracture mechanics (Atkinson 1987), which is beyond the scope of this book. As noted by Timco and Weeks (2010), the use of the rigourous fracture mechanics methods in sea ice problems is still rare, though some progress in this area has been observed in recent years.

3.4.1 Tensile Fracture

When an initially crack-free sample of polycrystalline ice is subjected to tensile stress, its initial stages of deformation, as described earlier (see Fig. 3.7 on p. 30), are due to three ductile behaviour mechanisms, namely the instantaneous elastic response ε^e, the delayed elastic strain ε^d, and the viscous creep strain ε^v. It is a generally accepted view (Sanderson 1988) that microcracks start to nucleate after some *critical strain* has developed in the material. Experimental evidence (Schulson et al. 1984) indicates that in the range of low to moderate strain-rates the critical strain is equal to the elastic delayed strain ε^d (which can be evaluated by applying formulae given in Sect. 3.3.2), whereas at high strain-rates it appears that the total strain ε, as defined by Eq. (3.12), is a more appropriate quantity to use as the crack nucleation criterion. For columnar S2 ice of a typical grain diameter $d = 5$ mm, a generally approved value of the elastic delayed strain ε^d at which the nucleation process starts is 10^{-4} (0.01%), irrespective of the type of loading. For instance, by applying relation (3.21), with the elastic modulus determined by (3.3) and assuming the ice temperature to be $T = 263$ K ($T_c = -10\,°\text{C}$), it can be found that the maximum stress at which microcracks will never develop is equal to about 0.5 MPa. If, however, the stress exceeds this limit value, then cracks will eventually nucleate, provided that the sample is left under loading for sufficiently long time. For example, for the stress level of 1 MPa the time needed for cracks to appear is about 1600 s, whereas for the tensile stress magnitude of 2 MPa this time decreases to about 110 s. For comparison, in the case of more finely-grained ice, with a mean diameter $d = 1$ mm, the threshold stress magnitude below which no cracks are nucleated is about 1.0–1.2 MPa, while the propagation of cracks starts at the stress levels exceeding about 1.2–2.0 MPa (Sanderson 1988).

The problem of crack formation and propagation under tensile stresses has been investigated experimentally by Schulson et al. (1984), who tested isotropic polycrystalline ice at temperatures ranging from −20 to −5 °C. The authors have concluded that the nucleation process gives rise to the formation of cracks of lengths which are of the order of an average grain diameter, and it is the largest, dominant flaw developed in the material that ultimately leads to the fracture of a specimen. A peak tensile stress at which the fracture occurs has been measured as a function of grain size, and it has turned out that the stress needed for fracture to occur, whether it is driven by nucleation or propagation of cracks, obeys relations which are very similar in form to each other. Accordingly, the dependence for the nucleation-controlled failure is expressed by the formula (Schulson et al. 1984; Schulson 2001):

$$\sigma_N = \sigma_0 + K_N\, d^{-1/2}, \tag{3.30}$$

which relates tensile stress, σ_N, needed to nucleate cracks, to the grain diameter d in terms of two empirically determined material constants, σ_0 and K_N. Both σ_0 (in MPa) and K_N (in MPa m$^{1/2}$) are temperature-dependent, and Schulson et al. (1984) give them, respectively, the following number values: 0.70 and 0.016 at −20 °C, 0.60 and 0.020 at −10 °C, and 0.55 and 0.020 at −5 °C. The tensile stress needed to induce the crack propagation is, in turn, described by the relation

$$\sigma_P = K_P\, d^{-1/2}, \tag{3.31}$$

which expresses stress σ_P, needed to propagate a crack, in terms of the grain size d by means of only one empirical constant, K_P. The value of the latter does not vary with temperature, and is equal to 0.044 MPa m$^{1/2}$ (Schulson et al. 1984).

Equations (3.30) and (3.31) determine a grain diameter at which the stresses σ_N and σ_P are equal; that is, the stresses required to nucleate cracks and propagate them are in equilibrium. This particular value of grain diameter, denoted by d_c, is given by

$$d_c = \left(\frac{K_P - K_N}{\sigma_0}\right)^2. \tag{3.32}$$

Since the constants K_N and σ_0 are temperature-dependent, one could expect that also d_c has this property. Simple calculations with the values of material parameters specified in the text after Eq. (3.30) show, however, that Eq. (3.32) yields nearly the same values of d_c for the three temperatures considered: $d_c = 1.6$ mm at $T_c = -20\,°\mathrm{C}$, $d_c = 1.6$ mm at $T_c = -10\,°\mathrm{C}$, and $d_c = 1.9$ mm at $T_c = -5\,°\mathrm{C}$. This feature is illustrated in Fig. 3.12, showing the variation of σ_N and σ_P as a function of d. The diameter d_c defines a *critical grain size*, at which the stress required to nucleate cracks equals the stress needed to propagate them. Following Schulson et al. (1984), the diameter $d_c = 1.6$ mm is adopted as a universal value. The plots in Fig. 3.12 indicate that for the grain size $d > d_c$ the fracture process is controlled by nucleation, while for $d < d_c$ the fracture is controlled by crack propagation. In other words, the critical grain size d_c marks a transition from a brittle (nucleation-controlled) to a less-brittle (propagation-controlled) behaviour of ice. Since both sea ice and polar glacier

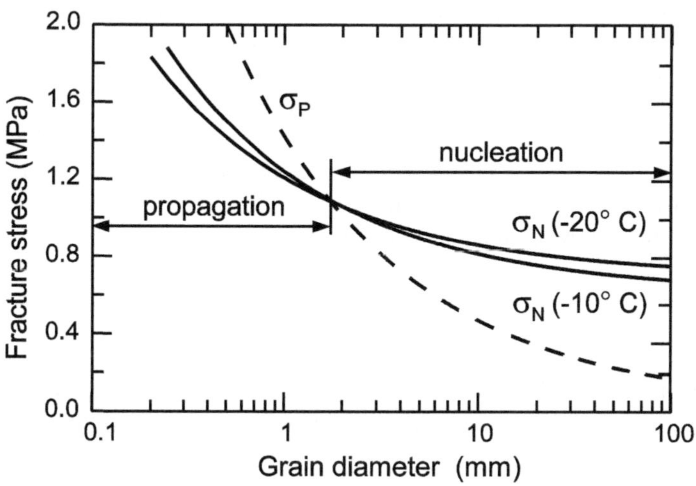

Fig. 3.12 Tensile fracture stresses needed for nucleation (σ_N) and propagation (σ_P) of cracks as a function of mean grain diameter d

ice are usually composed of grains exceeding 1.6 mm in diameter, this means that once cracks have developed in ice, they propagate immediately. There are practical consequences of this situation. Namely, if a sample of ice is initially crack-free, then its *tensile fracture strength* equals the stress needed to nucleate cracks. If, however, there are already flaws in ice, and their length exceeds a few grain diameters, then the ice failure is governed by crack propagation (as requiring, for $d > 1.6$ mm, smaller stress than that needed for crack nucleation).

As can be seen from Fig. 3.12, the typical tensile fracture strength of solid sea ice (without brine or air pockets), for mean grain diameters $d > 1$ mm, is equal to about 1 MPa, and the latter value is commonly used in engineering applications as the standard tensile strength parameter. In some problems of practical importance, though, in which a sea ice sheet is bent, for instance during ice-breaking, *flexural strength* seems to be a more appropriate measure of the brittle material capability of sustaining tensile stresses. It turns out that the flexural strength of ice is typically by a factor of about 1.7 larger than its tensile strength (Timco and O'Brien 1994). This is because the largest cracks which govern the fracture mechanism are generally inside an ice cover, not near its surfaces (Schulson and Duval 2009). The tensile and flexural strength properties of sea ice are strongly dependent on the porosity of ice (caused by the brine inclusions in its structure). Timco and O'Brien (1994) proposed, on the basis of experimental measurements, the following approximate formula describing the reduction of the flexural strength σ_f with increasing brine content ϕ_b:

$$\sigma_f = 1.76 \exp\left(-5.88\sqrt{\phi_b}\right) \quad [\text{MPa}]. \tag{3.33}$$

This relation predicts, for the brine content $\phi_b = 0.05$ (a typical value for first-year sea ice), the value of $\sigma_f = 0.47$ MPa; this means a significant 73% reduction in the flexural strength of ice. One can expect a very similar reduction in the axial tensile strength of sea ice with its increasing porosity as well.

In order to evaluate the tensile strength of the material more accurately, one has to examine what is the stress magnitude that makes an existing crack propagate. The basic criterion is provided by the linear elastic fracture mechanics (LEFM), see, for example, Atkinson (1987). This criterion states that a sharp crack of length $2a$ will propagate (or 'open') in a material under a far-field uniform tensile stress σ (see Fig. 3.13) if the following inequality is satisfied:

$$\sigma > \frac{K_{IC}}{\sqrt{\pi a}}. \tag{3.34}$$

The parameter K_{IC} in (3.34) is called the *fracture toughness* of the material (more precisely, it is 'mode-I fracture toughness', which means that the crack surfaces are opening in the direction perpendicular to the crack faces, as opposed to the modes of crack sliding and crack tearing). The parameter K_{IC} is a material property and can be easily measured for elastic engineering materials (Sanderson 1988). In the case of polycrystalline ice, however, it is considerably more difficult to experimen-

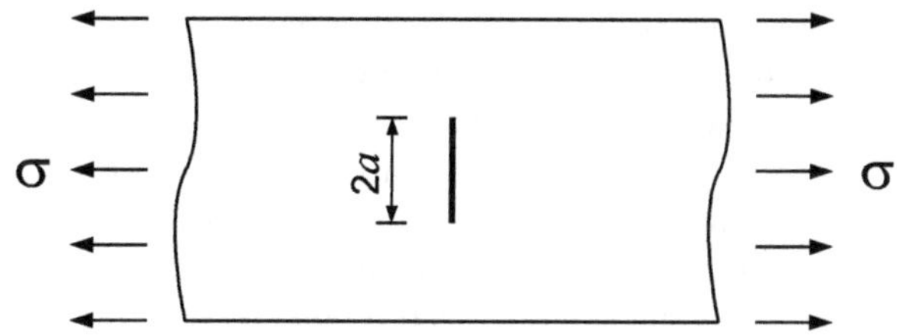

Fig. 3.13 Simple configuration for propagation of a crack of length $2a$ under uniform tensile stress σ

tally determine K_{IC} because of the creeping properties of the material. In general, the measurements should be carried out at sufficiently high strain-rates, otherwise improper values (greater than actual) can be obtained.

The fracture toughness of pure ice is approximately 0.115 MPa $m^{1/2}$. Practically, it can be treated as a temperature-independent parameter (it increases very slightly with decreasing temperature of ice). The fracture toughness decreases with increasing grain size. This can be expressed by the relation (Schulson and Duval 2009)

$$K_{IC} = K_{IO} + \phi_d \, d^{-1/2}, \tag{3.35}$$

where $K_{IO} = 0.58$ MPa $m^{1/2}$ and $\phi_d = 0.042$ MPa $mm^{1/2}$. This relation gives the values K_{IC}=0.10 MPa $m^{1/2}$ for the grain size d=1 mm and $K_{IC} = 0.071$ MPa $m^{1/2}$ for $d = 10$ mm. Obviously, the fracture toughness decreases with ice porosity; therefore, the value of this parameter for sea ice is smaller than for pure ice. As a first approximation, one can assume that the toughness decreases linearly with increasing porosity (Rist et al. 2002). For simplicity, though, one can adopt the value of $K_{IC} = 0.1$ MPa $m^{1/2}$ (Sanderson 1988) as adequate for typical sea ice applications.

3.4.2 *Compressive Fracture*

As already noted, the process of ice failure due to crack nucleation and propagation under compressive loading is a more complex physical phenomenon than that under tensile stress (Schulson and Duval 2009). In compression, cracks may develop not only due to the accumulation of dislocations at grain boundaries, which is the case under tensile loading as well, but also due to their nucleation across ice crystals. Furthermore, the mechanism directly leading to the failure is associated now with the linkage of a large number of cracks and the formation and evolution of so-called wing and comb cracks (Schulson 2001), whereas in tension the final failure may be caused by a single, sufficiently large crack that propagates through the material.

Due to the above-mentioned points, the formal treatment of compressive failure is difficult, and in order to be done properly, application of elaborate theories of fracture mechanics would be expedient. However, following a simplified engineering approach, a nucleation criterion similar to that used for tensile fracture is applied. Hence, it is assumed that nucleation of cracks under compression begins when the lateral tensile strain resulting from the Poisson expansion reaches the same magnitude as the corresponding strain occurring during the tensile crack formation. Since under uniaxial compressive stress the lateral expansion of the material is the axial compression times the Poisson ratio ν, and, further, the elastic response of ice is roughly independent of direction, one can deduce that the axial compressive stress has to be $1/\nu$ times larger in magnitude than the critical tensile stress necessary for the tensile crack nucleation. Accordingly, by analogy to relation (3.30), the nucleation criterion for cracks to develop under compressive stress can be expressed by

$$\sigma_N^c = -\frac{1}{\nu}\left(\sigma_0 + K_N\, d^{-1/2}\right), \tag{3.36}$$

where σ_N^c (negative in compression) is the stress at which cracks nucleate, and the remaining constants have already been defined for Eq. (3.30). Since $\nu \sim 1/3$, it means that the stress needed to produce cracks under compression is about three times as large as the corresponding stress for tension. As it follows from (3.36), the compressive fracture of coarsely-grained ice ($d \gtrsim 5$ mm) occurs at the stress level of about 3 MPa.

Once the process of crack nucleation under compression has started, a sample of ice contains a wide variety of cracks of different sizes and orientations. Experimental investigations have shown (Sanderson 1988) that: (1) the average length of nucleated cracks, $2a$, is roughly proportional to the grain diameter d, with an approximate relation $2a \sim 0.65d$; (2) the crack orientations are clustered around the axis of compression, so that about 90% of cracks are aligned within the angle of 45° to the direction of the principal compressive stress axis; (3) for coarsely-grained ice ($d \gtrsim 5$ mm) a mean crack density is about one crack per grain.

At this stage (that is, after the crack nucleation), ice still remains in a stable state, and its brittle failure can occur only after the existing cracks merge with one another, which can be accomplished by increasing the compressive stress. At some critical loading level *wing cracks* develop in the material and the process of crack propagation starts (see Fig. 3.14). As the compressive stress is applied, both sides of

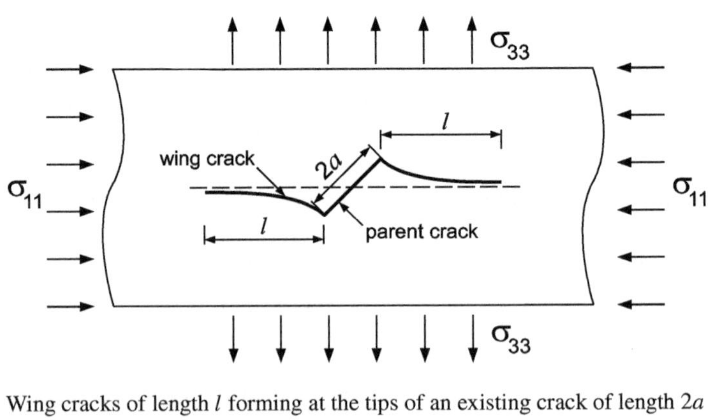

Fig. 3.14 Wing cracks of length l forming at the tips of an existing crack of length $2a$

the crack slide relative to each other due shear stresses and, as a result, tensile stress regions develop around the tips of the cracks. These tensile zones give rise to wing cracks that initially propagate roughly perpendicularly to the parent crack ends, but subsequently they tend to align along the principal axis of compression. The length of the wing crack, which usually develops in a stable manner, is a function of the current stress conditions and the mechanical properties of ice. This problem has been analysed in two dimensions by Ashby and Hallam (1986); the details of the analysis are omitted here and only the final results of this work are presented.

Assume that an initial crack of length $2a$ is oriented at some angle to the direction of the principal compressive stress σ_{11}, (Fig. 3.14). It is possible then to relate the length l of the wing crack by the formula

$$K_{IC} = -\frac{\sigma_{11}\sqrt{\pi a}}{\sqrt{3}\,(1+L)^{3/2}}\left[Lc + (1+L)^{-1/2}\right] \times \\ \times \left[1 - \xi - \lambda(1+\xi) - \sqrt{3}\,L\xi/c\right]. \tag{3.37}$$

In this formula, $L = l/a$, c is a coefficient, $\xi = -\sigma_{33}/\sigma_{11}$ is the ratio of confining lateral stress to compressive stress, and λ is the coefficient of friction across the crack. As an illustration, let consider unconfined uniaxial compression conditions, that is, assume that $\sigma_{33} = 0$, and hence $\xi = 0$. In this simplified case, Eq. (3.37) defines the compressive failure stress $\sigma_{11} = \sigma_P^c$ as follows:

$$\sigma_P^c = -\frac{\sqrt{3}\,K_{IC}\,(1+L)^{3/2}}{(1-\lambda)\sqrt{\pi a}}\left[Lc + (1+L)^{-1/2}\right]^{-1}. \tag{3.38}$$

The latter equation relates the compressive stress to the length of wing cracks. What is, however, a mean length of wing cracks when ice fails due to crack propagation, still remains an open question, and a number of approaches are possible to answer it (Ashby and Hallam 1986). Here it is supposed that the process of linkage of wing cracks, resulting in a catastrophic reduction in the ice strength, occurs when the length l of cracks is equal to half of the grain diameter d. This is consistent with the empirical data indicating that immediately after the nucleation of cracks, their density is approximately one crack per grain. Thus, one can expect that if $d = 2l$, then the initially separated cracks coalesce (since they all tend to align in the same direction), and hence the fracture process starts. Now, recalling that the length of newly nucleated cracks is given by the relation $2a = 0.65d$, it follows that $L = l/a \sim 1.53$. By adopting the values $c = 0.4$ and $\lambda = 0.3$ (suggested by Sanderson 1988), we obtain from Eq. (3.38) an approximate expression for the *compressive fracture strength* of ice:

$$\sigma_P^c \approx -7.97\,K_{IC}\,d^{-1/2}. \tag{3.39}$$

Taking $K_{IC} = 0.115$ MPa m$^{1/2}$ as the value of ice fracture toughness, Eq. (3.39) predicts, for ice of the mean grain size $d = 5$ mm, the compressive failure strength of about 13 MPa. According to Sanderson (1988), this is a reasonable estimate, and for many engineering purposes the approximate formula (3.39) seems to be adequate, in spite of a number of uncertainties and crude simplifications involved in its derivation.

One should keep in mind that relation (3.39) holds for pure ice. In the case of sea ice, the compressive fracture strength is obviously smaller because of the factors (ice porosity and salinity) discussed earlier in this chapter. For instance, Schulson et al. (2006) give the value of (5.0 ± 0.2) MPa at a temperature of $-10\,°\mathrm{C}$ for S2 columnar first-year Arctic ice, of grain size $d = (3.9 \pm 0.4)$ mm and salinity of 8 ppt. The above magnitude of 5 MPa is often used in sea ice applications as the parameter describing the strength of sea ice under horizontal compression.

As seen above, the theoretical analysis of ice fracture phenomena is complicated even in the case of plane problems. Therefore, the extension of the analysis to three dimensions proves a very challenging task. Nevertheless, a number of three-dimensional constitutive models for ice failure mechanism have been already developed. One of such general theories is due to Sjölind (1987), who treated ice as an orthotropic viscoelastic brittle material. The model is based on an assumption supported by observation that microcracks nucleate and grow in planes that are perpendicular to the directions of positive principal stresses. For each possible crack propagation direction, a so-called damage vector is defined, the magnitude of which is related to the average volume density of cracks. These damage vectors constitute a set of internal state variables which describe the macroscopic behaviour of the material. In order to derive evolution equations governing the growth-rates of the microcrack fields, the principles of irreversible thermodynamics have been applied. The model has been tested for uniaxial, compressive and tensile stress configurations, with the material properties pertinent to S2 columnar ice. The model predictions obtained numerically have been consistent with empirical data over a wide range of strain-rates. However, it appears that due to its complexity, it is difficult to apply this theory to realistic three-dimensional engineering problems.

In the fracture theories for ice, the mechanism of crack nucleation is usually assumed to result entirely from dislocation pile-ups at grain boundaries. Cole (1988) has proposed a theory based on the assumption that the elastic anisotropy of individual ice crystals can also contribute to the process of crack formation. This has been demonstrated by applying a model, in which the interaction of only two isolated crystals has been considered. A drawback of Cole's approach, apparently due its simplicity, is that it predicts equal tensile and compressive stresses at which the first cracks nucleate, which seems unrealistic. Shyam Sunder and Wu (1990a) have significantly extended the theory of crack nucleation due to the elastic anisotropy of ice grains, by applying a method developed by Eshelby (1957). The model results have shown that the stress magnitude required to nucleate the first crack under uniaxial compression is about 2.5 times as large as that under uniaxial tension, and the magnitude of the stress needed to form a crack in compression is strongly dependent on the orientation of crystals. The predicted range of the tensile nucleation stresses as a function of grain size has agreed well with the experimental data of Schulson et al. (1984).

In another attempt, Nixon (1996) has formulated a constitutive theory which describes ice failure due to the mechanism of unstable wing crack propagation. In contrast to Ashby and Hallam (1986) and Sanderson (1988), the author has considered a possibility that the formation of new cracks, rather than the propagation of wing cracks from those already developed, may result in the brittle fracture of ice. The analytical relations derived by Nixon resemble those obtained by Ashby and Hallam (1986); that is, they relate ice failure strength to grain diameter and a set of parameters that can be measured in experiments. The author has compared the predictions of his six different models with the available test data, and concluded that the brittle compressive failure of ice cannot be described solely by means of wing crack propagation, therefore other micro-mechanisms need to be accounted for as well.

3.5 Thermal Properties of Ice

All the processes associated with the formation and melting of ice on Earth are related to the climate. Further, most of the physical (and, in particular, the mechanical) properties of ice strongly depend on its temperature. Therefore, the phenomena of phase changes (freezing of water and melting of ice), and the processes of heat fluxes through ice and heat exchange between ice, water and air, are all of paramount importance to ice mechanics.

The amount of energy due to the solar irradiance, incoming to a plane perpendicular to the rays, and measured per unit area and unit time, is defined by the *solar constant*. Its value, measured at the top of Earth's atmosphere, is equal to 1361 $\mathrm{W\,m^{-2}}$. Due to the absorption of solar radiation by water vapour, ozone and carbon dioxide in the atmosphere, its absorption and reflection by clouds, and its scattering back to space by air, etc., only about 51% (Paterson 1994) of the total solar energy arriving at the outer layers of the atmosphere reaches the surface of Earth.

Of the energy which ultimately arrives at the ocean or land (ice) top surface, one part is absorbed by underlying medium and contributes to its heating, while the other part is reflected back to the atmosphere. How much of the solar energy is reflected at a given surface is defined by a dimensionless parameter known as *albedo* (or *whiteness*), assuming the values from zero to unity. A value of zero means that the body perfectly absorbs the incoming energy, whereas a value of unity means that the body perfectly reflects the entire energy. The albedo of ice is much higher than the albedos of other Earth surfaces. Its exact value depends on the nature of ice (for instance, its surface roughness) at a given location, and varies from about 0.35 to about 0.50 for clean ice, and, respectively, from 0.15 to 0.25 for dirty ice (Paterson 1994). The presence of snow significantly increases the surface albedo: it can be as high as 0.87 for newly fallen snow (Weeks 2010). Hence, compared to bare ice, the amount of solar energy absorbed by snow-covered ice can be reduced by a factor of about three. For comparison, a typical albedo of ocean surface can be as low as 0.06 (meaning that 94% of incoming energy heats the underlying water).

The solar energy that falls on the ice, depending on particular local conditions (such as the air temperature), can drive a number of thermodynamic processes. In the first place, melting of ice can occur on its surface. The amount of energy that is used during this process is defined by the *latent heat of melting*, which for water ice has the value of 3.34×10^5 J kg^{-1}. On the other hand, some energy can be released from ice if water evaporation occurs at the ice surface; the amount of this energy is described by the *latent heat of vaporization*, which has the value of 2.8×10^6 J kg^{-1}. Part of the incoming energy that is not involved it the ice phase change processes is transferred (conducted) into the ice interior. This heat transfer occurs at the rate defined by the *thermal conductivity* parameter, which for pure ice at 272 K has the value of 2.22 W m^{-1} K^{-1}. For comparison, water at the freezing point has the conductivity of 0.56 W m^{-1} K^{-1}, which is a value by a factor of four smaller than that for ice. The amount of energy which is added to (or released from) a body as a result of a temperature change is described by the *specific heat* parameter, (or *specific heat capacity*) which for ice at 263 K has the value of 2.11×10^3 J kg^{-1} K^{-1}. For comparison, the specific heat capacity of air-free water is 4.18×10^3 J kg^{-1} K^{-1}.

As ice is heated it expands, as nearly all natural materials (a rare exception is water near its freezing point, since it contracts with an increase in temperature). The mechanism of thermal expansion, which is usually neglected when dealing with ice on geophysical scales, may be important in the context of civil engineering applications, when stresses exerted by ice on structures need to be evaluated. The rate of expansion is defined by the *thermal expansion coefficient*, which for ice at 263 K is equal to 5.2×10^{-5} K^{-1}.

There are many more thermodynamic phenomena which take place in and between ice, water and air, and which need to be accounted for when analysing in detail the energy balances for sea and polar ice. These phenomena, however, which are often very complex and difficult to describe formally, are beyond the scope of this book and, therefore, are nor treated here.

References

Alley RB (1992) Flow-law hypotheses for ice-sheet modelling. J Glaciol 38(129):245–256

Ashby MF, Duval P (1985) The creep of polycrystalline ice. Cold Reg Sci Technol 11(3):285–300

Ashby MF, Hallam SD (1986) The failure of brittle solids containing small cracks under compressive stress-states. Acta Metall 34(3):497–510

Atkinson BK (ed) (1987) Fracture mechanics of rock. Academic Press, London

Baral DR, Hutter K, Greve R (2001) Asymptotic theories of large-scale motion, temperature, and moisture distribution in land-based polythermal ice sheets: a critical review and new developments. Appl Mech Rev 54(3):215–256

Barnes P, Tabor D, Walker JCF (1971) The friction and creep of polycrystalline ice. Proc R Soc Lond A 324(1557):127–155

Budd WF, Jacka TH (1989) A review of ice rheology for ice sheet modelling. Cold Reg Sci Technol 16(2):107–144. https://doi.org/10.1016/0165-232X(89)90014-1

Colbeck SC, Evans RJ (1973) A flow law for temperate glacier ice. J Glaciol 12(64):71–86

Cole DM (1988) Crack nucleation in polycrystalline ice. Cold Reg Sci Technol 15(1):79–87

Dantl G (1969) Elastic moduli of ice. In: Riehl N, Bullemer B, Engelhardt H (eds) Physics of ice. Plenum Press, New York, pp 223–230
Doake CSM, Wolff EW (1985) Flow law for ice in polar ice sheets. Nature 314(6008):255–257
Duval P (1981) Creep and fabric of polycrystalline ice under shear and compression. J Glaciol 27(95):129–140
Duval P, Ashby MF, Anderman I (1983) Rate-controlling processes in the creep of polycrystalline ice. J Phys Chem 87(21):4066–4074
Duval P, Lorius C (1980) Crystal size and climatic record down to the last ice age from Antarctic ice. Earth Planet Sci Lett 48:59–64
Eshelby JD (1957) The determination of the elastic field of an ellipsoidal inclusion, and related problems. Proc R Soc Lond A 241:376–396
Findley WN, Lai JS, Onaran K (1976) Creep and relaxation of nonlinear viscoelastic materials. North-Holland, Amsterdam
Flügge W (1967) Viscoelasticity. Blaisdell, Toronto
Gammon PH, Kiefte H, Clouter MJ, Denner WW (1983) Elastic constants of artificial and natural ice samples by Brillouin spectroscopy. J Glaciol 29(103):433–460
Glen JW (1955) The creep of polycrystalline ice. Proc R Soc Lond A 228(1175):519–538
Gold LW, Sinha NK (1980) The rheological behaviour of ice at small strains. In: Tryde P (ed) Physics and mechanics of ice, proceedings of the IUTAM Symposium, Copenhagen 1979. Springer, Berlin, pp 117–128
Goodman DJ, Frost HJ, Ashby MF (1981) The plasticity of polycrystalline ice. Philos Mag A 43(3):665–695
Gow AJ, Meese DA, Alley RB, Fitzpatrick JJ, Anandakrishnan S, Woods GA, Elder BC (1997) Physical and structural properties of the Greenland Ice Sheet Project 2 ice core: a review. J Geophys Res 102(C12):26559–26575. https://doi.org/10.1029/97JC00165
Green AE, Zerna W (1992) Theoretical elasticity. Dover, Mineola, New York
Hawkes I, Mellor M (1972) Deformation and fracture of ice under uniaxial stress. J Glaciol 11(61):103–131
Hill R (1952) The elastic behaviour of a crystalline aggregate. Proc Phys Soc A 65(389):349–354
Hill R (1965) Continuum micro-mechanics of elastoplastic polycrystals. J Mech Phys Solids 13(2):89–101
Hobbs PV (2010) Ice physics. Oxford University Press, Oxford
Hutter K (1975) Floating sea ice plates and the significance of the dependence of the Poisson ratio on brine content. Proc R Soc Lond A 343(1632):85–108
Hutter K (1983) Theoretical glaciology. Material science of ice and the mechanics of glaciers and ice sheets. Reidel, Dordrecht
Jacka TH (1984) The time and strain required for development of minimum strain rates in ice. Cold Reg Sci Technol 8(3):261–268. https://doi.org/10.1016/0165-232X(84)90057-0
Kamb WB (1961) The glide direction in ice. J Glaciol 3(30):1097–1106
Le Gac H, Duval P (1980) Constitutive relations for the non-elastic deformation of polycrystalline ice. In: Tryde P (ed) Physics and mechanics of ice, proceedings of the IUTAM symposium, Copenhagen 1979. Springer, Berlin, pp 51–59
Lliboutry L (1969) The dynamics of temperate glaciers from the detailed viewpoint. J Glaciol 8(53):185–205
Lliboutry L, Duval P (1985) Various isotropic and anisotropic ices found in glaciers and polar ice caps and their corresponding rheologies. Ann Gheophys 3(2):207–224
Mellor M (1980) Mechanical properties of polycrystalline ice. In: Tryde P (ed) Physics and mechanics of ice, proceedings of the IUTAM symposium, Copenhagen 1979. Springer, Berlin, pp 217–245
Mellor M, Cole DM (1982) Deformation and failure of ice under constant stress or constant strain-rate. Cold Reg Sci Technol 5(3):201–219
Mellor M, Cole DM (1983) Stress/strain/time relations for ice under uniaxial compression. Cold Reg Sci Technol 6(3):207–230

Mellor M, Testa R (1969a) Creep of ice under low stress. J Glaciol 8(52):147–152
Mellor M, Testa R (1969b) Effect of temperature on the creep of ice. J Glaciol 8(52):131–145
Morland LW (1993) The flow of ice sheets and ice shelves. In: Hutter K (ed) Continuum mechanics in environmental sciences and geophysics. Springer, Wien, pp 403–466
Morland LW (1996) Dynamic impact between a viscoelastic ice floe and a rigid structure. Cold Reg Sci Technol 24(1):7–28
Morland LW (2001) Influence of bed topography on steady plane ice sheet flow. In: Straughan B, Greve R, Ehrentraut H, Wang Y (eds) Continuum mechanics and applications in geophysics and the environment. Springer, Berlin, pp 276–304
Nakawo M, Sinha NK (1981) Growth rate and salinity profile of first-year ice in the high Arctic. J Glaciol 27(96):315–330
Nanthikesan S, Shyam Sunder S (1994) Anisotropic elasticity of polycrystalline ice Ih. Cold Reg Sci Technol 22(2):149–169
Nixon WA (1996) Wing crack models of the brittle compressive failure of ice. Cold Reg Sci Technol 24(1):41–55
Paterson WSB (1994) The physics of glaciers, 3rd edn. Butterworth-Heinemann, Oxford
Rist MA, Sammonds PR, Oerter H, Doake CSM (2002) Fracture of Antartctic shelf ice. J Geophys Res 107(B1):ECV 2-1–ECV 2-13. https://doi.org/10.1029/2000JB000058
Sanderson TJO (1988) Ice mechanics. Risks to offshore structures, Graham and Trotman, London
Schulson EM (2001) Brittle failure of ice. Eng Fract Mech 68(17–18):1839–1887
Schulson EM, Duval P (2009) Creep and fracture of ice. Cambridge University Press, Cambridge
Schulson EM, Fortt AL, Iliescu D, Renshaw CE (2006) Failure envelope of first-year Arctic sea ice: the role of friction in compressive fracture. J Geophys Res 111:C11S25. https://doi.org/10.1029/2005JC003234186
Schulson EM, Lim PN, Lee RW (1984) A brittle to ductile transition in ice under tension. Philos Mag A 49(3):353–363
Schwarz J, Weeks WF (1977) Engineering properties of sea ice. J Glaciol 19(81):499–531
Shyam Sunder S, Wu MS (1989a) A differential flow model for polycrystalline ice. Cold Reg Sci Technol 16(1):45–62
Shyam Sunder S, Wu MS (1989b) A multiaxial differential model of flow in orthotropic polycrystalline ice. Cold Reg Sci Technol 16(3):223–235
Shyam Sunder S, Wu MS (1990a) Crack nucleation due to elastic anisotropy in polycrystalline ice. Cold Reg Sci Technol 18(1):29–47
Shyam Sunder S, Wu MS (1990b) On the constitutive modeling of transient creep in polycrystalline ice. Cold Reg Sci Technol 18(3):267–294
Sinha NK (1978a) Rheology of columnar-grained ice. Exp Mech 18(12):464–470
Sinha NK (1978b) Short-term rheology of polycrystalline ice. J Glaciol 21(85):457–473
Sinha NK (1979) Grain boundary sliding in polycrystalline materials. Philos Mag A 40(6):825–842
Sinha NK (1983) Creep model of ice for monotonically increasing stress. Cold Reg Sci Technol 8(1):25–33
Sinha NK (1989) Elasticity of natural types of polycrystalline ice. Cold Reg Sci Technol 17(2):127–135
Sjölind SG (1985) Viscoelastic buckling analysis of floating ice sheets. Cold Reg Sci Technol 11(3):241–246
Sjölind SG (1987) A constitutive model for ice as a damaging visco-elastic material. Cold Reg Sci Technol 14(3):247–262
Smith GD, Morland LW (1981) Viscous relations for the steady creep of polycrystalline ice. Cold Reg Sci Technol 5(2):141–150
Spring U, Morland LW (1983) Integral representations for the viscoelastic deformation of ice. Cold Reg Sci Technol 6(3):185–193
Thorsteinsson T, Kipfstuhl J, Miller H (1997) Textures and fabrics in the GRIP ice core. J Geophys Res 102(C12):26583–26599. https://doi.org/10.1029/97JC00161

Timco GW, O'Brien S (1994) Flexural strength equation for sea ice. Cold Reg Sci Technol 22(3):285–298. https://doi.org/10.1016/0165-232X(94)90006-X

Timco GW, Weeks WF (2010) A review of the engineering properties of sea ice. Cold Reg Sci Technol 60(2):107–129. https://doi.org/10.1016/j.coldregions.2009.10.003

Treverrow A, Budd WF, Jacka TH, Warner RC (2012) The tertiary creep of polycrystalline ice: experimental evidence for stress-dependent levels of strain-rate enhancement. J Glaciol 58(208):301–314. https://doi.org/10.3189/2012JoG11J149

Weeks WF (2010) On sea ice. University of Alaska Press, Fairbanks

Weeks WF, Assur A (1967) The mechanical properties of sea ice. USA, U.S, Army Cold Regions Research and Engineering Laboratory, Hanover, NH

Weertman J (1983) Creep deformation of ice. Annu Rev Earth Planet Sci 11:215–240

Zhan C, Evgin E, Sinha NK (1994) A three dimensional anisotropic constitutive model for ductile behaviour of columnar grained ice. Cold Reg Sci Technol 22(3):269–284

Chapter 4
Sea Ice in Civil Engineering Applications

In civil engineering applications, in which an interaction between a structure and a floating ice cover occurs, the main interest is in the evaluation of forces which the ice exerts on the structure, with the particular importance of their maximum magnitudes, since these determine design loads for a given engineering object. As already discussed in Chap. 3, a variety of deformation mechanisms can be observed in ice, depending on stress, strain and strain-rate to which the material is subjected. Typically, at the beginning of an ice–structure interaction process, elastic strains develop in ice, but these are small in magnitude compared to other modes of deformation. When the ice is in good contact with the structure (is frozen to its walls), so that the forces (induced by winds and water currents) which drive the ice change slowly in time, and when the sea waves are small, then the material deforms in a continuous way, by creep. If, however, the stresses in ice are large, and/or the sea wave action breaks the ice cover, then the ice–structure interaction has a dynamic character, and the ice behaves in a typical brittle manner.

In what follows in this chapter, the three above-mentioned types of the ice mechanical behaviour, that is, the elastic, creep and brittle responses of the material, will be analysed. Hence, several problems of the interaction between a coherent floating ice cover and an engineering structure are discussed. First, in Sect. 4.1, the problem of purely elastic response of ice during its short-time (measured in seconds) interaction with a rigid vertical structure is investigated, with the aim to evaluate the magnitudes of the maximum horizontal forces exerted on the structure; these forces are assumed to be those which lead to an elastic buckling failure of an ice plate under compressive and transverse loadings. Then, in Sect. 4.2, quasi-static ice–structure interaction events lasting for hours and days are investigated, in which the deformations of ice are dominated by its creep. Hence, rheological models describing the creep of sea ice are first discussed, and these models are used to analyse the mechanism of creep buckling of a floating ice sheet, and then the interaction problems involving rigid structures of rectangular and cylindrical planar cross-sections are considered.

R. Staroszczyk, *Ice Mechanics for Geophysical and Civil Engineering Applications*,
GeoPlanet: Earth and Planetary Sciences, https://doi.org/10.1007/978-3-030-03038-4_4

Further, on the basis of the results of numerical simulations, the effects of different sea ice rheologies on the predicted magnitudes of the ice–structure interaction forces are examined. Finally, Sect. 4.3 is devoted to the problem of a dynamic impact of an ice floe on a structure, during which the ice behaves in a brittle manner.

4.1 Elastic Interaction of Ice with a Rigid Wall

The purely elastic behaviour of ice is rarely observed in floating ice, which is due to the fact that at typical stress levels in sea ice (1–5 MPa), creep strains overtake elastic ones within a period of seconds after the application of loading forces. Despite this, the problem of evaluating the forces in ice during the very short period of its elastic response is of practical importance, since these forces may reach the values which exceed the magnitudes of forces in ice at later, creep or fracture, modes of deformation. Hence, the elastic response forces need to be taken into account when determining design loads on an engineering structure.

It is a common assumption (Sanderson 1988) that the maximum values of the elastic response forces exerted on a structure are bounded by the magnitude of a force which is required to cause an elastic buckling of the part of an ice cover which directly interacts with the structure. Usually, the floating ice sheet that interacts with a structure is supposed to have, in the horizontal plane, a shape of a truncated wedge of a finite or semi-infinite length. Such a geometry reflects the conditions frequently occurring in the field, when radial cracks propagating from the vertical edges of the structure develop, bounding thus the domain of the ice cover which effectively interacts with the structure.

When analysing the behaviour of coherent floating ice, it is usually assumed that the ice cover can be treated as a continuous plate. The problem of the elastic buckling of a wedge-shaped plate on an elastic foundation (the underlying water) has been investigated in a number of papers, for instance, by Kerr (1978), Nevel (1980) and Sanderson (1988), in which approximate estimates for the buckling forces, derived analytically, are given. Some relevant analytical results can be also found in the work by Kerr and Palmer (1972), and experimental data on the elastic buckling of ice have been reported by Sodhi et al. (1983). The results presented here have been obtained by the author Staroszczyk (2002) by applying a finite-element method (FEM). These results are compared with those predicted by approximate solutions proposed by Kerr (1978), and it will be shown that the approximations of the latter author lead to a significant overestimation of the buckling forces that a wedge-shaped elastic plate can sustain. Moreover, some inconsistency in the analytical results by Kerr (1978) has been discovered. Therefore, by fitting to the FEM results obtained, new approximate formulae enabling simple, but reliable, calculations of the buckling forces in floating wedge-shaped elastic plates have been proposed for the use by engineers.

4.1.1 Interaction Problem Formulation

The behaviour of a continuous ice cover floating on the free surface of water is considered. The ice is assumed to be driven by horizontal drag forces due to the wind and water currents action. As the ice cover deforms, not only in the horizontal direction but also transversely, it undergoes vertical loading resulting from the reaction of the underlying water. In this analysis, the ice cover is treated as an elastic plate that floats on a liquid foundation, and is subject to the combined action of the in-plane as well transverse (out-of-plane) forces. The definitions of the internal forces (axial and shear forces and bending moments) and external loadings, together with the adopted frame of rectangular coordinates, are shown in Fig. 4.1a. The plate of ice is assumed to be of uniform thickness, denoted by h, and to be in perfect contact with the underlying water (that is, there are no air pockets between the ice and the water), see Fig. 4.1b. Due to the possible variation of the ice porosity and temperature with depth, the mechanical properties (such as the Young modulus and the ice viscosity) may change across the ice cover (usually, the ice is weaker near its base). The consequence of this is the plate inhomogeneity along the vertical direction, which implies that, in general, the neutral and middle surfaces in the plate do not coincide.

The vertical z-axis, directed downwards, is chosen in such a way that $z = 0$ at the top surface of the plate, and $z = h$ at its bottom surface. The plate transverse

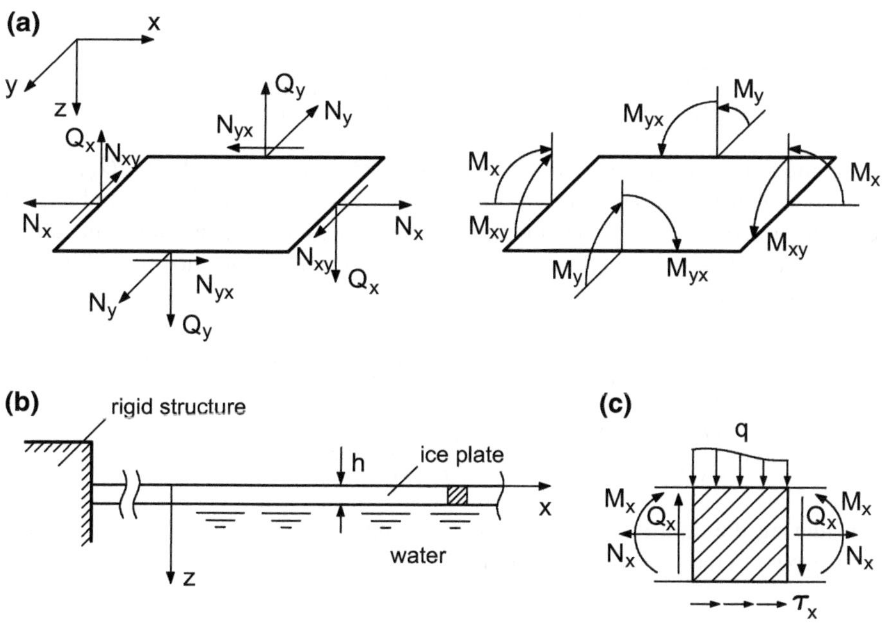

Fig. 4.1 **a** Adopted coordinate system and definitions of internal forces acting on a plate element, **b** and **c** vertical cross-sections of the plate

displacement along the z-axis is denoted by w. The ice is assumed to be in contact with an engineering structure and to exert forces on it. For simplicity, the structure is modelled as a rigid body which interacts with the floating ice along vertical walls. The prime objective is to evaluate the forces which the wind-driven ice exerts on the structure. In particular, the magnitude of the horizontal compressive force in ice under which the ice plate buckles in an elastic manner will be determined, as this buckling force defines the maximum horizontal load exerted on the structure.

In order to solve the problem, the classical Kirchhoff–Love theory of thin plates (Timoshenko and Woinowsky-Krieger 1959) is applied, which is based on the assumptions that (1) the plate thickness is small compared to its characteristic lengths, (2) the plate deflections are small, that is, not exceeding its thickness, (3) the effects due to shear stresses are neglected, so (4) the plate cross-sections which are normal to the middle plane prior to bending remain plane and normal to the middle surface in the deformed state, and (5) the normal stresses in the direction transverse to the plate surfaces are disregarded.

In the horizontal plane Oxy, the internal loads acting in the plate are the axial forces N_x and N_y (considered positive in tension) and the shear forces $N_{xy} = N_{yx}$, all measured per unit length. Apart from them, there are tangential forces (tractions) acting over the top and bottom surfaces of the plate, caused by the wind stress and water drag. The two components of these forces, defined per unit area of the middle plane of the plate, are denoted by τ_x and τ_y (see Fig. 4.1c). The equilibrium balances of the forces along the x and y axes, ignoring inertia forces due to small horizontal velocities of ice, are expressed by

$$\frac{\partial N_x}{\partial x} + \frac{\partial N_{xy}}{\partial y} + \tau_x = 0, \qquad \frac{\partial N_{xy}}{\partial x} + \frac{\partial N_y}{\partial y} + \tau_y = 0. \tag{4.1}$$

Along the z-axis, the plate is subject to the vertical shear forces, Q_x and Q_y, and the transverse distributed load q. In a deformed state, also the forces N_x, N_y and N_{xy}, all acting in directions tangential to the deflection surface $w(x, y)$, have relevant vertical components. Hence, neglecting the own weight of ice, the projection of all forces on the z-axis direction gives

$$\begin{aligned} &\frac{\partial Q_x}{\partial x} + \frac{\partial Q_y}{\partial y} + q + N_x \frac{\partial^2 w}{\partial x^2} + \frac{\partial N_x}{\partial x}\frac{\partial w}{\partial x} + N_y \frac{\partial^2 w}{\partial y^2} + \frac{\partial N_y}{\partial y}\frac{\partial w}{\partial y} + \\ &+ 2N_{xy}\frac{\partial^2 w}{\partial x \partial y} + \frac{\partial N_{xy}}{\partial x}\frac{\partial w}{\partial y} + \frac{\partial N_{xy}}{\partial y}\frac{\partial w}{\partial x} = 0. \end{aligned} \tag{4.2}$$

Considering the equilibrium of moments acting on an infinitesimal plate element with respect to the y and x axes, we find that

$$\frac{\partial M_x}{\partial x} - \frac{\partial M_{xy}}{\partial y} - Q_x = 0, \qquad \frac{\partial M_y}{\partial y} - \frac{\partial M_{xy}}{\partial x} - Q_y = 0, \tag{4.3}$$

where M_x and M_y are the bending moments, and $M_{xy} = M_{yx}$ are the twisting moments, all defined per unit length. The only transverse load that is exerted on the plate comes from the reaction of the underlying water, when the plate is either lifted or depressed from its floating equilibrium state. It is assumed that the reaction of the water is purely elastic and is proportional to the plate deflection w (thus, the liquid base can be regarded as the Winkler–Zimmerman-type foundation). Accordingly,

$$q = -\varrho_w g w, \tag{4.4}$$

where ϱ_w is the density of water and g is the acceleration due to gravity.

It is useful to eliminate the vertical shearing forces Q_x and Q_y from the equilibrium balances (4.2) and (4.3), thus reducing the number of equations to be solved. Accordingly, by inserting in (4.2) the expressions for Q_x and Q_y defined by (4.3), and then using relations (4.1) and (4.4) in the resulting equation, one arrives at the equilibrium equation of the form:

$$\begin{aligned} &\frac{\partial^2 M_x}{\partial x^2} - 2\frac{\partial^2 M_{xy}}{\partial x \partial y} + \frac{\partial^2 M_y}{\partial y^2} + N_x\frac{\partial^2 w}{\partial x^2} + 2N_{xy}\frac{\partial^2 w}{\partial x \partial y} + N_y\frac{\partial^2 w}{\partial y^2} + \\ &- \varrho_w g w - \tau_x\frac{\partial w}{\partial x} - \tau_y\frac{\partial w}{\partial y} = 0. \end{aligned} \tag{4.5}$$

This equation involves the bending moments and in-plane loads, the plate deflection w and its spatial derivatives, and the driving forces τ_x and τ_y.

The internal forces in Eq. (4.5) can be expressed in terms of the stresses σ_{xx}, σ_{yy} and σ_{xy} which act in the transverse cross-sections of the plate. Hence, the in-plane axial and shear forces are given by

$$N_x = \int_0^h \sigma_{xx} \mathrm{d}z, \quad N_y = \int_0^h \sigma_{yy} \mathrm{d}z, \quad N_{xy} = \int_0^h \sigma_{xy} \mathrm{d}z, \tag{4.6}$$

and the bending and twisting moments are defined by

$$\begin{aligned} &M_x = \int_0^h \sigma_{xx}(z - z_0)\mathrm{d}z, \quad M_y = \int_0^h \sigma_{yy}(z - z_0)\mathrm{d}z, \\ &M_{xy} = -\int_0^h \sigma_{xy}(z - z_0)\mathrm{d}z. \end{aligned} \tag{4.7}$$

In relations (4.7), z_0 is the position of the neutral surface in the plate in its undeformed state.

To proceed further, one needs specific constitutive laws describing the material response of ice; that is, the equations which relate stresses to strains (in the case of the elastic response), strain-rates (in the case of the creep response), or both (in the case of the viscoelastic response of ice or more complex material responses).

The equilibrium equation (4.5) describes the two-dimensional behaviour of a plate floating on the water surface. In order to solve this equation, even for simple two-dimensional geometries, one has to resort to one of discrete methods, since no analytical solutions are available. As already mentioned, in typical ice–structure interaction phenomena, due to the propagation of cracks in the ice cover, the domain of ice which actively interacts with an engineering object has, to a good approximation, the horizontal shape of a truncated wedge, with the ice which is outside the wedge playing only a passive role. The geometry of the problem is illustrated in Fig. 4.2a, showing the planar view of the wedge-shaped plate being in contact with a flat, vertical wall at $x = 0$.

Before solving the elastic plate buckling problem, the latter is simplified by assuming that the plate wedge, defined by the wall width b_0 and the angle $\alpha \geq 0$, is symmetric about the x-axis, and extends to infinity. Further, it is also assumed that the forces induced by the wind and water drag action are symmetric about the x-axis as well, so that the plate is pushed towards the structure along the negative direction of the x-axis. Due to the above assumptions, another simplification is introduced, by supposing that the plate deflection w and all the loads and forces acting on the plate are functions of only one horizontal coordinate, x. Thus, $w = w(x), q = q(x)$, $M_x = M_x(x)$, etc., which implies that there is no bending of the plate in the lateral

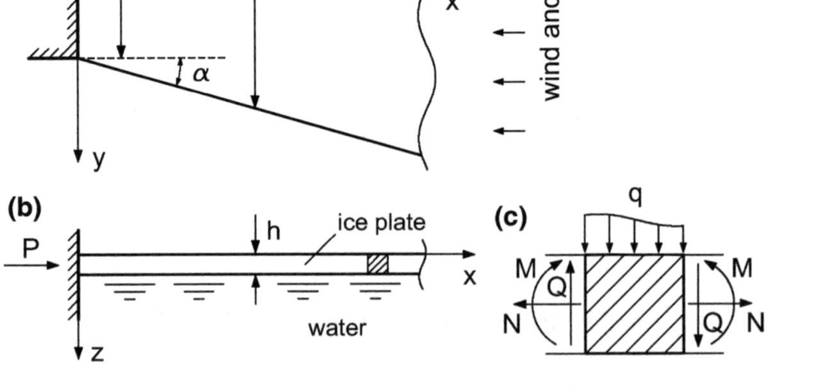

Fig. 4.2 Geometry of a wedge-shaped plate of floating ice interacting with a rigid structure of width b_0: **a** plane view, **b** plate cross-section, **c** definition of internal forces

y-direction; that is, the plate is bent cylindrically in the vertical plane Oxz. Accordingly, the problem is in fact reduced to that of a beam of thickness h and variable width $b(x)$ floating on water, and subjected to bending and axial compression (however, the dependence of the plate flexural rigidity on the Poisson ratio, in a form typical of plates, will be retained to account for the constraints in the lateral y-axis direction). In spite of the considerable simplifications, it is believed that the results obtained, at least for small wedge angles α, will not differ significantly from those that can be obtained by solving a fully two-dimensional plate problem, and therefore they will prove useful for engineering practice.

Since the problem is treated as a one-dimensional, we omit henceforth the subscripts in the notations of relevant internal forces, as shown in Fig. 4.2c. The equilibrium equation (4.5) now becomes

$$\frac{\mathrm{d}^2 M}{\mathrm{d}x^2} + N\,\frac{\mathrm{d}^2 w}{\mathrm{d}x^2} = \varrho_w g w + \tau \frac{\mathrm{d}w}{\mathrm{d}x}, \tag{4.8}$$

where $\tau = \tau_x$. The plane cross-section assumption implies a linear variation of axial strains $\epsilon_{xx}(x, z)$ and $\epsilon_{yy}(x, z)$ with z, with zero strains at the neutral surface. Supposing that the axial strain $\epsilon_{xx}(x, z)$ across the plate is proportional to the curvature κ_x of the middle surface of the deformed plate, one can express ϵ_{xx} as

$$\epsilon_{xx} = \kappa_x (z - z_0). \tag{4.9}$$

By Hooke's law, the axial strains are related to axial stresses by

$$\epsilon_{xx} = \frac{1}{E}\,(\sigma_{xx} - \nu\sigma_{yy}), \quad \epsilon_{yy} = \frac{1}{E}\,(\sigma_{yy} - \nu\sigma_{xx}), \tag{4.10}$$

in which the Young modulus is a function of z. To ensure the plate continuity in the lateral y-direction, the constrain $\epsilon_{yy} = 0$ is introduced. Due to this constraint, relations (4.10) supply the following expression for the axial stress σ_{xx} due to bending:

$$\sigma_{xx} = \frac{E(z)}{1 - \nu^2}\,\kappa_x (z - z_0), \tag{4.11}$$

which shows that the axial stress distribution across the plate is not linear in z because of E being a function of z. Using the latter expression for σ_{xx} in the bending moment definition (4.7), on obtains the relation

$$M = \kappa_x D_p\,, \tag{4.12}$$

in which D_p is the flexural rigidity of the plate defined by

$$D_p = \frac{1}{1 - \nu^2} \int_0^h (z - z_0)^2\, E(z)\mathrm{d}z. \tag{4.13}$$

The position of the neutral surface, z_0, can be determined from the condition that the resultant axial force N due to bending, obtained by integrating over $0 \leq z \leq h$ the stress σ_{xx} defined by (4.11), is zero. This yields

$$\int_0^h (z - z_0)\, E(z)\mathrm{d}z = 0. \tag{4.14}$$

For practical purposes, it can be assumed that Young's modulus varies linearly with the depth of ice. Let $E = E_0$ at the upper surface of ice ($z = 0$), and $E = \beta E_0$ at the ice base ($z = h$), with $0 \leq \beta \leq 1$. Then, for the linear variation of $E(z)$ between the two limit values, relation (4.14) gives

$$z_0 = h\, \frac{1 + 2\beta}{3(1 + \beta)}. \tag{4.15}$$

With z_0 defined by (4.15), the plate flexural rigidity (4.13) becomes

$$D_p = \frac{E_0 h^3}{12(1 - \nu^2)} \left[\frac{1 + 4\beta + \beta^2}{3(1 + \beta)} \right]. \tag{4.16}$$

Obviously, for $\beta = 1$ (E is uniform across the plate thickness), (4.15) yields $z_0 = h/2$, and the plate rigidity (4.16) is equal to $D = E_0 h^3/[12(1 - \nu^2)]$. On the other hand, for the limit case of $\beta = 0$, the neutral surface position is $z_0 = h/3$, and the plate flexural rigidity D_p reduces to 1/3 of that for the homogeneous ice plate.

Now, let us return to the plate equilibrium equation (4.8). Due to the small deflection assumption, implying $(dw/dx)^2 \ll 1$, the plate middle surface curvature (considered positive if the deformed plate is convex downward) can be approximated by $\kappa_x = -d^2w/dx^2$. By substituting the latter expression for κ_x in the bending moment definition (4.12), then using the resulting expression for M in (4.8), and finally by multiplying both sides of the ensuing equation by the plate width $b(x)$, one arrives at the following fourth-order differential equation for the plate deflection $w(x)$:

$$D_p\, b(x)\, \frac{\mathrm{d}^4 w}{\mathrm{d}x^4} + P\, \frac{\mathrm{d}^2 w}{\mathrm{d}x^2} + \varrho_w g b(x) w = 0, \quad 0 < x < \infty, \tag{4.17}$$

where the varying width of the wedged-shaped plate is defined by

$$b(x) = b_0 + 2x \tan\alpha. \tag{4.18}$$

In Eq. (4.17), $P = -Nb$ is the total compressive load acting on the whole cross-section of width b of the plate wedge (see Fig. 4.2). The load P is assumed to be independent of x in the region adjacent to the rigid structure located at $x = 0$; that is, in the region in which, for $\alpha > 0$, elastic buckling of the plate is expected to occur. Moreover, in derivation of (4.17), the horizontal traction τ has been neglected. Such a simplification seems to be permitted, since a typical magnitude of the buckling force P is much larger than the resultant tangential load due to wind and water drag stresses applied over a relatively small area of the ice cover in the immediate vicinity of the structure.

The differential equation (4.17) must be supplemented by boundary conditions at the ice–structure contact area at $x = 0$. Two types of these conditions are considered. The first type describes the case of a simply-supported edge of the plate, with zero deflection w and bending moment M, whereas the other type corresponds to the case of a rigidly-supported (clamped) end, with zero deflection and zero slope at $x = 0$. According to Sanderson (1988), the first case of the simply-supported plate at $x = 0$ is more realistic in practice, since the perfect contact between floating ice and a structure is rarely observed in the field. These two types of boundary conditions are expressed, respectively, by the relations

$$\text{simply-supported end:} \quad w(0) = 0, \quad \frac{d^2 w}{dx^2}(0) = 0, \tag{4.19}$$

$$\text{rigidly-supported end:} \quad w(0) = 0, \quad \frac{dw}{dx}(0) = 0. \tag{4.20}$$

Besides the boundary conditions (4.19) and (4.20), the regularity condition of the plate deflection being bounded at $x \to \infty$ has to be satisfied.

4.1.2 *Finite-Element Solution of the Problem*

The fourth-order differential equation (4.17), supplemented by the boundary conditions, either (4.19) or (4.20), describes an eigenvalue problem from which the buckling force P can be calculated. Because of the presence of the variable coefficient $b(x)$ in (4.17), no exact closed-form analytical solution is available for the general case of a wedge-shaped plate defined by $\alpha > 0$. However, in a particular case of $\alpha = 0$, corresponding to the case a plate of uniform width $b(x) = b_0$, Eq. (4.17) simplifies to that with constant coefficients. For such an equation, an analytical solution can be obtained in a straightforward manner, and has the form

$$P_0 = 2b_0\sqrt{\varrho_w g D_p}\,, \tag{4.21}$$

which is valid for both simply-supported and clamped boundary conditions at $x = 0$ (Kerr 1978). The corresponding buckling mode, described by $w(x) = \sin(k_0 x)$, yields the buckling half-wave length L_0 given by

$$L_0 = \pi/k_0\,, \quad \text{with} \quad k_0^4 = \frac{\varrho_w g}{D_p}\,. \tag{4.22}$$

An attempt to construct a semi-analytical solution of the problem defined by Eqs. (4.17), (4.19) and (4.20) for the case of $\alpha > 0$ was made by Kerr (1978).

However, it has proved that the approximate relations proposed in the latter paper are erroneous (Staroszczyk 2002), since these relations are inconsistent with the solution (4.21) for $\alpha = 0$, and they significantly overestimate the magnitudes of buckling forces for the case of a rigidly-supported plate; this will be demonstrated in Sect. 4.1.4.

In order to solve the eigenvalue problem defined by Eqs. (4.17), (4.19) and (4.20), one can apply a discrete method, for instance, a finite-difference method. It turns out, however, that then the discretization of the problem leads to the necessity of solving a generalized eigenvalue problem for non-symmetric matrices, which is much more difficult to solve numerically than a problem involving symmetric matrices. For this reason, a finite-element method (FEM) is applied, in which case all the matrices resulting from the discretization of the problem are symmetric, which significantly simplifies numerical calculations.

A weighted residual, or Galerkin, method (Zienkiewicz et al. 2005) is employed, in which the problem equation is satisfied in an integral mean sense. Following this method, the plate of variable width b is discretized along the x-axis by introducing one-dimensional finite elements. To ensure the continuity of both the plate deflection curve and its slope between elements, at each discrete nodal point two parameters are used to describe the plate deformations, namely w and $\mathrm{d}w/\mathrm{d}x$. Assuming that a given finite element is defined by two nodes i and j, located at x_i and x_j, respectively, we approximate the continuous function $w(x)$ within the element ij by means of four interpolation (shape) functions $\Phi_r(x)$ $(r = 1, \ldots, 4)$ as follows:

$$w(x) = w_i \Phi_1 + \theta_i \Phi_2 + w_j \Phi_3 + \theta_j \Phi_4\,, \tag{4.23}$$

where $\theta_i = (\mathrm{d}w/\mathrm{d}x)_i$ and $\theta_j = (\mathrm{d}w/\mathrm{d}x)_j$ are the nodal values of the plate slope. Let introduce a dimensionless local coordinate ξ defined by

$$\xi = \frac{x - x_c}{a}, \quad x_c = \frac{x_i + x_j}{2}, \quad -1 \le \xi \le 1, \tag{4.24}$$

where $2a$ is the length of the element ij. Then, the adopted shape functions are given by

$$\begin{aligned} \Phi_1 &= \frac{1}{4}\,(\xi - 1)^2(2 + \xi), \quad \Phi_2 = \frac{a}{4}\,(\xi - 1)^2(\xi + 1), \\ \Phi_3 &= \frac{1}{4}\,(\xi + 1)^2(2 - \xi), \quad \Phi_4 = \frac{a}{4}\,(\xi + 1)^2(\xi - 1). \end{aligned} \tag{4.25}$$

By multiplying Eq. (4.17), in turn, by a set of weighting functions, which in the Galerkin method are identical with the interpolation functions Φ_r, and then integrating the resulting relations over the plate length $x \ge 0$ and applying in the process Green's theorem (Zienkiewicz et al. 2005) to reduce by one the order of differentiation, one obtains a system of $2N$ linear algebraic equations, with N being the number of discrete nodes. This system of equations can be expressed in matrix form as

$$(\boldsymbol{K} + P\boldsymbol{B} + \boldsymbol{C})\,\boldsymbol{w} = \boldsymbol{O}, \tag{4.26}$$

where the vector $\boldsymbol{w} = (w_1, \theta_1, \ldots, w_i, \theta_i, w_j, \theta_j, \ldots, w_N, \theta_N)^T$ contains the values of the plate deflections and slopes at all nodal points of the discrete system. The plate stiffness matrix $\boldsymbol{K}$ and the matrices $\boldsymbol{B}$ and $\boldsymbol{C}$ are aggregated from respective element matrices $\boldsymbol{K}^e$, $\boldsymbol{B}^e$ and $\boldsymbol{C}^e$ in a way typical of the finite-element method (Zienkiewicz et al. 2005). The element matrices, each of the dimension 4×4, have the entries which for the element ij are defined by the following integrals:

$$
\begin{aligned}
K^e_{rs} &= D_p \int_{x_i}^{x_j} b(x) \frac{\mathrm{d}^2 \Phi_r}{\mathrm{d}x^2} \frac{\mathrm{d}^2 \Phi_s}{\mathrm{d}x^2} \mathrm{d}x, \quad B^e_{rs} = \int_{x_i}^{x_j} \Phi_r \frac{\mathrm{d}^2 \Phi_s}{\mathrm{d}x^2} \mathrm{d}x, \\
C^e_{rs} &= \varrho_w g \int_{x_i}^{x_j} b(x)\, \Phi_r \Phi_s \mathrm{d}x, \quad (r, s = 1, \ldots, 4; \;\; i, j = 1, \ldots, N),
\end{aligned}
\tag{4.27}
$$

in which the shape functions involved are given by (4.25).

Equation (4.26) defines a generalized eigenvalue problem from which the value of the buckling force P, the lowest eigenvalue of the problem, can be calculated, together with the associated eigenvector $\boldsymbol{w}$. To accomplish this, the matrix $\boldsymbol{B}$ is first decomposed into a product of the lower and upper triangular matrices, and then, by matrix inversions and multiplications, the general eigenvalue problem is reduced to a standard problem for a real and symmetric matrix. The latter problem can be solved by using standard numerical tools.

4.1.3 Numerical Simulations

The finite-element model presented in Sect. 4.1.2 was applied to simulate the interaction od a wedge-shaped ice cover with a rigid wall. The model included 200 finite elements of the uniform length for plates thinner than $h = 0.2$ m, and 100 elements otherwise, and the length of each element was assumed to be equal to $3h$. Thus, the length of the plate adopted to approximate the behaviour of a semi-infinite ice cover was equal to either $600h$ or $300h$. The material constants were taken to be those pertinent to isotropic granular T1 ice at temperature $-5\,^\circ$C. Hence, on account of relations (3.2) and (3.7), the Young modulus was $E = 8.99$ GPa, and the Poisson ratio was $\nu = 0.308$. The water density was assumed to be $\varrho_w = 10^3\,\mathrm{kg\,m^{-3}}$, and $g = 9.81\,\mathrm{m\,s^{-2}}$.

The results of simulations illustrating the dependence of the elastic buckling load P on the ice cover thickness h and the wedge angle α, for a rigid structure of the width $b_0 = 10$ m, are plotted in Fig. 4.3. The solid lines in the figure show the results obtained for a simply-supported edge of the plate at $x = 0$, and the dashed lines correspond to the case of a rigidly-supported (clamped) edge. The labels by the curves indicate the ice cover thickness in metres. The values of the buckling forces are normalized by the magnitude of the load P_0 defining the buckling force

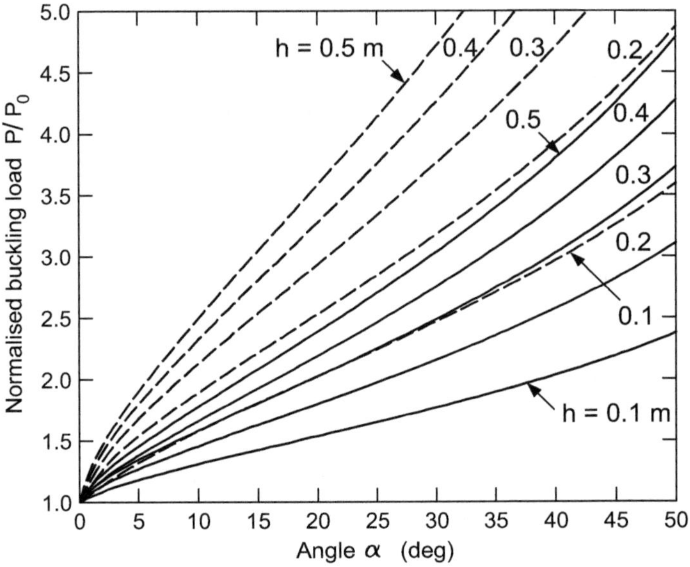

Fig. 4.3 Dependence of the normalized buckling load P/P_0 on the wedge angle α and the ice plate thickness h for simply-supported (solid lines) and rigidly-supported (dashed lines) edge conditions at $x = 0$, for a structure width $b_0 = 10$ m

of a parallel-sided plate of width b_0 and the respective thickness h, see Eq. (4.21); accordingly, the ratios P/P_0 are plotted in the graph.

Figure 4.4 presents the average pressures, defined by $P/(b_0 h)$, which are exerted on a structure by the floating ice cover. Again, the rigid wall is assumed to be 10 m long. The dependence of contact pressures on the plate geometry described by the angle α for different ice thicknesses h is illustrated for a plate which is simply-supported at its edge at $x = 0$. The horizontal dashed-dotted line in the figure indicates a pressure level at which ice fails by crushing. The value of the latter limit pressure, corresponding to the compressive strength of ice (see p. 51) was adopted as 5 MPa. Above this value, elastic buckling is unlikely to occur in the ice cover since the ice strength is exceeded earlier than the critical magnitude of compressive load P needed to buckle the ice is reached. The plots in Fig. 4.4 show that, for the range of most realistic wedge angles $30° \lesssim \alpha \lesssim 45°$, elastic buckling can occur only for the ice which is thinner than about 0.2 m. In such a case, the average contact stresses at the structure wall are smaller than the ice compressive strength of 5 MPa. For thicker ice, in turn, the stresses in ice at its buckling exceed the ice compressive strength. Hence, such thick ice crushes in a brittle manner, and the maximum contact pressures on the wall are equal to 5 MPa. When the plate is clamped at the wall, rather than simply-supported, then the limit ice thicknesses h above which no elastic buckling of ice can occur decreases to about 0.12 m.

Corresponding to the previous diagram is Fig. 4.5, showing the effect of weakening of ice with its increasing depth on the average contact pressures sustained by a

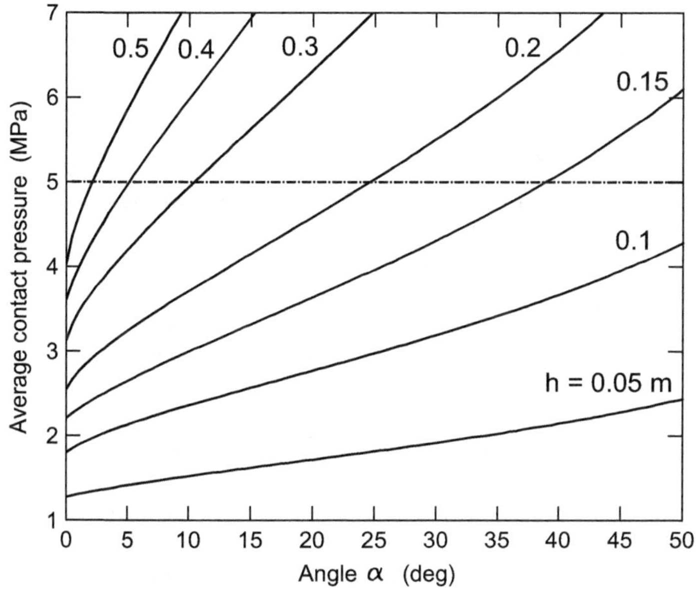

Fig. 4.4 Dependence of the average contact pressure on the wedge angle α and the ice sheet thickness h for simply-supported edge conditions at $x = 0$ and a structure width $b_0 = 10\,\mathrm{m}$. The dashed-dotted line indicates the compressive strength of ice

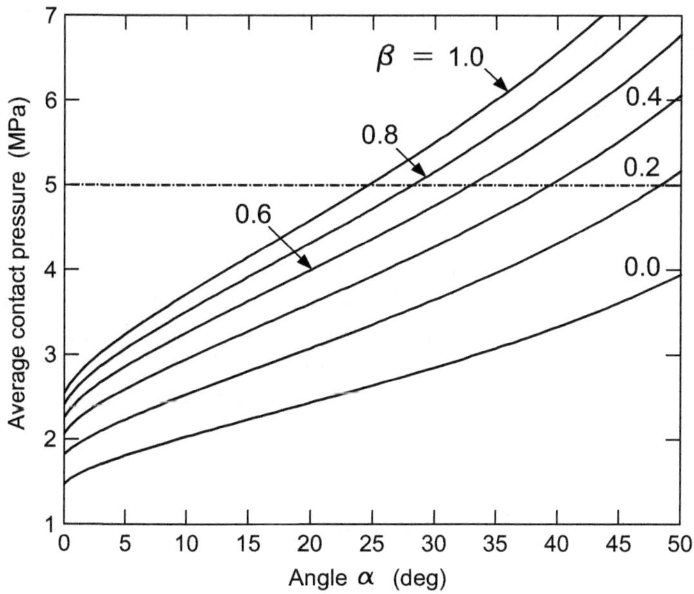

Fig. 4.5 Variation of the average contact pressure with angle α and the parameter β for the plate $h = 0.2\,\mathrm{m}$ thick and $b_0 = 10\,\mathrm{m}$ wide at $x = 0$, with a simply-supported edge. The dashed-dotted line indicates the compressive strength of ice

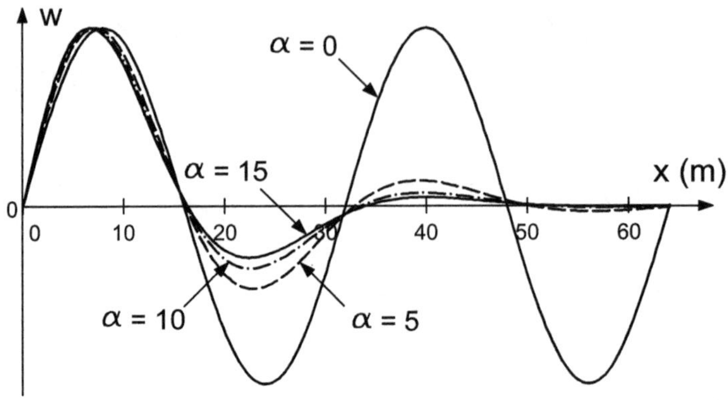

Fig. 4.6 Buckling deflection modes for various wedge angles α (given in deg), for the plate $h = 0.2\,\text{m}$ thick and $b_0 = 10\,\text{m}$ wide at $x = 0$, with a simply-supported edge

structure. The plots, presenting the predictions for the ice cover of thickness $h = 0.2\,\text{m}$ and the structure width $b_0 = 10\,\text{m}$, illustrate the dependence of the average contact pressure on the parameter β, the latter describing a linear variation of Young's modulus $E(z)$ from its maximum value E_0 to βE_0 between the top and the bottom surfaces of ice, respectively.

Figure 4.6 illustrates the shapes of buckling deflection modes for an ice plate of thickness $h = 0.2\,\text{m}$ and width $b_0 = 10\,\text{m}$, with simply-supported edge conditions at $x = 0$. The curves depict the plate deflections w for different wedge angles α, expressed in deg. For a plate of uniform width, defined by $\alpha = 0$, the fundamental buckling mode is described by the function $w(x) = \sin(k_0 x)$, with k_0 given by relation (4.22). For the adopted material constants, the latter parameter determines the buckling half-wave length $L_0 = \pi/k_0$ equal to 16.0 m. We note that the buckling mode length obtained by solving numerically the eigenvalue problem (4.26) agrees well with the length determined analytically. It can be also observed that for wedge angles $\alpha > 0$, even as small as 5°, the deflection of the ice plate in buckling attenuates rapidly with the distance x from the structure, showing that elastic buckling of the ice cover can take place only in a small region adjacent to the structure wall at $x = 0$. It also confirms that the length of the wedge-shaped plate adopted in the discrete model, equal to minimum $300h$ (i.e. 60 m for $h = 0.2\,\text{m}$) seems to be adequate, as the deflections w at the truncated edge of the buckled plate are negligibly small for $\alpha \geq 5°$.

4.1.4 Approximate Analytical Solution

As already noted earlier in this section, Kerr (1978) attempted to construct semi-analytical formulae that enable simple estimations of elastic buckling forces in wedge-shaped floating ice plates for the general case of $\alpha > 0$. The relations derived by Kerr are expressed by

$$\begin{aligned} &\text{simply-supported end:} \qquad P = c_1 k_0 D_p \, (k_0 b_0 + 2 \tan \alpha), \\ &\text{rigidly-supported end:} \qquad P = c_2 k_0 D_p \, (2 k_0 b_0 + 2 \tan \alpha), \end{aligned} \tag{4.28}$$

where k_0 is defined in (4.22), and $c_1 = 5.3$ and $c_2 = 8$ are constants. It can be easily proved, however, that for the angle $\alpha = 0$ (the case of a parallel-sided plate) relations (4.28) give $P/P_0 = c_1/4 = 1.325$ for a simply-supported plate, and $P/P_0 = c_2/2 = 4$ for a rigidly-supported plate, instead of unity in either case. Therefore, the above formulae by Kerr (1978) are apparently erroneous, with a particularly large error occurring in the case of a plate clamped at its edge at $x = 0$. For this reason, the finite-element results presented in the previous Sect. 4.1.3 have been used to construct alternative approximate relations to be used to estimate plate buckling forces with accuracy levels that are satisfactory for an engineer. Such an approximate relation, common for both simply and rigidly-supported plate ends, can be expressed in the following dimensionless form (Staroszczyk 2002):

$$\frac{P}{P_0} = 1 + \left(\frac{h}{h^*}\right)^{r_1} \left(\frac{b_0}{b_0^*}\right)^{-r_2} \left(r_3 \, \alpha + r_4 \, \alpha^2 + r_5 \, \alpha^3\right), \tag{4.29}$$

in which P_0 is defined by (4.21), and r_i $(i = 1, \ldots, 5)$ are coefficients. In Eq. (4.29), two characteristic scales are introduced: $h^* = 0.1\,\text{m}$ for the plate thickness, and $b_0^* = 10\,\text{m}$ for the width of a structure interacting with floating ice.

The coefficients r_i have been determined by correlating relation (4.29) with the finite-element results by using the method of least-squares. The correlations have been carried out for ice thicknesses h ranging from 0.05 to 0.5 m and structure widths b_0 ranging from 5 to 50 m, separately for the two types of boundary conditions at the ice–structure interface, defined by (4.19) and (4.20). The best results have been achieved with the two sets of the coefficients r_i listed in Table 4.1.

Figure 4.7 compares the predictions of the finite-element method (solid lines) with the approximations (4.28) (dotted lines) and (4.29) (dashed lines), on an example of the plate $h = 0.2\,\text{m}$ thick and $b_0 = 10\,\text{m}$ wide. It is immediately seen that, for the whole range of wedge angles $\alpha \leq 50°$, the results obtained by Kerr (1978), and subsequently repeated by Sanderson (1988), significantly overestimate elastic buckling forces for a plate which is clamped at $x = 0$; in the case of a simply-supported plate edge the FEM results and Kerr's estimates differ by about 20–30%. On the other hand, a good agreement between the results given by the proposed analytical approximation (4.29) and the finite-element results, for both simply-supported and clamped plate boundary conditions, is observed. For the adopted parameters ($h = 0.2\,\text{m}$ and $b_0 = 10\,\text{m}$), the maximum relative discrepancies between the FEM results and those determined by (4.29) are equal to 2.8% for the simply-supported plate and 4.0% for the clamped plate.

Table 4.1 Coefficients r_i for two types of boundary conditions at the edge $x = 0$

	r_1	r_2	r_3	r_4	r_5
Simple support	0.630	0.840	2.002	−1.959	1.681
Rigid support	0.590	0.786	4.276	−4.698	3.678

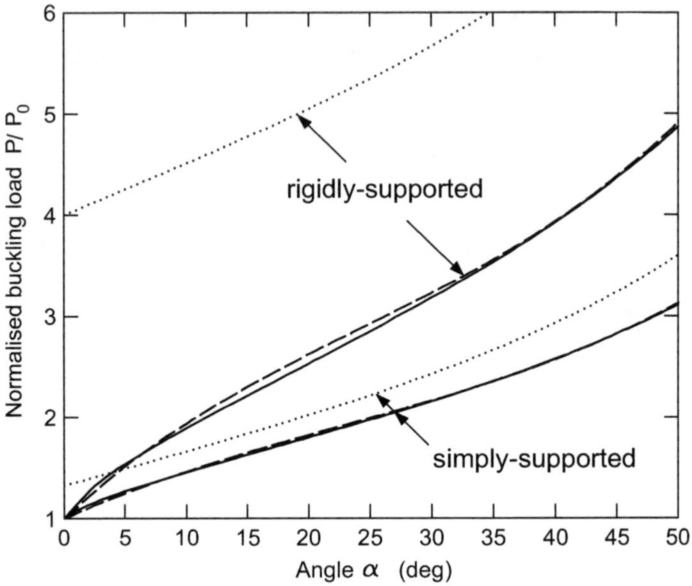

Fig. 4.7 Comparison of the finite-element results (solid lines) with the estimates given by Eq. (4.29) (dashed lines) for a floating ice sheet $h = 0.2$ m thick and $b_0 = 10$ m wide, for simply-supported and rigidly-supported edge conditions. Dotted lines show the results predicted by Eq. (4.28) proposed by Kerr (1978)

For other combinations of h and b_0 than that illustrated in Fig. 4.7 (the ranges $5\,\mathrm{m} \leq b_0 \leq 50\,\mathrm{m}$, $0.05\,\mathrm{m} \leq h \leq 0.5\,\mathrm{m}$, and $0^\circ \leq \alpha \leq 50^\circ$ have been explored), the maximum relative discrepancies between the FEM and approximate analytical predictions (4.29) are of similar order. As the results presented by Staroszczyk (2002) demonstrate, in the case of narrow plates ($b_0 = 5$ m), these discrepancies do not exceed 6.6% for simply-supported plates, and 9.5% for clamped plates. For much wider plates ($b_0 = 50$ m), in turn, the maximum relative differences are smaller, and equal to 3.8% and 5.7% for simply-supported and clamped plates, respectively. This shows that the proposed approximate formula (4.29) provides the predictions which can be regarded as sufficiently accurate for civil engineering applications.

4.2 Interaction of Creeping Ice with a Structure

As has already been pointed out on a few occasions in this book, the creep behaviour of sea ice is a dominant mechanism of its deformation in a wide range of stress, strain and strain-rate regimes. This section starts with the discussion of rheological models describing the creep behaviour of ice. Then, creep buckling of a floating ice plate is investigated, and key features of this mechanism are analysed in the context of the elastic buckling phenomenon considered in Sect. 4.1. Further in this section, based on the equations formulated in Sect. 4.1.1, a two-dimensional finite-element model is presented for an ice–structure interaction problem, in which the creep rheology is implemented. This model is applied to calculate the forces exerted by sea ice on a rigid structure of a rectangular planar cross-section; the results are shown in Sect. 4.2.4. Finally, in Sect. 4.2.5, the interaction of sea ice with a cylindrical structure is investigated. Hence, the ice equilibrium equations are formulated, and solved numerically, in cylindrical polar coordinates, and the results obtained for different rheological models for ice are compared to examine their effect on the predicted forces acting on the structure walls.

4.2.1 Rheological Models for Sea Ice

Various forms of the constitutive laws describing the rheology of creeping ice have been formulated, tested in numerical models and verified by field observations. Essentially, these laws include, among others, non-linearly viscous, viscous-plastic and elastic-viscous-plastic rheologies, all constructed with the purpose to describe as well as possible the complex mechanical response of sea ice to loading. It appears that two major classes of the rheological models are of the greatest significance to the sea ice modelling; these are: non-linearly viscous fluid models, which are formally and computationally simpler, but less realistic, and more realistic, and more complex, non-linearly viscous-plastic models, which describe distinct material responses below and above a critical level of a strain-rate invariant. The second class of models also comprises constitutive formulations which describe the failure of ice by other mechanisms than plasticity, for instance by brittle fracture.

The non-linearly viscous fluid models are usually based on a Reiner-Rivlin constitutive equation. The first model of this kind was introduced to the sea ice applications by Smith (1983), and was followed by Overland and Pease (1988) and a few other authors, including Gray and Morland (1994), Schulkes et al. (1998) and Morland and Staroszczyk (1998).

The viscous-plastic rheological models originate from the work by Hibler (1979). The main idea of these models is to use an elliptic plastic yield curve which restricts permissible stress states in the floating ice pack. For strain-rates below a critical value the stresses in ice are determined by a viscous flow relation, while for those above the critical value the stresses lie on a yield curve. To eliminate some drawbacks of the original formulation, various modifications of the model were subsequently proposed (Ip et al. 1991; Hibler and Ip 1995; Tremblay and Mysak 1997; Hibler 2001), including one in which non-physical elasticity was introduced (Hunke and Dukowicz 1997) in order to improve the numerical performance of the model. A simplified form of the viscous-plastic rheological model is the cavitating fluid model proposed by Flato and Hibler (1992). Comparisons between the predictions of the above two classes of constitutive laws are made by Schulkes et al. (1998), who have applied four different ice rheologies to simulate the behaviour of an ice cover.

In what immediately follows, we focus on the viscous-fluid and viscous-plastic rheologies, since these rheologies will be implemented in numerical models for sea ice, the predictions of which will be discussed in the further part of this chapter, and also later in Chap. 5.

Viscous Fluid Model

Commonly, viscous fluid (VF) rheological models are constructed as forms of a general, frame-indifferent Reiner-Rivlin constitutive law (Chadwick 1999), which expresses the Cauchy stress tensor $\boldsymbol{\sigma}$ in terms of the strain-rate tensor and its invariants. When applied to the sea ice pack, which is assumed to be stress-free in diverging flow (that is, when adjacent ice floes move away from each other), the Reiner-Rivlin law is expressed as

$$\boldsymbol{\sigma} = [\phi_1(\eta, \gamma)\boldsymbol{I} + \phi_2(\eta, \gamma)\boldsymbol{D}]\, H(-\eta). \tag{4.30}$$

In the above equation, $\boldsymbol{I}$ is the unit tensor, and $\boldsymbol{D}$ denotes the two-dimensional strain-rate tensor, the components of which are defined in terms of the components v_i of the ice velocity vector $\boldsymbol{v}$ by

$$D_{ij} = \frac{1}{2}\left(\frac{\partial v_i}{\partial x_j} + \frac{\partial v_j}{\partial x_i}\right) \quad (i, j = 1, 2). \tag{4.31}$$

In (4.30), η and γ are two invariants of $\boldsymbol{D}$, defined for the two-dimensional deformation by

$$\eta = \operatorname{tr} \boldsymbol{D}, \quad \gamma^2 = \tfrac{1}{2}\operatorname{tr}(\hat{\boldsymbol{D}}^2), \tag{4.32}$$

where $\operatorname{tr}(\cdot)$ denotes the trace of a tensor, and $\hat{\boldsymbol{D}}$ is the deviatoric strain-rate tensor given by

$$\hat{\boldsymbol{D}} = \boldsymbol{D} - \tfrac{1}{2}\eta\boldsymbol{I}. \tag{4.33}$$

In strain-rate components, the two invariants, the dilatation-rate η and the shear-rate invariant γ, are expressed by

$$\eta = D_{11} + D_{22}, \quad \gamma^2 = D_{12}^2 + \tfrac{1}{4}\left(D_{11} - D_{22}\right)^2. \tag{4.34}$$

The invariant η is positive in diverging flow, and negative in converging flow.

The function $H(\cdot)$, entering Eq. (4.30), denotes the Heaviside unit step function, which is zero for the negative values of its argument, and is unity for the positive values of the argument; hence,

$$H(-\eta) = \begin{cases} 1 & \text{for} \quad \eta < 0, \quad \text{i.e. in converging flow,} \\ 0 & \text{for} \quad \eta > 0, \quad \text{i.e. in diverging flow.} \end{cases} \tag{4.35}$$

In view of the above definition, the constitutive equation (4.30) gives $\boldsymbol{\sigma} = \mathbf{0}$ for the diverging flow ($\eta > 0$).

The two response functions ϕ_1 and ϕ_2, appearing in the general flow law (4.30), describe the material behaviour of the medium under applied deformation-rates. In order to express the viscous fluid constitutive law in a more conventional way, that is, in terms of viscosities, let first introduce the deviatoric stress $\boldsymbol{S}$, defined by

$$\boldsymbol{S} = \boldsymbol{\sigma} + p\boldsymbol{I}, \quad p = -\tfrac{1}{2}\,\mathrm{tr}\,\boldsymbol{\sigma}, \tag{4.36}$$

where p denotes a mean pressure in ice. Then, by taking the spherical and deviatoric parts of both sides of Eq. (4.30), one can determine the bulk and shear responses of the material, the former response associated with the action of pressure, and the latter with the action of the deviatoric stress (Morland and Staroszczyk 1998; Staroszczyk 2005). By introducing standard definitions of the bulk and shear viscosities, ζ and μ respectively, the viscous behaviour of ice can be described by

$$p = -\zeta\eta, \quad \boldsymbol{S} = 2\mu\hat{\boldsymbol{D}}, \tag{4.37}$$

and the material response functions can then be expressed as

$$\phi_1 = (\zeta - \mu)\eta, \quad \phi_2 = 2\mu. \tag{4.38}$$

With the above relations, the viscous flow law (4.30) becomes

$$\boldsymbol{\sigma} = [(\zeta - \mu)\eta\boldsymbol{I} + 2\mu\boldsymbol{D}]\, H(-\eta). \tag{4.39}$$

Note that this is a constitutive law for an ice pack, that is, a material consisting of a large number of floes and, therefore, having little ability to sustain tensile stresses on the scale of a pack. The latter feature is accounted for by the inclusion of the Heaviside function term in (4.39). When only a single, continuous, floe is considered, then this term should be omitted in the law.

Since the viscosities μ and ζ are the functions of the current deformation-rate invariants η and γ, Eq. (4.39) represents, in general, a non-linear constitutive law. In a particular case of $\zeta = \mu$, relation (4.39) simplifies to the form

$$\boldsymbol{\sigma} = 2\mu\boldsymbol{D}H(-\eta), \tag{4.40}$$

with a single viscosity measure. While this is not a realistic rheological model for the sea ice pack, it may have some value for numerical testing (Schulkes et al. 1998; Staroszczyk 2003).

An example of the rheological model based on the general Reiner-Rivlin viscous fluid law (4.30) is the model proposed by Overland and Pease (1988), sometimes referred to as the OP-rheology. In this constitutive model, the isotropic stress in ice (described by the response function ϕ_1) is assumed to depend on the ice thickness h, whereas the deviatoric stress (defined by the function ϕ_2) depends also on the shear-rate invariant γ (that is, there is no dependence on the dilatation-rate η). This specific form of the rheological law was investigated by Schulkes et al. (1998), who used it in finite-element simulations of a large sea ice pack behaviour under the action of wind. Another, more complex, form of the non-linearly viscous fluid rheological model was developed by Morland and Staroszczyk (1998). In their model, possible stress states in sea ice are assumed to lie within an envelope in the principal stress plane, which makes it a little similar to viscous-plastic rheological models considered further in this section.

Viscous-Plastic Model

In the viscous-plastic rheological models, often referred to as the VP-models, introduced to the sea ice dynamics by Hibler (1979), it is assumed that floating ice has zero tensile strength, and when it is subject to compressive stresses, which occur during converging flow of the ice pack, the ice can behave in two ways, depending on the current rate of its deformation. Below a certain critical level of strain-rate, the ice behaves as viscous fluid, whereas above that critical strain-rate it deforms by plastic yield. The limit stress state in the ice in its plastic flow is defined by a yield curve, the shape of which prescribes admissible stresses in the material. When viewed on the principal stress plane, the stress states on the yield curve occur during plastic flow, while those inside the yield curve occur during viscous flow.

Various shapes of the yield envelopes have been proposed and used in simulations so far; some of them, plotted in principal stress axes (σ_1, σ_2), with positive values denoting tension, are presented in Fig. 4.8. Usually, an elliptic curve has been adopted (Hibler 1979; Ip et al. 1991), though other shapes, including tear-drop curves (Rothrock 1975; Morland and Staroszczyk 1998) have also been tried. For comparison, the straight lines, representing the Coulomb-Mohr rheology widely applied for granular media, are also plotted in the figure.

In the original formulation of the viscous-plastic model (Hibler 1979) zero-tensile strength of ice was postulated. It turned out, however, that such an assumption gives rise to some problems with numerical stability during simulations, since an arbitrarily small change in the divergence rate through zero results in a large change in the creep response of the medium. For this reason, to remove this source of numerical instability, the original viscous-plastic model has been modified in such a way that a small tensile strength of ice is allowed in diverging flow regime (Staroszczyk 2006). Hence, two strength parameters: $P_1 > 0$ for compression and $P_2 > 0$ for tension, with $P_1 \gg P_2$, are used in the modified viscous-plastic model to describe the properties of sea ice. In this manner, without introducing any non-physical diffusive terms in the flow equations (which is a common practice in numerical modelling of sea ice), the stability of a numerical method is significantly improved. The adopted elliptic yield curve, plotted in the two-dimensional principal stress space, is illustrated in Fig. 4.9 (the solid line). The tensile stress states are those in the first quadrant (near point B) of the (σ_1, σ_2) plane.

The elliptic yield curve presented in Fig. 4.9 is specified by the equation

$$F(\sigma_1, \sigma_2) = (\sigma_1 + \sigma_2 + P_1 - P_2)^2 + e^2(\sigma_1 - \sigma_2)^2 - (P_1 + P_2)^2 = 0, \quad (4.41)$$

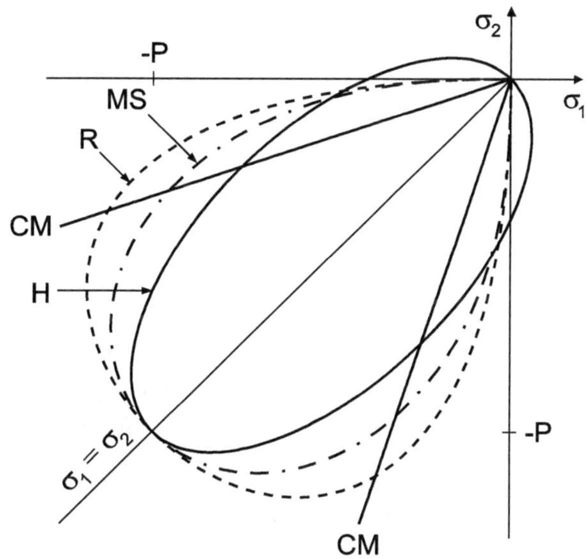

Fig. 4.8 Comparison of different yield curves for viscous-plastic rheology: H—elliptic curve by Hibler (1979), R—tear-drop curve by Rothrock (1975), MS—teardrop curve by Morland and Staroszczyk (1998), CM—Coulomb–Mohr straight lines. *P* is the compressive strength of ice

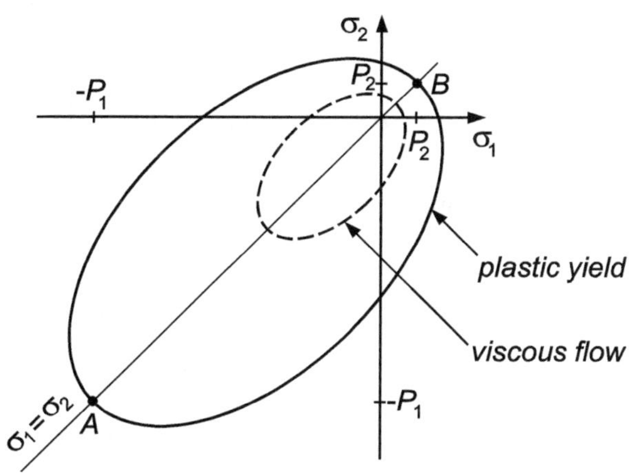

Fig. 4.9 Elliptic yield curve (solid line) with the ice compressive strength P_1 (stress point A) and the small tensile strength P_2 (point B), and a smaller ellipse (dashed line) describing stress states in viscous flow

where σ_1 and σ_2 are the principal stress components, and $e \geq 1$ defines the ellipse eccentricity (the ratio of the major to the minor axis lengths of the ellipse). In physical terms, e defines the ratio of the maximum shear yield stress in the material to the maximum mean pressure: the larger value of the parameter e, the smaller is the shear resistance of ice, with $e \to \infty$ describing the cavitating fluid rheology.

Following Hibler (1977), it is assumed that sea ice during its yield (when the stress lies on the yield curve) obeys a normal flow rule, implying that the principal strain-rate vector is normal to the yield curve $F(\sigma_1, \sigma_2)$. Hence, an associated flow law is applied, expressed in the form:

$$D_{ij} = \lambda \frac{\partial F(\sigma_{ij})}{\partial \sigma_{ij}} \Bigg|_{F=0}, \quad i, j = 1, 2, \quad \lambda > 0, \tag{4.42}$$

where λ is a function of strain-rate. It can be shown (Staroszczyk 2006) that the latter function is given by the relation

$$\lambda = \frac{\Delta}{4(P_1 + P_2)}, \tag{4.43}$$

with

$$\Delta^2 = \eta^2 + 4\gamma^2/e^2, \quad \Delta \geq 0. \tag{4.44}$$

Since Δ is a function of two invariants η and γ, already defined by formulae (4.32) or (4.34), it itself is an invariant of the strain-rate tensor $\boldsymbol{D}$ (recall that the dimensionless rheological parameter e is a constant).

Substitution of the definition (4.43) for λ into the plastic flow rule (4.42), with $F(\sigma_1, \sigma_2)$ defined by the formula (4.41), gives expressions for the strain-rates in terms of the stresses. Inversion of the latter expressions prescribes the stresses σ_{ij} in terms of the strain-rates D_{ij}. These relations, when set in the tensor form, give the following frame-indifferent flow law:

$$\boldsymbol{\sigma} = 2\mu \boldsymbol{D} + \left[(\zeta - \mu)\eta - \frac{1}{2}(P_1 - P_2) \right] \boldsymbol{I}, \tag{4.45}$$

where the parameters ζ and μ are defined by

$$\zeta = \frac{P_1 + P_2}{2\Delta}, \quad \mu = \frac{\zeta}{e^2} = \frac{P_1 + P_2}{2\Delta e^2}. \tag{4.46}$$

Comparison of the viscous-plastic flow law (4.45) with the viscous fluid flow relation (4.39) shows that the parameters ζ and μ can be identified as the bulk and shear viscosities of ice, respectively, both being now functions of the ice strength parameters P_1 and P_2 and the strain-rate invariant Δ.

The law (4.45), in conjunction with the viscosity definitions (4.46), describes the behaviour of ice during plastic yield. The latter is assumed to occur when the strain-rate invariant Δ reaches a certain critical value, denoted by Δ_c; that is, plastic deformations take place when $\Delta \geq \Delta_c$. Below the critical value, when $\Delta < \Delta_c$, ice is supposed to undergo viscous deformations, with constant (that is, independent of the current strain-rates) viscosities ζ and μ. Following Hibler (1979), the magnitudes of the latter two parameters are set to be equal to the viscosities at the onset of plastic yield. Hence, on account of (4.46), they are defined by

$$\zeta_m = \frac{P_1 + P_2}{2\Delta_c}, \quad \mu_m = \frac{P_1 + P_2}{2\Delta_c e^2}. \tag{4.47}$$

The above parameters ζ_m and μ_m can be considered as the upper bounds on the viscosities of sea ice. The critical level Δ_c of the strain-rate invariant can be inferred from in situ observations of the sea ice behaviour. A typical value is $\Delta_c = 2 \times 10^{-9}\ \mathrm{s}^{-1}$ (Hibler 1979; Schulkes et al. 1998).

When sea ice pack deforms in the viscous regime, which occurs for $\Delta < \Delta_c$, then, following the idea of Ip et al. (1991), the ice strength parameters P_1 and P_2 must be scaled down by a factor Δ/Δ_c as follows:

$$P_1 \to \frac{\Delta}{\Delta_c} P_1, \quad P_2 \to \frac{\Delta}{\Delta_c} P_2, \quad \Delta < \Delta_c\,, \tag{4.48}$$

to avoid a physically unsound response, in which isotropic non-zero stress arises in the ice pack in the absence of any deformation-rate. Therefore, two distinct relations must be used to describe two distinct deformation regimes, plastic and viscous, depending on the current strain-rate magnitude relative to its critical level:

$$\boldsymbol{\sigma} = \begin{cases} 2\mu \boldsymbol{D} + \left[(\zeta - \mu)\eta - \frac{1}{2}(P_1 - P_2)\right] \boldsymbol{I} & \text{for} \quad \Delta \geq \Delta_c\,, \\ 2\mu_m \boldsymbol{D} + \left[(\zeta_m - \mu_m)\eta - \frac{1}{2}\frac{\Delta}{\Delta_c}(P_1 - P_2)\right] \boldsymbol{I} & \text{for} \quad \Delta < \Delta_c\,. \end{cases} \tag{4.49}$$

It can be shown that in the viscous flow, when $\Delta < \Delta_c$, the stresses predicted by the second equation (4.49) lie on an ellipse, the centre of which approaches the stress origin, and the major and minor axes decrease monotonically to zero, as $\Delta \to 0$. One such an ellipse is plotted in Fig. 4.9 (see the dashed line).

The constitutive model represented by Eq. (4.49) has four free parameters: P_1 and P_2 defining, respectively, compressive and tensile strength of ice, Δ_c prescribing the critical strain-rate at which plastic yield starts, and e defining, through (4.46) and (4.47), the ratio of the shear to bulk viscosities of ice.

4.2.2 Creep Buckling of Floating Ice

In Sect. 4.1 the mechanism of elastic buckling of a floating ice plate is considered. It has been shown that this failure mechanism is possible to occur only in relatively thin ice sheets, of thicknesses usually not exceeding 0.3–0.4 m, depending on the geometry of the plate and the type of boundary conditions at the ice–structure contact zone (see Fig. 4.4 on p. 71). There is, however, a vast field evidence (Sanderson 1988) that under certain conditions, in particular at very low horizontal velocities of the floating ice cover, ice sheets significantly thicker than 0.5 m, and sometimes even more than 1.0 m, are also susceptible to the out-of-plane buckling. Typically, during late Arctic spring, when ice becomes softer and undergoes thermal expansion, buckles form in floating ice sheets over the periods of up to several days, until tensile cracks develop at the surface of ice, leading to its gradual failure. Similar buckle features occur when ice is pushed against a vertical structure at very low loading levels.

Certainly, the reason for such a behaviour of floating ice is its creep, which is substantial comparing with other materials encountered in civil engineering. At typical stress levels of 1 MPa, the time required for creep strains to exceed elastic strains in ice is about one minute (Mellor 1980; Sanderson 1988). This clearly indicates that not only elastic, but also, and first of all, creep (viscous) effects in ice must be taken into account to properly determine realistic contact forces between floating ice and an engineering structure. That is, the maximum forces occurring in ice during its creep buckling must be found, as these are the forces which can be exerted by ice on a structure. Relatively little research has been devoted to this topic so far (Sjölind 1985; Sanderson 1988), and this is limited to the problems of plates of a uniform width.

Here an extension of the analyses carried out in the two latter papers is presented, in which a floating plate having in the horizontal plane a shape of a truncated wedge is considered. This is the same geometry as that adopted in Sect. 4.1 for analysing the mechanism of the elastic buckling of ice (see Fig. 4.2 on p. 64). Accordingly, the same Eqs. (4.1) to (4.7) from Sect. 4.1 are used to describe the equilibrium balances of forces acting on the plate in the horizontal and vertical planes. The significant difference is the constitutive description of the material: instead of the equations of elasticity (4.10), the viscous flow law (4.39) is applied to express stresses in ice in terms of deformation-rates. An important factor now is temperature, since the viscosity of ice is strongly temperature-dependent, making the material mechanically inhomogeneous across the plate thickness.

The equation of equilibrium of the ice sheet floating on water, under the combined action of axial compression and bending, can be derived by the method analogous to that described Sect. 4.1 for the case of the elastic response of ice. A simplified form of this equation for a wedge-shaped plate is obtained by treating the latter as a beam of its width varying with x. For the viscous behaviour of ice, the equilibrium equation has the form (Staroszczyk and Hedzielski 2004):

$$Rb(x)\,\frac{\partial^4 \dot{w}}{\partial x^4} + P\,\frac{\partial^2 w}{\partial x^2} + \varrho_w g b(x) w = 0, \quad 0 < x < \infty, \tag{4.50}$$

where $P = -Nb$ is the compressive force acting along the x-axis, and $b(x)$ is the varying plate width in the lateral y-direction. The quantity R describes the flexural viscous behaviour of the ice plate and is defined by

$$R = \int_0^h \zeta_a(z)\, z\, (z - z_0) \mathrm{d}z, \tag{4.51}$$

where $\zeta_a = \mu + \zeta$ denotes the axial viscosity of ice, with z_0 denoting the position of a neutral plane in the vertical cross-section of the plate. Comparison of Eq. (4.17) on p. 66 with (4.50) shows that they differ only in the first term, in which, instead of the elastic plate rigidity D_p, the parameter R is used, and instead of the plate deflection w its time-derivative $\dot{w}$ now appears.

The temperature-dependence of the viscous properties of sea ice is described by the three Eqs. (3.18) to (3.20) on p. 40, defining a dimensionless factor $a(T)$ (Smith and Morland 1981) that is used to scale the viscosities with the ice temperature. For instance, the change of ice temperature from -1 to $-5\,°\mathrm{C}$ increases the ice viscosity by a factor of about 3.5. Since temperature variations of such a magnitude are quite usual in sea ice due to diurnal (24 h) cycles of heating and cooling, this clearly indicates how substantially the creep properties of ice can change over relatively short time scales. The solution of the heat conduction equation

$$\frac{\partial T}{\partial t} = k \frac{\partial^2 T}{\partial z^2}, \tag{4.52}$$

where $k = 1.15 \times 10^{-6}\,\mathrm{m}^2\,\mathrm{s}^{-1}$ is the thermal diffusivity coefficient for ice, shows that a free surface temperature perturbation during 24-h temperature cycles is attenuated by a factor of 10 at a depth of ice of about 0.4 m (more rapid temperature variations decay faster). Hence, it may be assumed that, typically, the daily temperature changes affect only the upper layer of thick sea ice.

Before proceeding further, it is useful to realize an essential difference between the mechanisms of elastic and creep buckling. Elastic buckling occurs instantly after a critical load has been reached, and is followed by unstable failure of ice. In contrast, creep buckling is a rather slow time-dependent process which occurs at any load level, and leads to the failure of ice only if sufficiently large strain-rates (and hence stresses reaching the flexural strength of ice) develop in the medium.

Analytical Results for a Uniform-Width Plate

In general, the solution of the ice plate equilibrium equation (4.50) for the plate deflection $w(x, t)$, with the variable coefficient $b(x)$, is possible only by an approximate method. An analytic solution is possible only in a particular case of a plate of uniform width $b(x) = b_0$, when (4.50) simplifies to the equation with constant coefficients

$$Rb_0 \frac{\partial^4 \dot{w}}{\partial x^4} + P \frac{\partial^2 w}{\partial x^2} + \varrho_w g b_0 w = 0. \tag{4.53}$$

Unlike elastic buckling, creep buckling requires an initial perturbation in the plate deflection w; this perturbation will subsequently evolve under applied loading. However, not any initial buckle $w(x, 0)$ will grow with time under a given load level P. In order to prove this, let re-write Eq. (4.53) in the form

$$Rb_0 \frac{\partial^4 \dot{w}}{\partial x^4} = q(x, t), \tag{4.54}$$

where

$$q(x, t) = -\left(P \frac{\partial^2 w}{\partial x^2} + \varrho_w g b_0 w \right). \tag{4.55}$$

The expression $q(x,t)$ in (4.54) can be treated as a transverse load depending on the axial force P and the current plate deflection $w(x,t)$. The existing deflection will grow with time only if $q(x,t) > 0$, and, reversely, it will decay with time if $q(x,t) < 0$; a stationary state, with w not evolving, occurs for $q(x,t) = 0$. Assuming that (4.55) can be solved by the method of separation of variables, and adopting the boundary conditions $w(0,t) = 0$ and $\partial^2 w/\partial x^2(0,t) = 0$ (a simply-supported plate at its edge $x = 0$), a general solution for the plate deflection w can be expressed in the form

$$w(x,t) = A(t)\sin(\pi x/L), \tag{4.56}$$

where $A(t)$ is a time-dependent buckle amplitude, and L is an arbitrary half-wavelength of a buckle. By substituting (4.56) into (4.55), a critical length of a buckle half-wave, denoted by L_c, can be determined as:

$$L_c = \pi\sqrt{\frac{P}{\varrho_w g b_0}}\,. \tag{4.57}$$

This critical length L_c determines the longest buckling half-wave, the amplitude of which can increase with time. Any existing buckles of lengths $L > L_c$ will decrease with time, as long as P is not increasing.

The length of a buckle (for $L < L_c$) affects the rate of growth of its current amplitude. It is supposed here that the amplitudes $A(t)$ of creeping buckles increase in an exponential manner, that is

$$A(t) = w_0 \exp(t/\tau), \tag{4.58}$$

where w_0 is an initial small deflection amplitude of a given buckle, and τ is a time constant. On inserting relation (4.58) into (4.56), and then substituting the resulting expression for $w(x,t)$ into the differential equation (4.53), the following relation is obtained:

$$\frac{1}{\tau} = \frac{1}{R}\left[\frac{P}{b_0}\left(\frac{L}{\pi}\right)^2 - \varrho_w g\left(\frac{L}{\pi}\right)^4\right], \tag{4.59}$$

which describes the growth-rate parameter τ in terms of the buckle length L and the axial load P. From among all possible perturbations of different lengths L, the fastest growing is the one for which τ attains the minimum value. By differentiating (4.59) with respect to L and setting it to zero, one can find that τ is minimized for the buckle half-wavelength L_0 given by

$$L_0 = \pi\sqrt{\frac{P}{2\varrho_w g b_0}}\,. \tag{4.60}$$

The corresponding growth-rate parameter τ, obtained by substituting (4.60) into (4.59), is expressed by

$$\tau_0 = \frac{4R\varrho_w g b_0^2}{P^2}\,. \tag{4.61}$$

Like the critical buckle length L_c, the length L_0 of the fastest-growing buckle depends on the load magnitude P, but does not depend on the viscous properties of the ice plate described by the parameter R. As the creep deformation of the plate develops from its initial state with small perturbations of various lengths L, the buckle of the half-wavelength L_0, with the largest growth-rate defined by τ_0, gradually becomes the dominant buckling mode.

By comparing expressions (4.57) and (4.60) one can note that, independently of the loading level P, the critical and dominant buckle half-wavelengths remain always at a constant ratio given by

$$\frac{L_c}{L_0} = \sqrt{2}\,. \tag{4.62}$$

Numerical Results for a Wedge-Shaped Plate

The fourth-order in space, and first-order in time, partial differential equation (4.50), which describes the creep behaviour of a wedge-shaped floating ice sheet in response to the compressive load P, has been solved numerically by applying a finite-element method. Essentially, the same discretization scheme is used as that employed in Sect. 4.1.2 for solving the problem of elastic buckling of a wedge-shaped plate. Hence, linear finite elements are used for the discretization of the plate along the x-axis, with the approximation method described by Eq. (4.23), and the element shape functions given by (4.25). One significant difference is that now, instead of the stiffness matrix $\boldsymbol{K}$ with components depending on the elastic flexural rigidity D_p, as defined by Eq. (4.27), a similar in structure damping matrix is formed, with its components depending on the parameter R involving the ice viscosity. The details are omitted here; they will be addressed in the next Sect. 4.2.4, devoted to the finite-element solution of a fully two-dimensional problem of creep behaviour of a floating ice sheet.

In numerical simulations, 400 finite elements of the same length, equal to $1.5\,h$, were used. Thus, the behaviour of a semi-infinite plate was approximated by the plate of the finite length of $600\,h$. The axial viscosity of ice at the melting point was adopted to be $\zeta_a = 1 \times 10^{11}\,\mathrm{kg\,m^{-1}\,s^{-1}}$. The results presented below were obtained for ice temperature equal to $-2\,°\mathrm{C}$ at the top surface, and $0\,°\mathrm{C}$ at the bottom surface of the plate, with the ice viscosities adjusted accordingly across the plate depth to account for the temperature dependence of creep properties of ice. The elastic constants, Young's modulus E and Poisson's ratio ν, were equal to 9.0 GPa and 0.31, respectively (they were needed to evaluate the magnitude of the elastic buckling force, P_e, for the plate). The flexural strength of ice was assumed to be $\sigma_f = 0.2$ MPa, corresponding to the ice of about 10% porosity.

The initial perturbed small deflection of the plate was adopted as a sum of twenty harmonic components, given by

$$w_0(x) = \sum_{i=1}^{20} \pm w_0^{(i)} \sin\left(\frac{i\pi x}{L}\right), \tag{4.63}$$

where the signs ($\pm$) were selected at random, and all the component amplitudes $w_0^{(i)}$ were equal and such that the maximum initial deflection was $w_0 = 0.001$ m. L, defining the length of the longest initial perturbation, was chosen to be three times the length L_0 of the dominant buckle half-wavelength for a plate of uniform width. In this way, the initial deflection $w_0(x)$ includes two components which are longer than the critical half-wavelength L_c determined by (4.57). In the simulations, the value of the compressive load P exerted on the floating plate was normalized by the magnitude of the force P_e causing elastic buckling of the plate; the latter force was calculated by using the method presented in Sect. 4.1.

Figure 4.10 illustrates the time variation of the deflection $w(x, t)$ of the plate of a unit width and the thickness $h = 0.2$ m, subjected to the compressive axial force $P = 0.1\,P_e$. The plots show how the plate vertical displacements, plotted at the intervals of 1.25 h (hours), gradually evolve from the initial, random distribution of small perturbations, into a regular pattern which, with increasing time, is more and more dominated by the buckling mode of the length L_0 defined by (4.60). The evolution of the plate deflection $w(x, t)$ from its initial state, prescribed by (4.63), is followed up to the time t_c, called the critical time, at which the tensile stress at any point in the plate exceeds the value of the ice flexural strength σ_f and the plate begins to fail due to the propagation of tensile cracks. The deflection of the plate at the critical time $t = t_c = 10.55$ h is plotted in Fig. 4.10b by the solid line. We note that the maximum deflections $w(x, t_c)$ at the onset of the plate failure are equal to about $h/2$. For comparison, the results of the analytic solution, indicated by the solid circles, are also presented in Fig. 4.10b to demonstrate the accuracy of the finite-element solution.

Figure 4.11 illustrates the effect of the in-plane axial load magnitude P/P_e on the ice displacement at the failure times t_c. The results, obtained for the plate of a unit width and the thickness $h = 0.2$ m, show that while the maximum plate deflections $w(x, t_c)$ decrease by a factor of about two with a fourfold increase in the load level, the values of the critical time at which the ice cover starts to fail change with the normalized load very substantially, decreasing by a factor of about 18 for the same, fourfold increase in loading.

The values of the critical time t_c required to fail a floating ice sheet due to its creep deformation started from initial, small-amplitude imperfections, are plotted in Figs. 4.12 and 4.13 as functions of the angle α defining the in-plane geometry of the truncated wedge (see Fig. 4.2 on p. 64). Fig. 4.12 illustrates, for the structure width $b_0 = 10$ m and the ice cover thickness $h = 0.2$ m kept constant, the dependence of the critical time t_c on the normalized axial load P/P_e (the corresponding plate deflections for selected ratios P/P_e and $\alpha = 0$ are shown in Fig. 4.12).

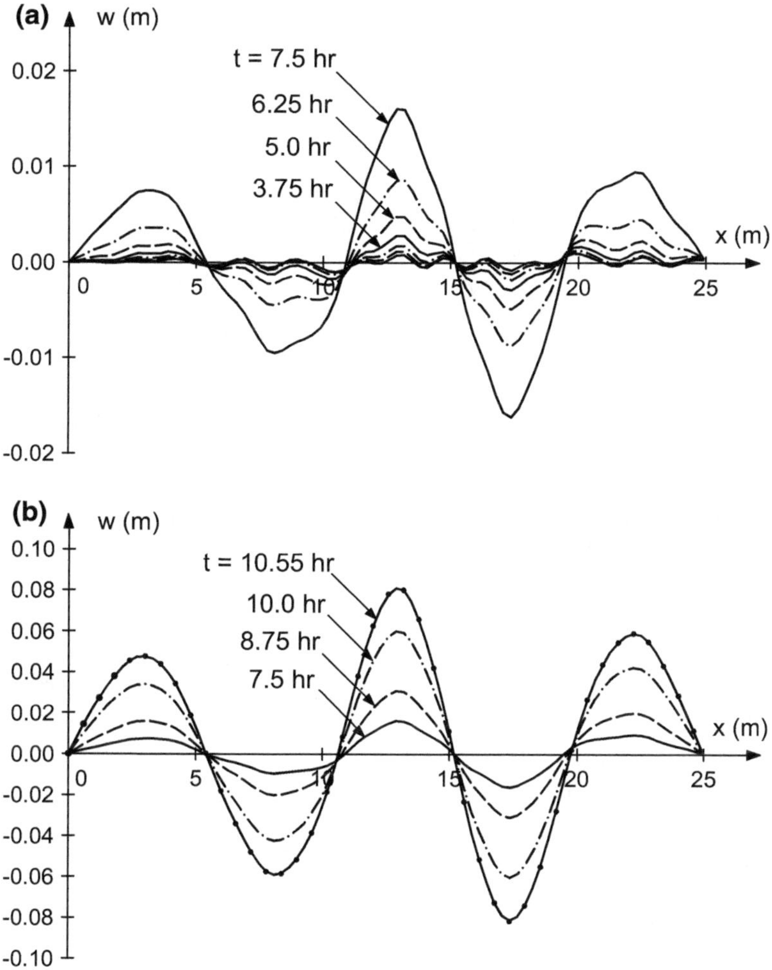

Fig. 4.10 Evolution of the deflection $w(x, t)$ of a uniform-width plate of thickness $h = 0.2$ m under the axial load $P/P_e = 0.1$: **a** for $t \leq 7.5$ h, **b** for $t \geq 7.5$ h. The solid circles in (**b**) show the results of the analytic solution for the critical time $t = t_c = 10.55$ h. Reprinted from Staroszczyk and Hedzielski (2004), Fig. 2, with permission of the Institute of Fundamental Technological Research of the Polish Academy of Sciences

Figure 4.13 displays, at the constant load $P/P_e = 0.1$, the variation of t_c for different plate thicknesses h. One can note that for thinner ice plates the values of the critical time initially slightly increase with the increasing angle α, while for thicker ice the values of t_c decrease monotonically with α.

More results obtained by the discrete method described above and illustrating the mechanism of ice creep buckling can be found in the paper by Staroszczyk and Hedzielski (2004).

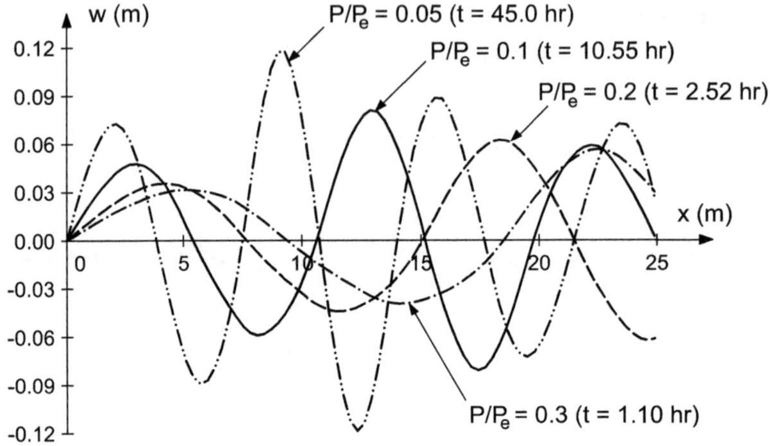

Fig. 4.11 Deflection of a uniform-width plate at the critical time $t = t_c$ (expressed in hours) as a function of the normalized load P/P_e, for the ice thickness $h = 0.2$ m. Reprinted from Staroszczyk and Hedzielski (2004), Fig. 3, with permission of the Institute of Fundamental Technological Research of the Polish Academy of Sciences

Fig. 4.12 Variation of the critical time t_c (given in hours) with the angle α and the normalized axial load P/P_e, for the ice thickness $h = 0.2$ m and the structure width $b_0 = 10$ m. Reprinted from Staroszczyk and Hedzielski (2004), Fig. 5, with permission of the Institute of Fundamental Technological Research of the Polish Academy of Sciences

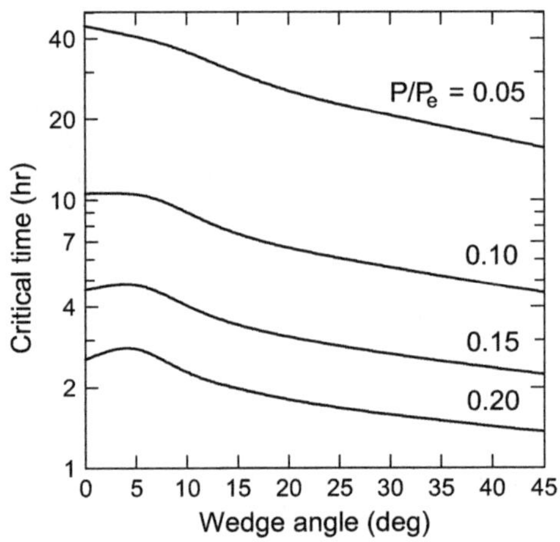

4.2.3 Ice Plate Failure Due to Its Thermal Expansion

In the preceding part of this section it was tacitly assumed that the in-plane axial forces N (and hence P) which caused the creep buckling of floating ice were generated by stresses arising on the ice surface due to the action of wind. Further, it was assumed that the temperature of ice did not change during the process of ice creep; that is, the properties of ice (in the first place its viscosity) which depend on temperature remained constant in time. Now we proceed to the problem in which the forces driving the creep buckling of floating ice are of a different origin—they are caused

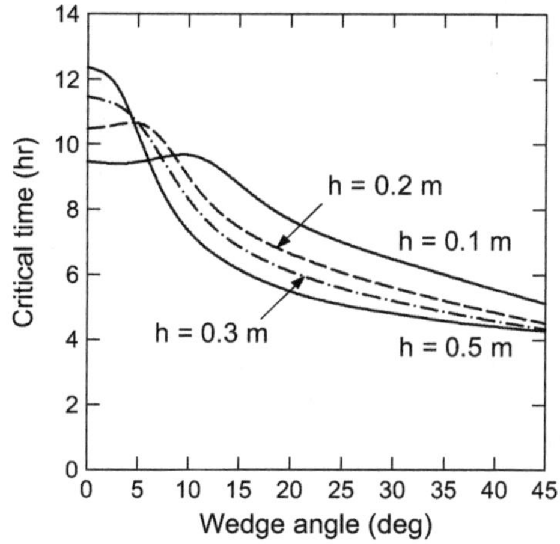

Fig. 4.13 Variation of the critical time t_c (given in hours) with the angle α and the ice thickness h, for the normalized load $P/P_e = 0.1$ and the structure width $b_0 = 10\,\text{m}$. Reprinted from Staroszczyk and Hedzielski (2004), Fig. 6. Copyright by the Institute of Fundamental Technological Research of the Polish Academy of Sciences

by the phenomenon of thermal expansion of ice due to its heating at the free surface of the ice cover. In order for such buckling force to develop in ice, the latter must be somehow constrained in the lateral (horizontal) directions. It is assumed here, that the lateral deformation of ice is prohibited by vertical, rigid walls representing elements of an engineering structure. Certainly, of the main interest are then the magnitudes of the forces exerted by ice on the constraining walls, and the evolution of these forces as the ice creep deformation progresses until the instant of the ice failure due to its flexural fracture. The evolution of the thermally-induced forces within the ice plate is not only due to the rise in temperature at the ice top surface, but also due to the vertical heat transfer through the ice cover from its top to the bottom. The latter process results in the variation, in time and space, of the elastic and viscous properties of the material; therefore, the plate of ice cannot be treated as a homogeneous, since its mechanical properties vary with depth.

Thermal Creep Plate Buckling Problem Formulation

The problem under consideration is sketched in Fig. 4.14. As previously, the floating ice cover is idealized by a plate of uniform thickness h. The lateral span of the plate is denoted by L, and is equal to the distance between the two constraining vertical walls at the ends of the plate. It is assumed that the top surface of ice is subjected to the action of varying in time temperature $T(t)$, with the ice at the base ($z = h$) being at the melting point temperature T_m at all time, and $T < T_m$ throughout the ice plate. It is also assumed that at the initial time $t = 0$ the plate is stress-free; that is, it is in equilibrium under an initial distribution of temperature in the plate. For simplicity, a plane-strain problem is analysed, so that the ice plate can be treated as a beam of uniform width, with its elastic flexural rigidity adjusted accordingly to account for the zero deformations in the direction normal to the plane Oxz.

All the equations describing the ice plate buckling problem are essentially those presented earlier in this section. Again, the viscous behaviour of ice is assumed to obey the Reiner-Rivlin-type constitutive law (4.39) on p. 77 (with the Heaviside

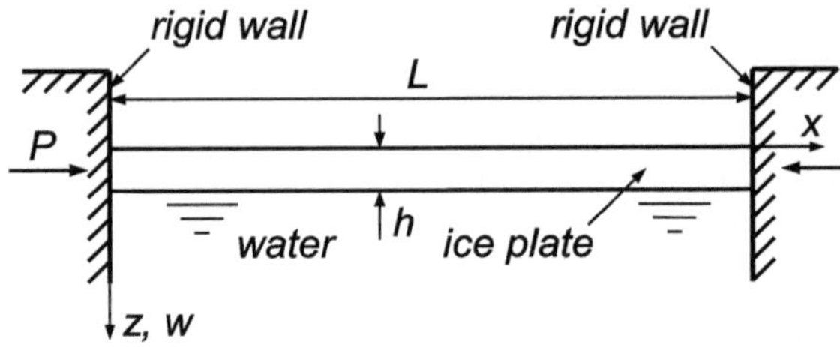

Fig. 4.14 Floating ice plate of thickness h and span L constrained by vertical rigid walls at its ends

function term $H(-\eta)$ omitted, since a single, continuous ice floe, not an ice pack, is now considered). Then, the equilibrium equation (4.53), for a plate of unit width b_0, becomes

$$R\,\frac{\partial^4 \dot{w}}{\partial x^4} + P\,\frac{\partial^2 w}{\partial x^2} + \varrho_w g w = 0, \tag{4.64}$$

where the parameter R, defining the viscous flexural 'rigidity' of the plate, is given by Eq. (4.51) which involves the axial viscosity $\zeta_a = \mu + \zeta$ being strongly sensitive to temperature. The above differential equation for the plate deflection curve $w(x, t)$ is solved with the boundary conditions at the plate edges at $x = 0$ and $x = L$ representing the case of a simply-supported plate, as being regarded (Sanderson 1988) as the most realistic conditions encountered in the field. Thus,

$$x = 0: \quad w = 0, \quad \frac{\partial^2 w}{\partial x^2} = 0; \quad x = L: \quad w = 0, \quad \frac{\partial^2 w}{\partial x^2} = 0. \tag{4.65}$$

The initial condition for the function $w(x, t)$ is prescribed by assuming that the plate deflection $w_0(x)$ at $t = 0$ consists of a number of small-amplitude, harmonic in x perturbations:

$$w_0(x) = \sum_{k=1}^{m} \pm A_k^0 \sin\left(\frac{k\pi x}{L}\right), \tag{4.66}$$

where the m component initial amplitudes A_k^0 and their signs ($\pm$) are selected at random. The number of components m is such that the shortest buckle half-wavelength is of order 1 m.

Besides the plate deflection evolution Eq. (4.64), also a heat conduction equation (4.52) on p. 83 which governs the evolution of the temperature field $T(z, t)$ in ice must be solved. The solution of that equation provides current vertical distributions of ice temperature, which in turn determine the current vertical distributions of elastic and viscous parameters of ice. The boundary conditions for the temperature field are prescribed at the top, $z = 0$, and the bottom, $z = h$, surfaces of the ice plate. At $z = 0$ a time-varying temperature distribution $T(0, t)$ is adopted to represent the ice heating conditions, whereas at $z = h$ it is assumed that the temperature of ice is constant and equal to the melting point, that is $T(h, t) = T_m$. Further, it is assumed that at the time $t = 0$ the temperature along the ice depth varies linearly between the initial top surface temperature $T_0(0)$ and the bottom surface temperature T_m.

In contrast to the problem considered in Sect. 4.2.2 in which the axial force P was treated as independent in time during the ice creep buckling process, now this force is time-dependent, since the elastic response of the constrained ice to the changing temperature field evolves as the ice heating progresses. The axial force $P(t)$ is defined by

$$P(t) = \alpha \int_0^h \Delta T(z, t) \frac{E(z, T)}{1 - \nu^2(z, T)} \mathrm{d}z, \tag{4.67}$$

where $\alpha = 5.2 \times 10^{-5}\,\mathrm{K}^{-1}$ is the thermal expansion coefficient for ice. The quantity $\Delta T(z, t)$ describes the difference between the current local temperature $T(z, t)$ and the initial local temperature $T_0(z)$ at the stable (stress-free) state of the plate. The force $P(t)$ given by (4.67) cannot exceed the force P_0 causing the instantaneous elastic buckling of the plate, which is given by Kerr (1978)

$$P_0 = 2\sqrt{\varrho_w g D_p}\,, \tag{4.68}$$

where the temperature-dependent plate elastic flexural rigidity D_p is

$$D_p = \int_0^h (z - z_0)^2 \frac{E(z, T)}{1 - \nu^2(z, T)} \mathrm{d}z. \tag{4.69}$$

In order to describe the temperature-dependence of the elastic constants, the Young modulus E and the Poisson ratio ν, the relations given in Chap. 3 are used. Hence, relation (3.2) on p. 34 is adopted to describe the function $E(T)$ for granular T1 ice, and relation (3.7) is employed for the function $\nu(T)$. The temperature-dependence of the ice viscous properties, μ and ζ, is, in turn, described by relations (3.18) to (3.20) on p. 40 due to Morland (1993, 2001); the same temperature scaling is applied to both shear and bulk viscosities. It is assumed here that the ice is porous. The degradation of the elastic properties with increasing ice porosity is described by formulae (3.9) on p. 36 due to Hutter (1983). Because of the lack of relevant data, it is supposed that the ice viscosity magnitudes μ and ζ are reduced for porous ice in the same manner as E; that is, by applying (3.9) with μ and ζ replacing E in the formulae. Finally, the weakening effect of ice porosity on the ice flexural strength σ_f is taken into account. For this purpose, relation (3.33) on p. 48 proposed by Timco and O'Brien (1994) is applied.

Results of Numerical Simulations

The differential equation (4.64) describing the evolution of the plate deflection surface $w(x, t)$ can be solved by either the analytical method for a uniform-width plate, or by a more general finite-element method for a wedge-shaped plate; both methods are described in the preceding Sect. 4.2.2. An enhanced analytical method is discussed by Staroszczyk (2018). In the calculations, the following material parameters were adopted: Young's modulus $E\ (T = 0\,°\mathrm{C}) = 8.93\,\mathrm{GPa}$, Poisson's ratio $\nu = 0.308$, ice porosity $\phi_b = 0.05$, ice viscosities $\mu = 1 \times 10^{11}$ Pa·s and $\zeta = 2 \times 10^{11}$ Pa·s, the ice flexural strength $\sigma_f = 0.47$ MPa (at $\phi_b = 0.05$) and the ice compressive strength $\sigma_c = 5$ MPa. The sea water density was assumed to be $\varrho_w = 1020\,\text{kg m}^{-3}$ and its freezing temperature was $T_m = -1\,°\text{C}$ (a compromise value between $T_m = 0\,°\text{C}$ for freshwater and $T_m = -1.9\,°\text{C}$ for ocean water of salinity 35 ppt). The ice thickness h and the plate length L were adopted of magnitudes typical of civil engineering applications.

The numerical simulations were run for a series of idealized sinusoidal ice surface temperature scenarios depicted in Fig. 4.15. The ice temperature T at the start of calculations was assumed to be equal to T_0, the maximum temperature reached during the daytime heating was T_{max} (supposed to be below the ice melting temperature T_m), and the minimum temperature at night was T_{min}. The results presented below were obtained for $T_0 \geq -5\,°\text{C}$ and $\Delta T_{max} = T_{max} - T_0 \leq 4\,°\text{C}$. For such moderate daily temperature increases, the maximum thermally-induced axial elastic force P determined from (4.67) was equal to around 1/3 of the magnitude of the elastic buckling force P_0 calculated from (4.68).

The temperature distribution along the ice plate depth was calculated by solving the heat conduction equation (4.52) by a finite-difference method, with the time integration performed by applying a Crank-Nicolson scheme. An example evolution of temperature depth profiles calculated by solving (4.52) for a plate of thickness $h = 0.3$ m is illustrated in Fig. 4.16. These profiles were obtained by adopting in the temperature scenario shown in Fig. 4.15 the values $T_0 = -5\,°\text{C}$ and $\Delta T_{max} = 4\,°\text{C}$, and for the day (heating period) lasting for 9 h. The solid lines in the figure illustrate the ice warming phase (increasing temperature period), and the dashed lines illustrate the cooling phase (decreasing temperature period), with the same colours corresponding to the same ice temperature at the upper surface of ice.

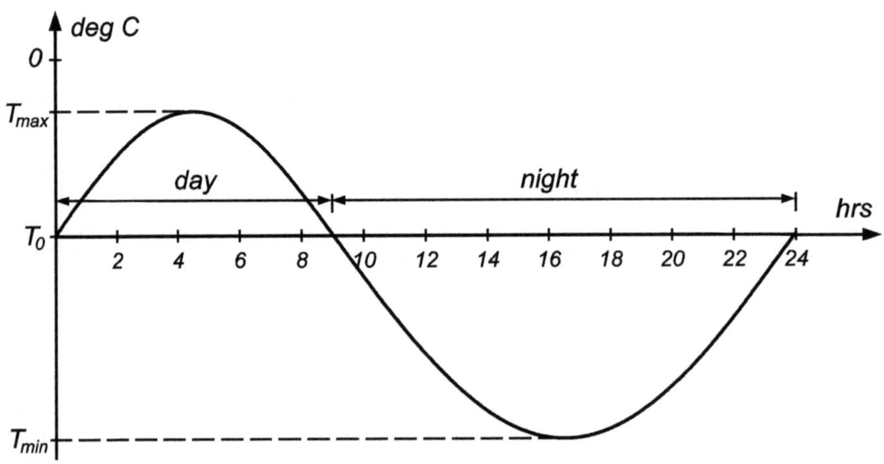

Fig. 4.15 Time variation of temperature $T(t)$ (time is given in hours) at the ice top surface $z = 0$. T_0 is the temperature at $t = 0$ (the start of simulations), T_{max} is the maximum temperature during the daytime, and T_{min} is the minimum temperature at night

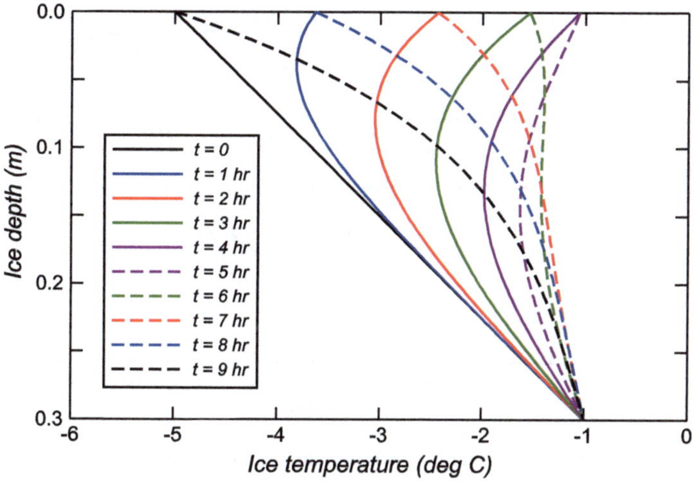

Fig. 4.16 Time variation of depth profiles of ice temperature $T(z)$ across a floating plate of thickness $h = 0.3$ m, for an initial temperature at the top surface of the plate $T_0 = -5\,°$C and a daily temperature increase $\Delta T_{max} = 4\,°$C. Reprinted with permission from Staroszczyk (2018), Fig. 3. Copyright 2018 by Elsevier

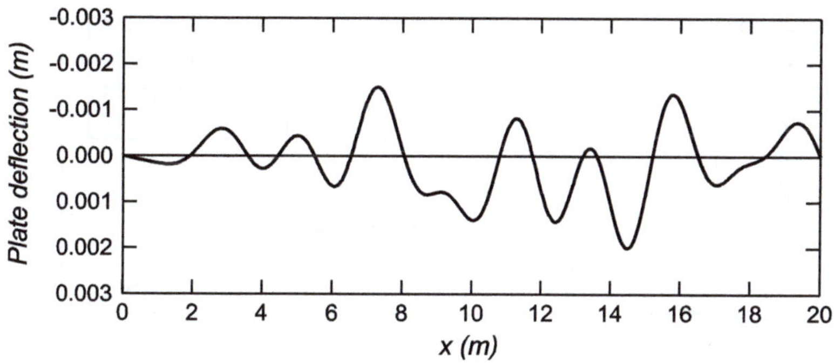

Fig. 4.17 Initial perturbed plate deflection curve $w_0(x)$ as a superposition of 20 harmonic components $A_0^{(k)}\,\sin(k\pi x/L)$ with random amplitudes, for the plate of thickness $h = 0.2$ m and length $L = 20$ m

The initial perturbed ice plate deflection curve, see Eq. (4.66), which was adopted for the plate of length $L = 20$ m and thickness $h = 0.2$ m is shown in Fig. 4.17. This curve was obtained by assuming $m = 20$ harmonic components, with the randomly selected component amplitudes scaled in such a way that the maximum initial plate deflection was equal to $1/100$ of the ice thickness.

The plots in Fig. 4.18 show the shapes of the plate deflection curves at the failure time for different maximum temperature amplitudes ΔT_{max}. The presented results illustrate the creep response of the plate of length $L = 20$ m and thickness $h = 0.2$ m, for $\Delta T_{max} = 3, 4$ and $5\,°$C; smaller values of ΔT_{max} do not generate sufficiently large

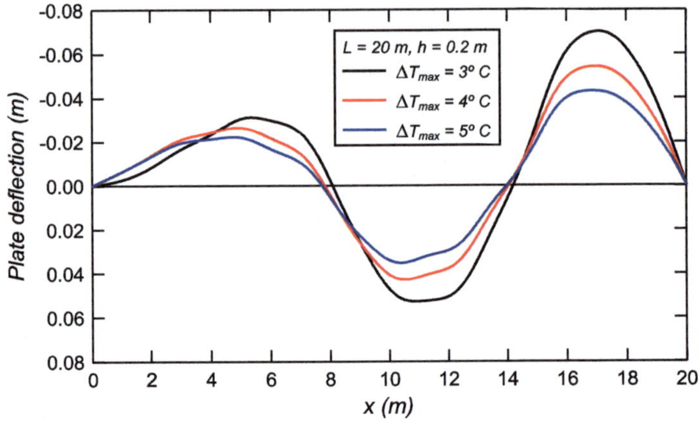

Fig. 4.18 Plate deflection curves $w(x)$ at corresponding failure times $t = t_f$ for different temperature amplitudes ΔT_{max}, for the plate of thickness $h = 0.2$ m and span $L = 20$ m. Reprinted with permission from Staroszczyk (2018), Fig. 7. Copyright 2018 by Elsevier

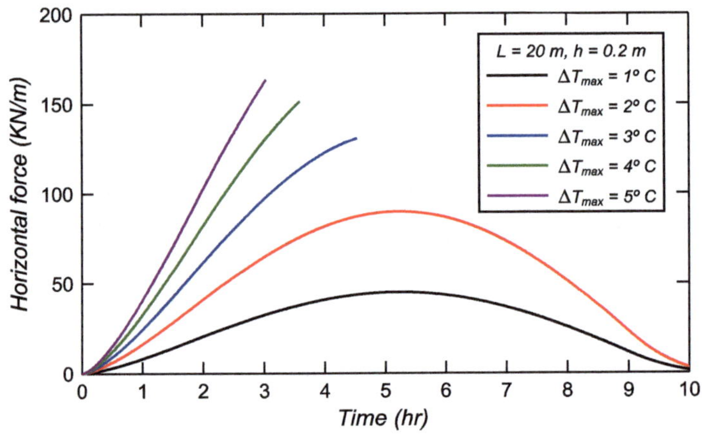

Fig. 4.19 Evolution of the horizontal force $P(t)$ exerted by ice on the rigid walls for different values of the maximum temperature increase ΔT_{max}, for the plate of thickness $h = 0.2$ m and span $L = 20$ m. Reprinted with permission from Staroszczyk (2018), Fig. 9. Copyright 2018 by Elsevier

stresses needed to fail the ice sheet. One can see that the creep behaviour of the plate, and in particular the length and the shape of the buckles, is similar for all the cases plotted in the figure, with the maximum plate deflections increasing monotonically with increasing temperature amplitudes. The corresponding values of the failure time ranged from about 4.5 h for $\Delta T_{max} = 3\,°$C to about 3 h for $\Delta T_{max} = 5\,°$C, see also the next figure.

The curves displayed in Fig. 4.19 show the dependence of the ice–wall reaction force $P(t)$ on the temperature amplitude ΔT_{max}, for the ice of thickness $h = 0.2$ m. The lines illustrating the evolution of $P(t)$ break at the corresponding ice failure

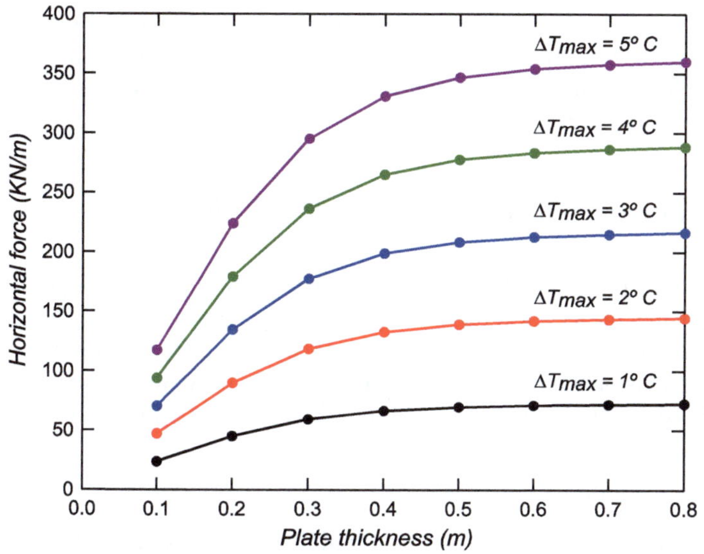

Fig. 4.20 Variation of the horizontal compressive force P with the plate thickness h for different temperature increase amplitudes ΔT_{max}, on the assumption that the plate does not fail. Reprinted with permission from Staroszczyk (2018), Fig. 10. Copyright 2018 by Elsevier

time instants, except the cases $\Delta T_{max} = 1$ and $2\,°\text{C}$ when the ice does not fracture during the heating period $0 < t < 9\,\text{h}$. As can be expected, an increase in the ice temperature-rate results in the ice failure occurring at earlier time. It can be noted that prior to the failure time the reaction force P increases in a smooth monotonic manner.

Finally, Fig. 4.20 illustrates the dependence of the maximum reaction forces P on the ice plate thickness h and the temperature increase amplitude ΔT_{max}, calculated on the assumption that no plate flexural failure occurs. The results plotted in Fig. 4.20 represent the maximum magnitudes of the compressive forces P which, at given h and ΔT_{max}, can theoretically develop in an ice sheet due to its heating. For safety reasons, the presented maximum forces should be adopted by an engineer as design loads on a structure that can be subjected to the action of floating ice. It is seen that the forces displayed in the figure are approximately proportional to the maximum temperature increase ΔT_{max}. Their increase with the plate thickness is most pronounced for $h \lesssim 0.5\,\text{m}$, whereas for thicker ice the thermally-induced axial forces caused by daily temperature variations do not significantly exceed those for $h \sim 0.5\,\text{m}$.

The results of calculations presented above show that a moderate temperature increase by a few degrees Celsius during a period of a few hours can lead to the fracture of a floating ice plate of thickness up to about 0.5 m. The thermally-induced forces exerted by floating ice on the walls constraining its lateral deformation can be of the magnitudes of several hundred kN per unit width of the ice plate. For instance, for the ice of thickness 0.5 m and a daily peak temperature increase by 5 °C, the maximum ice–structure reaction force predicted in the numerical simulations was equal to about 350 kN/m. More results illustrating the mechanism of floating ice creep buckling due to the phenomenon of thermal expansion of ice can be found in the paper by Staroszczyk (2018).

4.2.4 Plane Ice–Structure Interaction Problem

In Sect. 4.2.2, the creep behaviour of a floating ice sheet is analysed under the assumption that the ice sheet has a wedge-shaped geometry in the Oxy plane and the problem is symmetric with respect to the x-axis. This, in fact, simplifies the problem to a one-dimensional in the horizontal plane. Now a fully two-dimensional problem is considered, without introducing any simplifications regarding the planar geometry of an ice sheet.

Governing Equations

The equations of the floating ice plate equilibrium in the horizontal and vertical planes, together with the definitions of internal forces (in-plane axial and shear forces and bending and twisting moments) in terms of stresses have already been formulated in Sect. 4.1.1, see relations (4.1) to (4.7) (also refer to Fig. 4.1 on p. 61). In the case of the creep deformations of ice, the stresses are expressed in terms of strain-rates, see the constitutive relations discussed in Sect. 4.2.1. Here we focus on solving the plane problem in which the creep of ice is described by the viscous fluid rheological model, expressing the stresses in terms of strain-rates by the flow law (4.39) on p. 77. We note, however, that the methodology of constructing a solution for the viscous-plastic rheology, due to the formal similarities between the laws (4.39) and (4.49) on p. 81, is not much different from that described below for the viscous fluid rheology.

When expressed in components in the Cartesian coordinate plane Oxy, the viscous fluid flow law (4.39) takes the form:

$$\sigma_{ij} = \left[(\zeta - \mu)D_{kk}\delta_{ij} + 2\mu D_{ij}\right] H(-\eta) \quad (i, j, k = 1, 2), \tag{4.70}$$

where the summation convention applies for a repeated suffix. Recall that in (4.70) ζ and μ denote the bulk and shear viscosities, respectively, and D_{ij} are the components of the two-dimensional strain-rate tensor $\boldsymbol{D}$, see definition (4.31). The subscripts i and j stand for either x and y, with the equivalence $x_1 = x$ and $x_2 = y$.

Deformations in the ice cover can be expressed as a sum of the deformations in the neutral plane of the plate caused by the forces N_{ij}, and these can be regarded as functions of the horizontal coordinates x and y alone, and the deformations due to bending and twisting of the plate, which are functions of the depth z as well. Accordingly, the in-plane strain-rates are determined by using the horizontal velocity components $v_x(x, y)$ and $v_y(x, y)$ in (4.31), while the strain-rates due to bending and twisting of the plate are given in terms of the curvatures and twist of the deflection surface $w(x, y)$ as follows

$$
\begin{aligned}
D_{xx} &= \dot{\kappa}_x(z - z_0) = -\frac{\partial^2 \dot{w}}{\partial x^2}(z - z_0), \\
D_{yy} &= \dot{\kappa}_y(z - z_0) = -\frac{\partial^2 \dot{w}}{\partial y^2}(z - z_0), \\
D_{xy} &= -\dot{\kappa}_{xy}(z - z_0) = -\frac{\partial^2 \dot{w}}{\partial x \partial y}(z - z_0).
\end{aligned}
\tag{4.71}
$$

In these equations, κ_x and κ_y are the curvatures of the deflection surface along the x and y axes, respectively, κ_{xy} is the twist with respect to the x and y axes, and z_0 is the position of the neutral plane in the undeformed state. With the strain-rates given by (4.31) and (4.71), and the stresses determined through the constitutive law (4.70), the in-plane axial and shear forces (4.6) become

$$
\begin{aligned}
N_x &= (H_1 + H_2)\frac{\partial v_x}{\partial x} + (H_1 - H_2)\frac{\partial v_y}{\partial y}, \\
N_y &= (H_1 - H_2)\frac{\partial v_x}{\partial x} + (H_1 + H_2)\frac{\partial v_y}{\partial y}, \\
N_{xy} &= H_2\left(\frac{\partial v_x}{\partial y} + \frac{\partial v_y}{\partial x}\right),
\end{aligned}
\tag{4.72}
$$

and the bending and and twisting moments (4.7) are given by

$$
\begin{aligned}
M_x &= -\left[(R_1 + R_2)\frac{\partial^2 \dot{w}}{\partial x^2} + (R_1 - R_2)\frac{\partial^2 \dot{w}}{\partial y^2}\right], \\
M_y &= -\left[(R_1 - R_2)\frac{\partial^2 \dot{w}}{\partial x^2} + (R_1 + R_2)\frac{\partial^2 \dot{w}}{\partial y^2}\right], \\
M_{xy} &= 2R_2\frac{\partial^2 \dot{w}}{\partial x \partial y}.
\end{aligned}
\tag{4.73}
$$

In the above expressions for the internal forces, the parameters defining the plate viscous properties are given by

$$
H_1 = \int_0^h \zeta \mathrm{d}z, \quad H_2 = \int_0^h \mu \mathrm{d}z, \quad R_1 = \int_0^h \zeta z(z - z_0)\mathrm{d}z, \quad R_2 = \int_0^h \mu z(z - z_0)\mathrm{d}z.
\tag{4.74}
$$

For the sake of brevity of the notations, it is tacitly assumed henceforth in this chapter that the Heaviside unit step factor $H(-\eta)$ is included in the viscosity terms ζ and μ.

By substituting now the definitions (4.73) for the internal moments into the equilibrium relation (4.5), we obtain the following differential equation

$$R\left(\frac{\partial^4 \dot{w}}{\partial x^4}+2\frac{\partial^4 \dot{w}}{\partial x^2 \partial y^2}+\frac{\partial^4 \dot{w}}{\partial y^4}\right)=N_x\frac{\partial^2 w}{\partial x^2}+2N_{xy}\frac{\partial^2 w}{\partial x \partial y}+N_y\frac{\partial^2 w}{\partial y^2}+ \\ -\varrho_w g w-\tau_x\frac{\partial w}{\partial x}-\tau_y\frac{\partial w}{\partial y}, \tag{4.75}$$

with the definition

$$R=R_1+R_2=\int_0^h \zeta_a z(z-z_0)\mathrm{d}z, \tag{4.76}$$

where $\zeta_a=\mu+\zeta$ is the axial viscosity. Equation (4.75) describes the time and space variation of the plate deflection w in terms of the in-plane forces N_{ij} and the external driving forces τ_i $(i, j=1,2)$. The forces N_{ij}, the functions of the plate horizontal velocities v_x and v_y as given by (4.72), can be determined independently of (4.75) by solving the equations of the in-plane equilibrium (4.1).

The driving forces τ_i are due to wind and water drag on the top and bottom surfaces of ice. In this work, formulae that are quadratic in velocities are adopted to describe these drag forces in terms of the ice, wind and water velocities. Hence, the surface tractions are expressed in the following forms (Sanderson 1988) :

$$\boldsymbol{\tau}_a=C_a\varrho_a(\boldsymbol{u}_a-\boldsymbol{v})|\boldsymbol{u}_a-\boldsymbol{v}|,\quad \boldsymbol{\tau}_w=C_w\varrho_w(\boldsymbol{u}_w-\boldsymbol{v})|\boldsymbol{u}_w-\boldsymbol{v}|, \tag{4.77}$$

where ϱ_a and ϱ_w are, respectively, the air and water densities, and $\boldsymbol{u}_a$ and $\boldsymbol{u}_w$ are, respectively, the wind and ocean current velocity vectors. The parameters C_a and C_w in relations (4.77) denote dimensionless wind stress and water drag coefficients. On the basis of the data presented in the literature (Sanderson 1988; Kara et al. 2007; Lu et al. 2011), the values $C_a=2\times10^{-3}$ and $C_w=4\times10^{-3}$ have been adopted for numerical simulations.

Typical boundary conditions, with which the three Eqs. (4.75) and (4.1) are solved in sea ice applications, are those of a simply-supported plate edge at the contact region with the structure, with zero horizontal velocities in the direction normal to the ice-structure interface (so-called free-slip conditions). These conditions are expressed by

$$w=0,\quad M_n=0,\quad \boldsymbol{v}\cdot\boldsymbol{n}=0, \tag{4.78}$$

where n is the direction normal to the edge of the plate, defined by the outward unit vector $\boldsymbol{n}$, and M_n is the bending moment acting on the plate section normal to $\boldsymbol{n}$.

Finite-Element Formulation

The system of three differential equations for the ice plate deflection w and the horizontal velocities v_x and v_y, given by (4.75) and (4.1) with (4.72), is solved approximately by applying the finite-element method. The weighted residual, or Galerkin, version of the method is employed. The plate is discretized in the horizontal plane Oxy by using a mesh of triangular elements, with the unknown variables defined at the corner nodes. At each discrete node, apart from the two horizontal velocities, v_x and v_y, and the plate vertical displacement, w, also the two plate slopes, $\partial w/\partial x$ and $\partial w/\partial y$, are treated as unknown variables. Such an approach is typical of the plate theory and is applied in order to ensure the continuity of the plate deflection surface along the sides of the adjacent elements (Zienkiewicz and Taylor 2005). Thus, there are five discrete parameters to be calculated at each node, so altogether there are 15 degrees of freedom per each triangular element. The continuous functions w, v_x and v_y are approximated by the following representations:

$$\begin{aligned} w(x,y,t) &= \Phi_j^w(x,y)w_j(t), \quad (j=1,\ldots,9), \\ v_i(x,y,t) &= \Phi_j^v(x,y)v_{ij}(t), \quad (i=1,2;\; j=1,2,3), \end{aligned} \tag{4.79}$$

where w_j and v_{ij} are the unknown nodal parameters, the displacements and the velocities respectively, with the former including both the plate deflections and the plate slopes. Φ_j^w and Φ_j^v are shape functions, which are different for the displacement and the velocity fields. While the velocity field is interpolated by simple linear shape functions, for the plate deflection approximation fourth-order polynomials in both x and y are used, following the formulation due to Specht (1988).

By applying a typical finite-element procedure for so-called weak formulations of the problem equation, in which the latter are multiplied by weighting functions (which in the Galerkin method are identical to the shape functions Φ_j^w and Φ_j^v). Integration of the resulting relations (Staroszczyk 2003) reduces the problem defined by Eqs. (4.75), (4.1) and (4.72) to the solution of a system of first-order differential equations given in a matrix form by

$$\boldsymbol{C}\dot{\boldsymbol{w}} + \boldsymbol{K}\boldsymbol{w} = \boldsymbol{f}, \tag{4.80}$$

where the vector $\boldsymbol{w}$ includes the values of the plate deflections w_j, the plate slopes $(\partial w/\partial x)_j$ and $(\partial w/\partial y)_j$, and the velocities v_{xj} and v_{yj} at all nodal points j of the discrete system. We note that the matrix $\boldsymbol{K}$ depends on the horizontal velocities, so $\boldsymbol{K} = \boldsymbol{K}(\boldsymbol{w})$, which means that the system of Eqs. (4.80) is non-linear in $\boldsymbol{w}$. The matrices $\boldsymbol{C}$, $\boldsymbol{K}$ and the forcing vector $\boldsymbol{f}$ are aggregated from the respective element matrices and vectors in a way characteristic of the finite-element method. The element matrices, $\boldsymbol{C}^e$ and $\boldsymbol{K}^e$, each of size 15×15, are, in turn, composed of 9 submatrices of dimension 5×5 each. The non-zero entries in these component submatrices are given for the matrix $\boldsymbol{C}$ by

$$\begin{aligned} c_{rs}^{mn} = \int_A \Bigg[&(R_1+R_2)\left(\frac{\partial^2\Phi_i^w}{\partial x^2}\frac{\partial^2\Phi_j^w}{\partial x^2} + \frac{\partial^2\Phi_i^w}{\partial y^2}\frac{\partial^2\Phi_j^w}{\partial y^2}\right) + \\ &+ 4R_2\frac{\partial^2\Phi_i^w}{\partial x\partial y}\frac{\partial^2\Phi_j^w}{\partial x\partial y} + (R_1-R_2)\left(\frac{\partial^2\Phi_i^w}{\partial x^2}\frac{\partial^2\Phi_j^w}{\partial y^2} + \frac{\partial^2\Phi_i^w}{\partial y^2}\frac{\partial^2\Phi_j^w}{\partial x^2}\right)\Bigg]\,\mathrm{d}A, \end{aligned} \tag{4.81}$$

and for the matrix $\boldsymbol{K}$ they are

$$
\begin{aligned}
k_{rs}^{mn} &= \int\limits_A \Phi_i^w \left(\varrho_w g \Phi_j^w + \tau_x \frac{\partial \Phi_j^w}{\partial x} + \tau_y \frac{\partial \Phi_j^w}{\partial y} + \right. \\
&\quad \left. - N_x \frac{\partial^2 \Phi_j^w}{\partial x^2} - 2N_{xy} \frac{\partial^2 \Phi_j^w}{\partial x \partial y} - N_y \frac{\partial^2 \Phi_j^w}{\partial y^2} \right) \mathrm{d}A, \\
k_{rs}^{44} &= \int\limits_A \left[(H_1 + H_2) \frac{\partial \Phi_r^v}{\partial x} \frac{\partial \Phi_s^v}{\partial x} + H_2 \frac{\partial \Phi_r^v}{\partial y} \frac{\partial \Phi_s^v}{\partial y} \right] \mathrm{d}A, \\
k_{rs}^{45} &= \int\limits_A \left[(H_1 - H_2) \frac{\partial \Phi_r^v}{\partial x} \frac{\partial \Phi_s^v}{\partial y} + H_2 \frac{\partial \Phi_r^v}{\partial y} \frac{\partial \Phi_s^v}{\partial x} \right] \mathrm{d}A, \\
k_{rs}^{54} &= \int\limits_A \left[H_2 \frac{\partial \Phi_r^v}{\partial x} \frac{\partial \Phi_s^v}{\partial y} + (H_1 - H_2) \frac{\partial \Phi_r^v}{\partial y} \frac{\partial \Phi_s^v}{\partial x} \right] \mathrm{d}A, \\
k_{rs}^{55} &= \int\limits_A \left[H_2 \frac{\partial \Phi_r^v}{\partial x} \frac{\partial \Phi_s^v}{\partial x} + (H_1 + H_2) \frac{\partial \Phi_r^v}{\partial y} \frac{\partial \Phi_s^v}{\partial y} \right] \mathrm{d}A.
\end{aligned}
\tag{4.82}
$$

The indices in (4.81) and (4.82) are

$$
r, s, m, n = 1, 2, 3, \quad i = 3(r-1) + m, \quad j = 3(s-1) + n, \tag{4.83}
$$

and A denotes the plane domain of integration. The components of the forcing vector $\boldsymbol{f}$ are given by

$$
\begin{aligned}
f_r^m &= \oint\limits_\Gamma \Phi_i^w Q \mathrm{d}\Gamma, \\
f_r^4 &= \int\limits_A \Phi_r^v \tau_x \mathrm{d}A + \oint\limits_\Gamma \Phi_r^v T_x \mathrm{d}\Gamma, \\
f_r^5 &= \int\limits_A \Phi_r^v \tau_y \mathrm{d}A + \oint\limits_\Gamma \Phi_r^v T_y \mathrm{d}\Gamma,
\end{aligned}
\tag{4.84}
$$

where Γ denotes the boundary of the domain A. In the first of the above equations, Q is the vertical shear force acting on the boundary Γ, and T_x and T_y are, respectively, the x- and y-components of the in-plane traction vector $\boldsymbol{T}$ acting on Γ, and are defined by

$$
T_x = N_x n_x + N_{xy} n_y, \quad T_y = N_{xy} n_x + N_y n_y, \tag{4.85}
$$

with n_x and n_y being the components of the outward unit vector $\boldsymbol{n}$ normal to the boundary Γ.

The system of equations (4.80) is integrated in time by applying an implicit weighted residual θ-method (Zienkiewicz et al. 2005). Application of this method gives the relationship that connects the solution vectors $\boldsymbol{w}_n$ and $\boldsymbol{w}_{n+1}$ at two consecutive time levels, t_n and t_{n+1}:

$$(\boldsymbol{C} + \theta \Delta t \boldsymbol{K})\, \boldsymbol{w}_{n+1} = [\boldsymbol{C} - (1-\theta)\Delta t \boldsymbol{K}]\, \boldsymbol{w}_n + \Delta t \bar{\boldsymbol{f}}, \tag{4.86}$$

where $\Delta t = t_{n+1} - t_n$ is the time-step length. The vector $\bar{\boldsymbol{f}}$ is the time-averaged forcing vector which, assuming a linear variation of $\boldsymbol{f}$ from t_n to t_{n+1}, is defined by

$$\bar{\boldsymbol{f}} = (1-\theta)\boldsymbol{f}_n + \theta \boldsymbol{f}_{n+1}. \tag{4.87}$$

In numerical calculations, the value of $\theta = 0.6$ has been adopted, for which the method is unconditionally stable, and which guarantees that the time-discretization error is nearly of the order $(\Delta t)^2$.

Ice–Structure Interaction Simulations

Before applying the above-described finite-element model to simulate a plane ice–structure interaction event, the discrete model was tested on a one-dimensional problem, for which a closed-form analytical solution is available, as described in Sect. 4.2.2. Hence, the model was run for a uniform-width plate undergoing creep buckling under the action of a in-plane compressive horizontal force, with the initial plate deflection consisting of a number of small harmonic perturbation of various lengths and random amplitudes, see Eq. (4.63) on p. 86. It turned out (Staroszczyk 2003) that the finite-element predictions were in a very good agreement with the analytical results (the maximum relative error in the plate deflections given by the two methods was less than 3%).

After the successful verification of the accuracy of the discrete model in the one-dimensional configuration, a two-dimensional problem sketched in Fig. 4.21 was solved, in which the behaviour of a coherent floating ice cover interacting with a rigid structure, the horizontal cross-section of which has the shape of a rectangle of

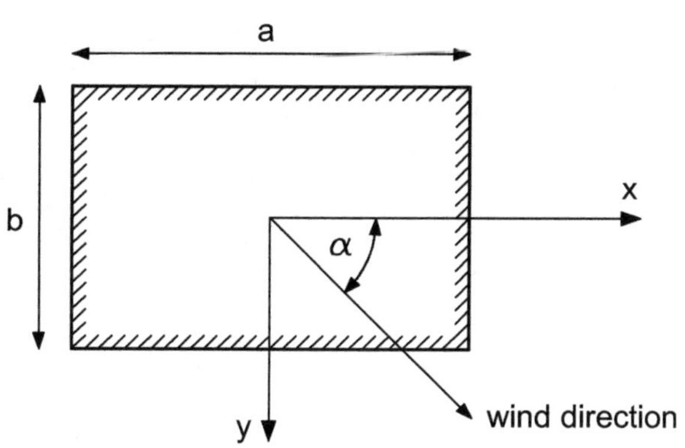

Fig. 4.21 A rectangular rigid structure of horizontal dimensions $a \times b$ interacting with sea ice driven by wind blowing at the angle α to the x-axis

dimensions defined by a and b. The ice cover was assumed to be driven towards the structure by air drag forces caused by a wind blowing in the direction defined by the angle α shown in the figure.

The simulations were carried out for a rigid structure situated at the centre of a rectangular in shape coherent ice field of the size 1 km × 1 km and the ice thickness $h = 0.5$ m. At the ice–structure interface, the plate was assumed to be simply-supported, and the free-slip boundary conditions, prescribed, by (4.78), were adopted. The wind had a speed $u_a = 30\,\mathrm{m\,s^{-1}}$, and its direction has been varied within the range $0 < \alpha < 90°$ in order to investigate how this affects the total loading exerted by the ice on the structure. Three particular cases of the structures of different shapes were considered, in which the width of the structure b was kept constant and equal to 10 m, and the length a was varied and equal to 20, 30 and 40 m, respectively. The results of numerical calculations, conducted with the mesh consisting of 4000 triangular finite elements and 10,400 degrees of freedom, are presented in Fig. 4.22. The results shown in the figure have been obtained for the ice viscosities $\zeta = \mu = 1.0 \times 10^9\ \mathrm{kg\,m^{-1}s^{-1}}$. The temperature at the top surface of the ice was assumed to be equal to $-2\,°\mathrm{C}$, and that at the bottom surface to be $0\,°\mathrm{C}$.

The plots illustrate the dependence of the magnitude of the total horizontal force F exerted by the ice on the structure on the wind direction angle α. Also shown are the components of the total force along the x and y axes, F_x and F_y respectively. The results obtained for the rectangle 20 m × 10 m are indicated by the solid lines, those for the rectangle 30 m × 10 m are given by the dashed lines, and those for the longest rectangle 40 m × 10 m are shown by the dashed-dotted lines. One can see in the figure that the geometry of the structure cross-section has a relatively small effect on the total force F sustained by the object during its interaction with creeping ice. Further, a rather small influence of the wind direction on the total force F is also noted. For the structure for which $a/b = 4$, the maximum and minimum forces, for $\alpha = 90°$ and $\alpha = 0°$ respectively, differ by about 20%, while for the structure, for which $a/b = 2$, the corresponding relative difference is about 10%. The results of a

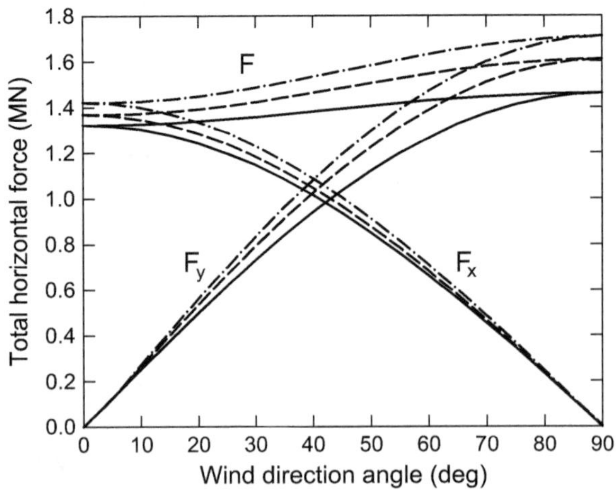

Fig. 4.22 Total horizontal forces F, together with their components F_x and F_y, exerted on the structure by the ice cover as a function of the wind direction angle α. Shown are the results for three rectangles of the same width $b = 10$ m and the lengths 20 m (solid lines), 30 m (dashed lines) and 40 m (dashed-dotted lines)

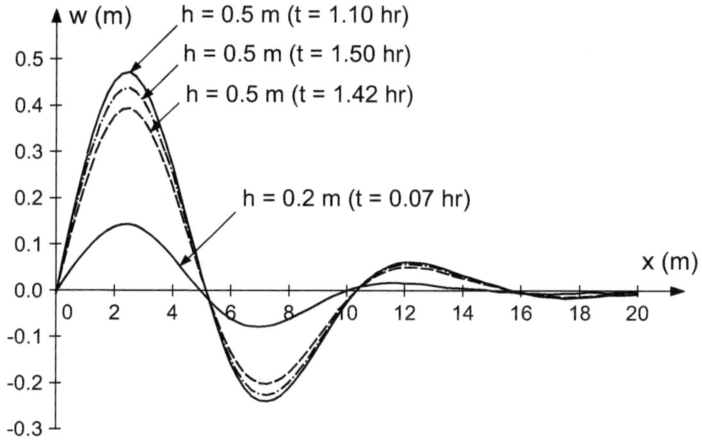

Fig. 4.23 Plate deflections w along the x-axis at the critical times t (given in hours) for the wind direction angle $\alpha = 180°$. Shown are the results for the plates of thickness $h = 0.2$ m and $h = 0.5$ m (solid lines); for the thicker plate also the deflections and corresponding critical times for colder ice (dashed line) and the ice with larger shear and bulk viscosities (dashed-dotted line) are presented

similar character, that is showing a relatively small effect of the wind angle α on the total force acting on a structure, have been also obtained for other thicknesses of the ice cover.

Figure 4.23 illustrates the ice plate deflections in the vicinity of the structure vertical wall. The results plotted in the figure have been obtained for the structure dimensions $a = 20$ m and $b = 10$ m, and for the wind blowing along the negative direction of the x-axis (that is for the angle $\alpha = 180°$, see Fig. 4.21). Plotted are the plate deflection curves along the positive x-axis at the critical times at which the process of flexural failure of ice starts (the origin of the x-axis is on the structure vertical wall; that is, it is shifted to the right by $a/2$ compared to Fig. 4.21). The two solid lines in the figure illustrate the plate deflections for two different thicknesses of the ice: $h = 0.2$ m and $h = 0.5$ m. It can be immediately noticed that the plate failure times for these two plates differ quite considerably: $t = 0.07$ h for the thinner ice and $t = 1.10$ h for the thicker ice. The dashed line displays, for $h = 0.5$ m, the plate deflection in the case of the top surface of the ice having the temperature -4 °C (compared to -2 °C for the ice represented by the respective solid line in the figure). Finally, the dashed-dotted line shows, for $h = 0.5$ m, the plate deflection for the ice of the viscosities ζ and μ increased by 30% with respect to the reference case plotted by the solid line; such a difference in viscosities occurs between isotropic ice and transversely isotropic columnar ice. It is seen that both the temperature and the type of ice anisotropy have quite a pronounced effect on the strength of ice, significantly increasing the values of the failure time (by about 30% in the presented example). On the other hand, the maximum plate deflections do not change much with the change of temperature and the type of ice.

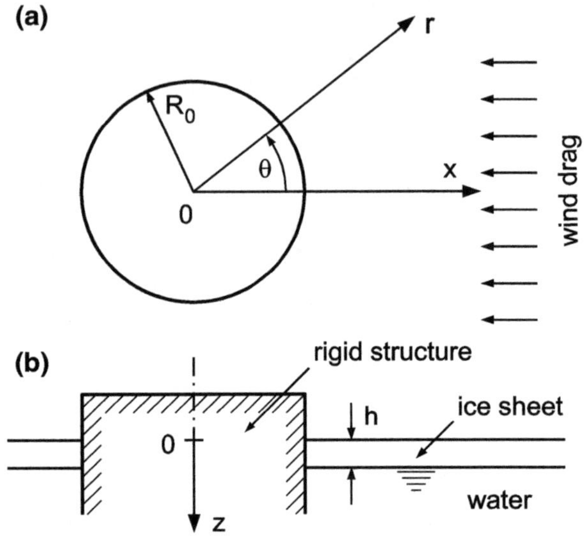

Fig. 4.24 Geometry of the problem and cylindrical polar coordinates: **a** plane view, **b** ice sheet cross-section

4.2.5 Interaction of Ice with Cylindrical Structures

In the previous Sect. 4.2.4 the plane problem of the interaction of floating ice with a rectangular in shape structure is discussed. Here a similar problem is considered, in which creeping ice interacts with a vertically-walled circular cylinder, see Fig. 4.24. Since the problem involves a single circular structure, cylindrical polar coordinates r, θ, z $(0 \leq \theta < 2\pi)$ are adopted, with the vertical z-axis coinciding with the axis of the rotational symmetry of the cylinder. As before, it is assumed that in the immediate vicinity of the structure the floating ice cover has a constant thickness, h. The z-axis, directed downwards, is chosen in such a way that $z = 0$ corresponds to the top surface of the ice sheet, and $z = h$ to its bottom. A circular cylinder, of radius R_0, is treated as a fixed rigid body that interacts with the ice sheet along its vertical walls at $r = R_0$. The purpose is to evaluate the values of the horizontal forces which the floating ice exerts on the structure during an interaction event.

Similar problems, of a circular cylinder interacting with sea ice, were previously investigated by Wang and Ralston (1983) and Sjölind (1985). In the first of these papers, the ice was treated as an elastic-plastic material, while in the second a viscoelastic rheology was adopted to describe the ice deformation. In this work, the forces acting on the structure are determined by adopting either the non-linearly viscous fluid rheology, or the viscous-plastic rheology, both discussed in Sect. 4.2.1.

Governing Equations

The definitions of internal forces acting on an infinitesimal plate element, with their components expressed in the adopted polar coordinates, are given in Fig. 4.25. Basically, all the equations describing the equilibrium of forces acting on a plate element in the horizontal and vertical planes are derived in a way analogous to that presented in Sect. 4.1.1. Hence, the balances of the forces acting in the horizontal plane $Or\theta$, involving the axial forces N_r and N_θ and the shear forces $N_{r\theta} = N_{\theta r}$, are expressed by

$$\begin{aligned} &\frac{\partial(rN_r)}{\partial r} - N_\theta + \frac{\partial N_{\theta r}}{\partial \theta} + rq_r = 0, \\ &\frac{1}{r}\frac{\partial(r^2 N_{r\theta})}{\partial r} + \frac{\partial N_\theta}{\partial \theta} + rq_\theta = 0, \end{aligned} \tag{4.88}$$

where q_r and q_θ denote the components of the external forces acting in the horizontal direction, which arise due to the wind stress and water current drag.

Along the z-direction, a plate element is subject to the vertical shear forces Q_r and Q_θ, and also to the transverse distributed load q_z coming from the underlying water. Since in our problem the in-plane forces N_r, N_θ and $N_{r\theta}$, all acting in the directions tangential to the deflection surface $w(r, \theta)$, can have magnitudes considerably larger than those of the vertical shear forces Q_r and Q_θ, we include the z-components of the former in the equilibrium balance. Accordingly, the projection of all forces on the vertical direction, with the own weight of ice neglected, gives

$$\begin{aligned} &\frac{\partial(rQ_r)}{\partial r} + \frac{\partial Q_\theta}{\partial \theta} + \frac{\partial}{\partial r}\left(rN_r\frac{\partial w}{\partial r}\right) + \frac{1}{r}\frac{\partial}{\partial \theta}\left(N_\theta\frac{\partial w}{\partial \theta}\right) + \\ &+ \frac{\partial}{\partial r}\left(N_{r\theta}\frac{\partial w}{\partial \theta}\right) + \frac{\partial}{\partial \theta}\left(N_{\theta r}\frac{\partial w}{\partial r}\right) + rq_z = 0. \end{aligned} \tag{4.89}$$

The above relations, apart from the internal forces, also involve the plate deflection spatial derivatives. The equilibrium of all moments (see Fig. 4.25b) acting on an

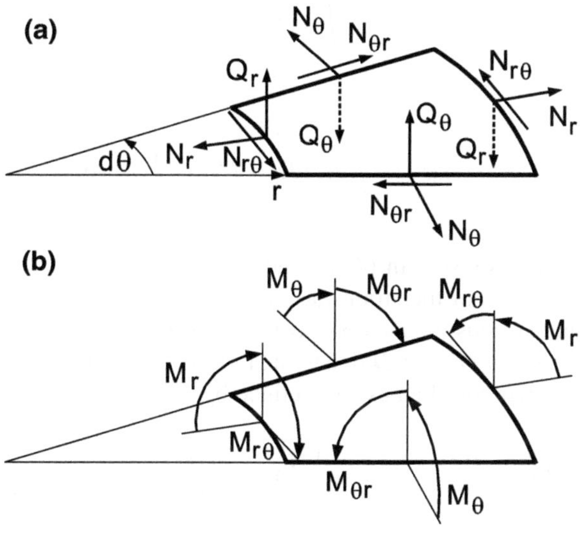

Fig. 4.25 Definitions of internal forces acting on a plate element: **a** axial and shear forces, **b** bending and twisting moments

infinitesimal plate element with respect to the radial (r) and circumferential (θ) directions yields the expressions

$$\begin{aligned}
\frac{\partial(rM_r)}{\partial r} - \frac{\partial M_{\theta r}}{\partial \theta} - rQ_r = 0, \\
\frac{\partial M_\theta}{\partial \theta} - \frac{\partial(rM_{r\theta})}{\partial r} - rQ_\theta = 0,
\end{aligned} \tag{4.90}$$

where M_r and M_θ are the bending moments, and $M_{r\theta} = M_{\theta r}$ are the twisting moments, all per unit width of the plate. Elimination of the shear forces Q_r and Q_θ from (4.89) by means of relations (4.90) gives the equilibrium equation

$$\begin{aligned}
&\frac{\partial^2(rM_r)}{\partial r^2} + \frac{1}{r}\frac{\partial^2 M_\theta}{\partial \theta^2} - \frac{\partial^2 M_{r\theta}}{\partial r \partial \theta} - \frac{1}{r}\frac{\partial^2(rM_{r\theta})}{\partial r \partial \theta} + \\
&+ rN_r\frac{\partial^2 w}{\partial r^2} + N_\theta\left(\frac{1}{r}\frac{\partial^2 w}{\partial \theta^2} + \frac{\partial w}{\partial r}\right) + 2rN_{r\theta}\frac{\partial^2}{\partial r \partial \theta}\left(\frac{w}{r}\right) + \\
&- rq_r\frac{\partial w}{\partial r} - q_\theta\frac{\partial w}{\partial \theta} + rq_z = 0.
\end{aligned} \tag{4.91}$$

The transverse distributed load q_z, resulting from the response of the underlying water, is assumed to be proportional to the plate deflection w:

$$q_z = -\varrho_w g w. \tag{4.92}$$

The internal forces in the vertical plate cross-sections are determined in terms of the axial, σ_{rr} and $\sigma_{\theta\theta}$, and shear, $\sigma_{r\theta}$, stresses by the integrals:

$$N_r = \int_0^h \sigma_{rr}\mathrm{d}z, \quad N_\theta = \int_0^h \sigma_{\theta\theta}\mathrm{d}z, \quad N_{r\theta} = \int_0^h \sigma_{r\theta}\mathrm{d}z, \tag{4.93}$$

and

$$M_r = \int_0^h \sigma_{rr} z\mathrm{d}z, \quad M_\theta = \int_0^h \sigma_{\theta\theta} z\mathrm{d}z, \quad M_{r\theta} = -\int_0^h \sigma_{r\theta} z\mathrm{d}z. \tag{4.94}$$

The stresses in (4.93) and (4.94), in the case of creeping behaviour of ice, are functions of the strain-rates and their invariants, as prescribed by the viscous fluid and viscous-plastic flow laws, (4.39) and (4.49) respectively. The components of the strain-rate tensor $\boldsymbol{D}$ due to the motion of ice in the horizontal plane, when expressed in polar coordinates, are defined by

$$D_{rr} = \frac{\partial v_r}{\partial r}, \quad D_{\theta\theta} = \frac{1}{r}\left(v_r + \frac{\partial v_\theta}{\partial \theta}\right), \quad D_{r\theta} = \frac{1}{2}\left[\frac{1}{r}\frac{\partial v_r}{\partial \theta} + r\frac{\partial}{\partial r}\left(\frac{v_\theta}{r}\right)\right], \tag{4.95}$$

and their invariants η and γ are given by

$$\eta = D_{rr} + D_{\theta\theta}, \quad \gamma^2 = D_{r\theta}^2 + \frac{1}{4}\left(D_{rr} - D_{\theta\theta}\right)^2 . \tag{4.96}$$

The strain-rates, developing in ice due to the bending and twisting of the plate, vary across its depth and are defined in terms of the time rates of the plate curvatures, κ_r and κ_θ, and the twist, $\kappa_{r\theta}$, of the deflection surface $w(r, \theta)$. Hence,

$$D_{rr} = \dot{\kappa}_r\,(z - z_0), \quad D_{\theta\theta} = \dot{\kappa}_\theta\,(z - z_0), \quad D_{r\theta} = -\dot{\kappa}_{r\theta}\,(z - z_0), \tag{4.97}$$

where z_0 denotes the position of the neutral plane in the undeformed state. In terms of the plate deflection function $w(r, \theta)$, the curvatures and the twist are given by

$$\kappa_r = -\frac{\partial^2 w}{\partial r^2}, \quad \kappa_\theta = -\frac{1}{r}\frac{\partial w}{\partial r} - \frac{1}{r^2}\frac{\partial^2 w}{\partial \theta^2}, \quad \kappa_{r\theta} = \frac{\partial}{\partial r}\left(\frac{1}{r}\frac{\partial w}{\partial \theta}\right). \tag{4.98}$$

So far in this chapter, when analysing the creep buckling of ice and the plane ice–structure interactions, the viscosities μ and ζ were assumed constant in the viscous fluid flow law (4.39). In the latter law, the Heaviside step function factor $H(-\eta)$ appears, the role of which is to ensure that no axial stresses develop in ice during its diverging flow (when $\eta < 0$), in order to model zero tensile strength of ice. Such an abrupt cut-off to zero stress during a change from converging to diverging flow, however, gives rise to instabilities in two-dimensional ice flow numerical models (Schulkes et al. 1998). For this reason, Morland and Staroszczyk (1998) proposed a replacement of the abrupt cut-off by a smooth transition to zero stress over a dilatation-rate range equal to approximately one-tenth to one-hundredth of the maximum convergence-rate typically appearing in sea ice flow problems, which significantly improved the stability of numerical algorithms. Hence, a scaling factor $\bar{H}(\eta)$ (Staroszczyk 2005) defined by

$$\bar{H}(\eta) = \begin{cases} 1 & \text{if } \eta < 0, \\ \exp[-(\eta/\eta_c)^2] & \text{if } \eta \geq 0, \end{cases} \tag{4.99}$$

is adopted to reduce ice viscosities in a narrow range of divergence-rates, which is unity at $\eta = 0$, tends to zero as $\eta \to \infty$, and has zero derivatives at $\eta = 0$. The free parameter $\eta_c > 0$ is a divergence-rate magnitude around which significant changes in viscosities occur. Accordingly, with the function $\bar{H}$, the flow law (4.39) is modified to take, in components, the form

$$\sigma_{ij} = \left[(\zeta - \mu)D_{kk}\delta_{ij} + 2\mu D_{ij}\right]\bar{H}(\eta) \quad (i, j, k = 1, 2), \tag{4.100}$$

with the indices i, j, k denoting either r or θ.

The stress tensor components given by the flow law (4.100), with the strain-rate tensor components expressed by (4.95), after their insertion into the definitions (4.93), yield the in-plane axial and shear forces in the forms

$$
\begin{aligned}
N_r &= (H_1 + H_2)\frac{\partial v_r}{\partial r} + (H_1 - H_2)\frac{1}{r}\left(v_r + \frac{\partial v_\theta}{\partial \theta}\right), \\
N_\theta &= (H_1 - H_2)\frac{\partial v_r}{\partial r} + (H_1 + H_2)\frac{1}{r}\left(v_r + \frac{\partial v_\theta}{\partial \theta}\right), \\
N_{r\theta} &= H_2\left[\frac{\partial v_\theta}{\partial r} + \frac{1}{r}\left(\frac{\partial v_r}{\partial \theta} - v_\theta\right)\right].
\end{aligned}
\tag{4.101}
$$

Similarly, the moment definitions (4.94), when combined with the relations (4.100), (4.97) and (4.98), express the bending and twisting moments as

$$
\begin{aligned}
M_r &= -\left[(R_1 + R_2)\frac{\partial^2 \dot{w}}{\partial r^2} + (R_1 - R_2)\frac{1}{r}\left(\frac{\partial \dot{w}}{\partial r} + \frac{1}{r}\frac{\partial^2 \dot{w}}{\partial \theta^2}\right)\right], \\
M_\theta &= -\left[(R_1 - R_2)\frac{\partial^2 \dot{w}}{\partial r^2} + (R_1 + R_2)\frac{1}{r}\left(\frac{\partial \dot{w}}{\partial r} + \frac{1}{r}\frac{\partial^2 \dot{w}}{\partial \theta^2}\right)\right], \\
M_{r\theta} &= 2R_2\frac{\partial^2}{\partial r \partial \theta}\left(\frac{\dot{w}}{r}\right).
\end{aligned}
\tag{4.102}
$$

The parameters H_1, H_2, R_1 and R_2 in Eq. (4.102) describe the viscous properties of the ice plate and are defined by relations (4.74) on p. 97, with the bulk and shear viscosities ζ and μ now replaced by $\zeta\bar{H}(\eta)$ and $\mu\bar{H}(\eta)$, accordingly.

Substitution of the moment expressions (4.102) into the equilibrium relation (4.91), with the distributed load q_z given by (4.92), yields the differential equation for the plate deflection function w in the form

$$
\begin{aligned}
R\,\nabla^2\nabla^2\dot{w} = {} & N_r\frac{\partial^2 w}{\partial r^2} + N_\theta\frac{1}{r}\left(\frac{\partial w}{\partial r} + \frac{1}{r}\frac{\partial^2 w}{\partial \theta^2}\right) + 2N_{\theta r}\frac{\partial^2}{\partial r \partial \theta}\left(\frac{w}{r}\right) + \\
& - q_r\frac{\partial w}{\partial r} - \frac{q_\theta}{r}\frac{\partial w}{\partial \theta} - \varrho_w g w,
\end{aligned}
\tag{4.103}
$$

where

$$
\nabla^2 = \frac{\partial^2}{\partial r^2} + \frac{1}{r}\frac{\partial}{\partial r} + \frac{1}{r^2}\frac{\partial^2}{\partial \theta^2}
\tag{4.104}
$$

is the Laplace operator expressed in polar coordinates, and $R = R_1 + R_2$, as defined by (4.76).

The above Eq. (4.101) for the in-plane forces, and (4.103) for the plate deflection evolution, have been derived for the ice creep behaviour described by the viscous fluid flow law (4.39), in its slightly modified version given by (4.100) with (4.99). When the viscous-plastic rheological model (4.49) is used instead to describe the creep of ice, then the ensuing equations are similar, which is due to the formal similarities between the flow laws (4.39) and (4.49). Both laws include the same two viscosity parameters, μ and ζ (though their physical meanings are different in the two laws), and additionally two ice strength parameters P_1 and P_2 enter the viscous-plastic flow relation (4.49). This makes the resulting equations more elaborate compared to (4.101) and (4.103), but the formal structure of the equations is retained.

Numerical Simulations

The system of two partial differential equations (4.88) with (4.93) for the unknown ice horizontal velocity components v_r and v_θ, and Eq. (4.103) for the unknown plate deflection w, was solved numerically by applying a finite-difference method in order to simulate the creep behaviour of a coherent ice cover interacting with a cylindrical structure (Staroszczyk 2005, 2006). The simulations were carried out for two sea ice rheological models: the non-linearly viscous fluid flow law (4.39) with its modification given by (4.100), and the viscous-plastic flow law described by (4.49). Owing to the symmetry of the problem with respect to the wind direction which was assumed to blow along the coordinate line $\theta = 0$, only the region $0 \leq \theta \leq \pi$ has been considered in the numerical model. In the radial direction, the ice domain was assumed to extend from the cylinder wall at $r = R_0$ to the free edge of the ice cover at $r = R_{max}$. The adopted computational mesh had 300 discrete nodes in the radial direction and 61 nodes in the circumferential direction, uniformly distributed along both r and θ ranges, so that there were 18,300 nodes in all, with 54,900 unknown values of the ice velocities and the plate deflections to be calculated.

At the ice–structure contact surface either no-slip (full bonding) or free-slip boundary conditions were assumed for the ice horizontal deformation, and the simply supported conditions for the ice plate bending. For a no-slip boundary these conditions are expressed by

$$r = R_0 : \quad \boldsymbol{v} = \boldsymbol{0}, \quad w = 0, \quad M_r = 0, \tag{4.105}$$

and for a free-slip boundary by

$$r = R_0 : \quad \boldsymbol{v} \cdot \boldsymbol{n} = 0, \quad N_{r\theta} = 0, \quad w = 0, \quad M_r = 0, \tag{4.106}$$

where $\boldsymbol{n}$ denotes the unit vector normal to the cylinder wall. The ice at the outer edge $r = R_{max}$ was assumed to be stress-free, that is,

$$r = R_{max} : \quad N_r = 0, \quad N_{r\theta} = 0. \tag{4.107}$$

Regarding the initial conditions, it was assumed that at the start of simulations the floating ice was undeformed and stress-free. The results presented below correspond to the ice flow stages when the magnitudes of the forces sustained by a structure attain their maximum values.

The simulations were carried out for a cylinder of the radius $R_0 = 10$ m, situated at the centre of a circular ice field extending to $R_{max} = 500$ m, with the thickness of the ice cover equal to $h = 0.2$ m. The ice was assumed to be driven onto the structure by a wind of a constant velocity, blowing along the coordinate line $\theta = 0$ in the negative direction of r. The dimensionless wind and drag coefficients appearing in (4.77) on p. 98 were adopted of the values $C_a = 2 \times 10^{-3}$ and $C_w = 4 \times 10^{-3}$, and the air and water densities were $\varrho_a = 1.3\,\text{kg}\,\text{m}^{-3}$ and $\varrho_w = 1.02 \times 10^3\,\text{kg}\,\text{m}^{-3}$.

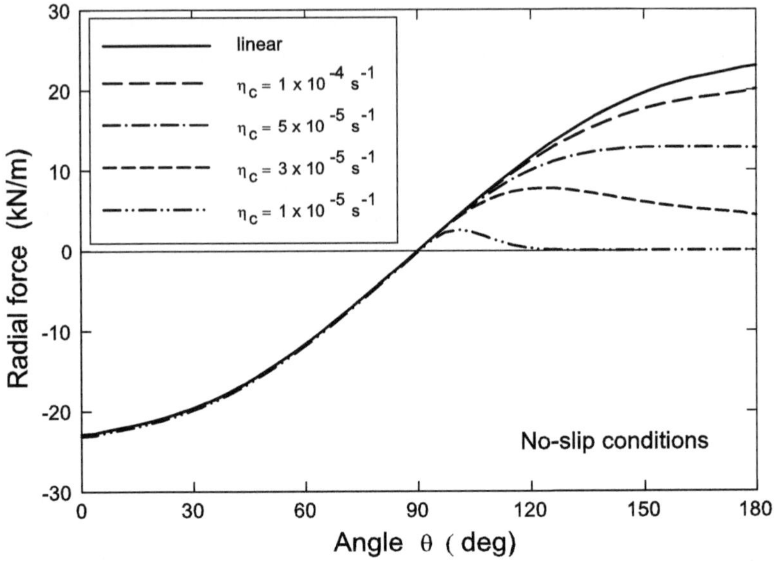

Fig. 4.26 Distribution of the radial forces N_r along the cylinder wall for no-slip boundary conditions and different magnitudes of the rheological parameter η_c

As first, the results obtained for the ice treated as a viscous fluid are presented. They have been obtained for the wind velocity $u_a = 30\,\mathrm{m\,s^{-1}}$ (such a wind generates a tangential stress $\tau_a \approx 2.3$ Pa on the ice surface). The viscous fluid rheological model involves three material parameters: two viscosities μ and ζ, and the critical strain-rate parameter η_c which describes the tensile ice strength reduction rate at the beginning of diverging flow. The adopted values of the viscosities were $\mu = 1.0 \times 10^9\ \mathrm{kg\,m^{-1}\,s^{-1}}$ and $\zeta = 2.0 \times 10^9\ \mathrm{kg\,m^{-1}\,s^{-1}}$, which can be regarded as typical viscosity magnitudes for floating ice. Since the maximum horizontal strain-rates occurring in the problem considered are of magnitudes equal to about $5 \times 10^{-5}\ \mathrm{s^{-1}}$, the parameter η_c has been adopted from within a range embracing the latter value. Thus, η_c has been chosen to vary from 5×10^{-6} to $1 \times 10^{-4}\ \mathrm{s^{-1}}$ to explore the effect of η_c on the magnitudes of the total contact forces exerted by ice on the cylinder.

The plots in Figs. 4.26 and 4.27 present the distribution of the forces exerted by the ice cover on the cylinder wall in the case of no-slip boundary conditions defined by (4.105). Illustrated is the dependence of the loads on the wall on the magnitude of the rheological parameter η_c; for comparisons, the contact forces generated in the case of a linearly viscous response of ice are also plotted. Figure 4.26 shows the variation of the radial force N_r with the angle θ. It can be noted that the effect of the ice viscosity reduction in diverging flow occurring at the leeward side of the cylinder ($90° < \theta \leq 180°$) is hardly observed on the opposite, windward part of the wall ($0° \leq \theta \leq 90°$), where the forces are practically insensitive to the value of η_c. In stark contrast, the radial forces on the leeward side decrease dramatically with decreasing η_c. The results in the plots suggest that for realistic modelling of the

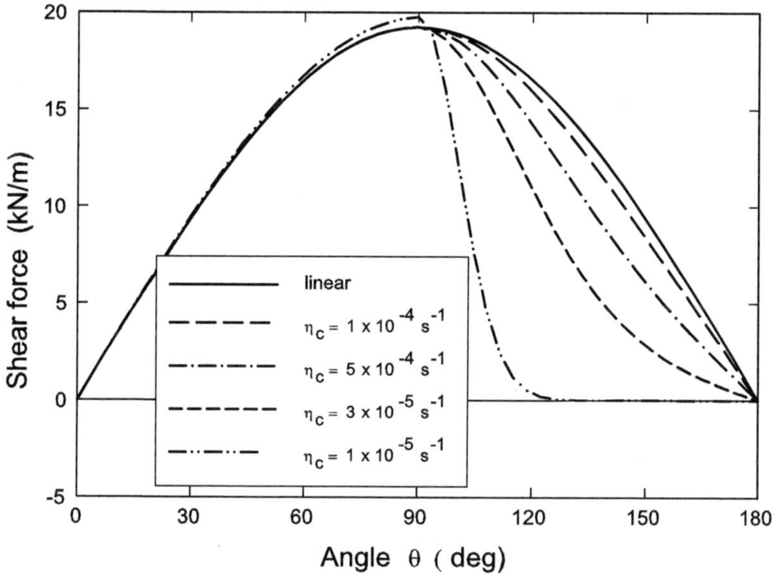

Fig. 4.27 Distribution of the shear forces $N_{r\theta}$ along the cylinder wall for no-slip boundary conditions and different magnitudes of the rheological parameter η_c

floating ice creep behaviour (small tensile strength of ice compared to its compressive strength), the magnitudes of the critical dilatation-rate η_c should be chosen of the order 1×10^{-5} s^{-1}.

A similar pattern is seen in Fig. 4.27, illustrating the variation of the contact shear forces $N_{r\theta}$ with the angle θ and the value of the rheological parameter η_c. Again, the tangential forces exerted on the walls on the windward side of the cylinder are roughly independent of η_c, while those on the leeward side rapidly approach zero values with η_c approaching the value 1×10^{-5} s^{-1}. Comparing this figure with the previous one, the change in the loading coming from the ice cover, for small values of η_c, seems even more dramatic.

Figure 4.28 displays the distribution of the normal forces N_r on the cylinder wall in the case of free-slip boundary conditions (4.106), when the tangential forces $N_{r\theta}$ are zero by definition. Due to $N_{r\theta} \equiv 0$, the whole loading from the ice is passed on the cylinder walls through the normal contact forces. For this reason, the normal forces at $\theta = 0$ (the wind direction) are by about 40 per cent larger than those in the case of no-slip conditions at the interface, see Fig. 4.26. Otherwise, qualitatively very similar features are observed in the plots for the free-slip and no-slip boundary conditions at the walls, with practically unchanged contact loading on the windward side, and a significant reduction of the N_r forces on the leeward side of the cylinder for the critical dilatation-rates $\eta_c \sim 1 \times 10^{-5}$ s^{-1}.

Of a particular interest to civil engineers are the magnitudes of total forces sustained by a cylindrical structure during its interaction with floating ice. These

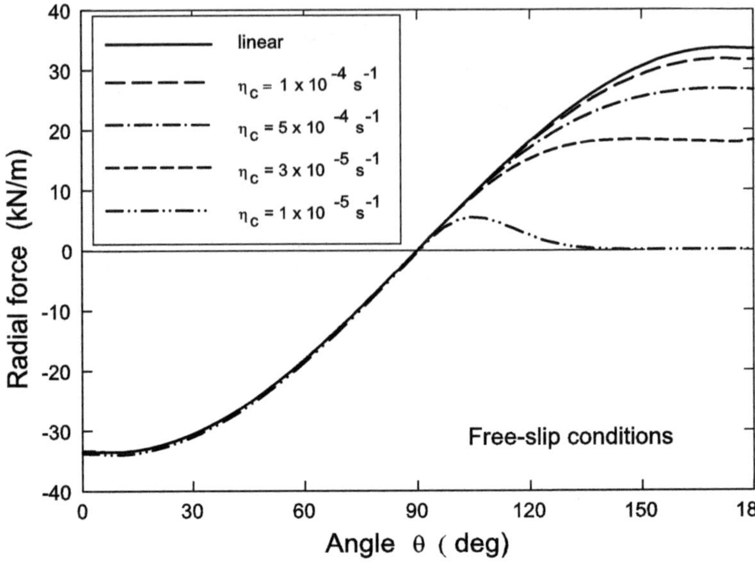

Fig. 4.28 Distribution of the radial forces N_r along the cylinder wall for free-slip boundary conditions and different magnitudes of the rheological parameter η_c

Table 4.2 Values of the total horizontal force F exerted on a cylindrical structure by the sea ice cover as a function of the rheological parameter η_c, for different cylinder diameters R_0 and boundary conditions on the wall ($\eta_c \to +\infty$ corresponds to the linearly viscous fluid solution). Results for the wind velocity of 30 m s^{-1}

η_c (s^{-1})	F (MN)					
	$R_0 = 5$ m		$R_0 = 10$ m		$R_0 = 20$ m	
	No-slip	Free-slip	No-slip	Free-slip	No-slip	Free-slip
$+\infty$	1.142	0.890	1.342	1.138	1.513	1.355
1×10^{-4}	1.026	0.859	1.296	1.114	1.497	1.343
5×10^{-5}	0.838	0.780	1.183	1.050	1.454	1.309
1×10^{-5}	0.669	0.490	0.792	0.615	0.989	0.828

magnitudes, obtained by integrating the radial and shear forces N_r and $N_{r\theta}$ along the whole perimeter of the cylinder, are listed in Table 4.2. Compared are the results for different cylinder diameters R_0, for different values of the critical dilatation-rate η_c, and for the two types (no-slip and free-slip) of the boundary conditions.

The viscous-plastic behaviour of sea ice predicted by the constitutive law (4.49) is determined by the values of the four constitutive parameters: P_1, P_2, e and Δ_c, from among which the first and the last, the ice compressive strength P_1 and the critical strain-rate invariant Δ_c, are most important in terms of quantitative results. Regarding the compressive strength of ice, P_1, there is no clarity in the literature as to its most proper magnitude. In the original formulation of the viscous-plastic model,

Hibler (1979) used the value 5×10^3 Pa for large-scale Arctic ice simulations, and the latter value was subsequently used by him and co-authors in a number of papers (Ip et al. 1991; Hibler and Ip 1995). Flato and Hibler (1992), in turn, applied a larger value, 2.75×10^4 Pa, also for describing the large-scale behaviour of ice. In our simulations, a value of $P_1 = 5 \times 10^4$ Pa which is slightly larger than the latter one was adopted, in belief that the strength of ice increases with decreasing spatial scales encountered in civil engineering applications, in accordance with the empirical data discussed in Sect. 3.4. Regarding the magnitude of the critical strain-rate invariant, a value of $\Delta_c = 2 \times 10^{-5}$ s^{-1} was used in the simulations. The latter value was adopted on the basis of the results presented above for the viscous fluid rheology, showing that the most realistic predictions were obtained for $\eta_c \sim 1 \times 10^{-5}$ s^{-1}, and also assuming that the strain-rate invariants η and γ are of comparable magnitudes in the relation (4.44) defining Δ_c. The rheological model parameter e, defining the shape of the yield curve, and hence the magnitude of the shear viscosity relative to the bulk viscosity, was commonly assumed (Hibler 1979) as 2 (implying $\mu/\zeta = 1/4$). In our simulations the range $1 \leq e \leq 3$ was explored. Finally, the remaining constitutive model parameter, P_2, used in the flow law (4.49) to define the tensile strength of ice, was adopted as a small fraction of the compressive strength P_1. Accordingly, a value $P_2 = 1 \times 10^3$ Pa, that is, $P_2 = P_1/50$ was used in the simulations. Recall that the small parameter $P_2 > 0$ was introduced in (4.49) to avoid numerical instabilities encountered in earlier viscous-plastic rheological models. Computational tests showed that this parameter (as long as it is small) has a very limited effect on the magnitudes of forces sustained by an engineering object, since most of the loading on structure walls comes from the ice that is under compression on the windward side of the structure.

The results plotted in Figs. 4.29 and 4.30 illustrate the distributions of the forces exerted by the ice on the structure walls in the case of no-slip boundary conditions (4.105). Shown is the dependence of the ice–structure contact forces on the rheological parameter e; that is, the effect of the ratio of the shear to bulk viscosities of ice is presented. The range of e varying from 1 to 3 corresponds to the viscosity ratios μ/ζ (or μ_m/ζ_m) decreasing from 1 (for $e = 1$) to $1/9 \sim 0.111$ (for $e = 3$), with the bulk viscosity ζ held constant for a given value of the strain-rate invariant Δ, as prescribed by relations (4.46) on p. 80. Figure 4.29 illustrates the variation of the radial force N_r with the polar angle θ. One can observe that the effect of the shear viscosity μ on the magnitude of N_r, for the no-slip boundary, is moderate, especially on the windward side of the structure. A little surprising is the prediction that most of the cylinder walls (for the no-slip conditions) is under the action of compressive contact forces.

The distribution of the shear forces $N_{r\theta}$ on the cylinder wall is shown in Fig. 4.30. It is seen that the shear forces exerted by the floating ice vary smoothly with the angle θ, with maximum values occurring at the angle $\theta \sim 60°$. Thus, the magnitudes of $N_{r\theta}$ on the windward side of the structure are larger than those on the leeward side, though the differences are not considerable, especially for smaller values of the shear viscosity (larger values of the parameter e).

Figure 4.31 illustrates the variation of the normal contact forces N_r with the angle θ and the rheological parameter e in the case of a free-slip boundary (4.106), when the

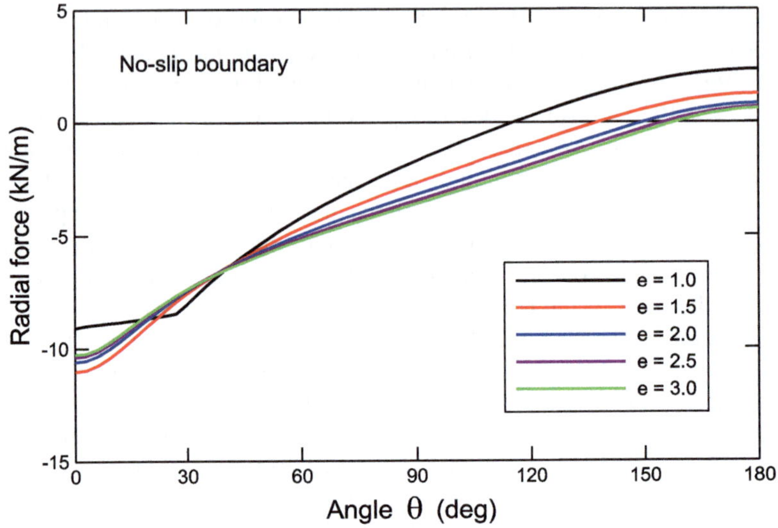

Fig. 4.29 Distribution of the radial forces N_r along the cylinder wall for no-slip boundary conditions and different values of the rheological parameter *e*

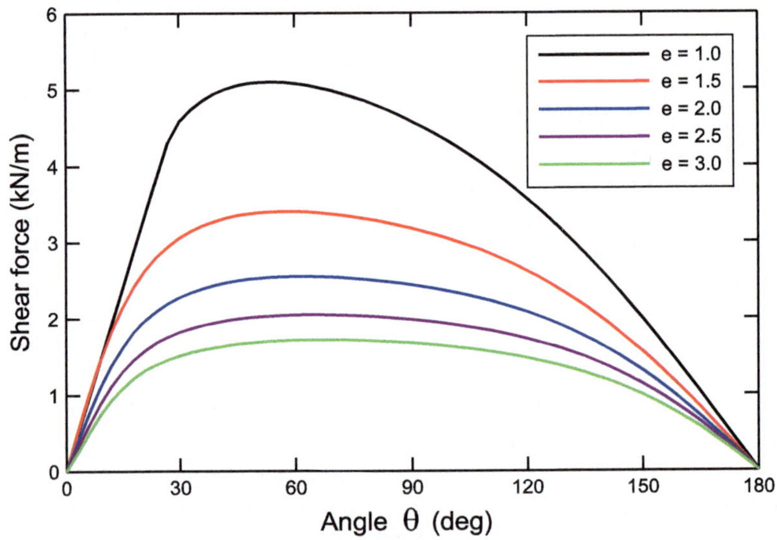

Fig. 4.30 Distribution of the shear forces $N_{r\theta}$ along the cylinder wall for no-slip boundary conditions and different values of the rheological parameter *e*

tangential forces $N_{r\theta}$ are identically zero. Comparing this figure with the analogous plots in Fig. 4.29 for the no-slip boundary conditions, one can note qualitatively distinct distributions of the radial forces along the cylinder walls. While in the no-slip case the forces N_r vary in a monotonic manner over the entire range of the

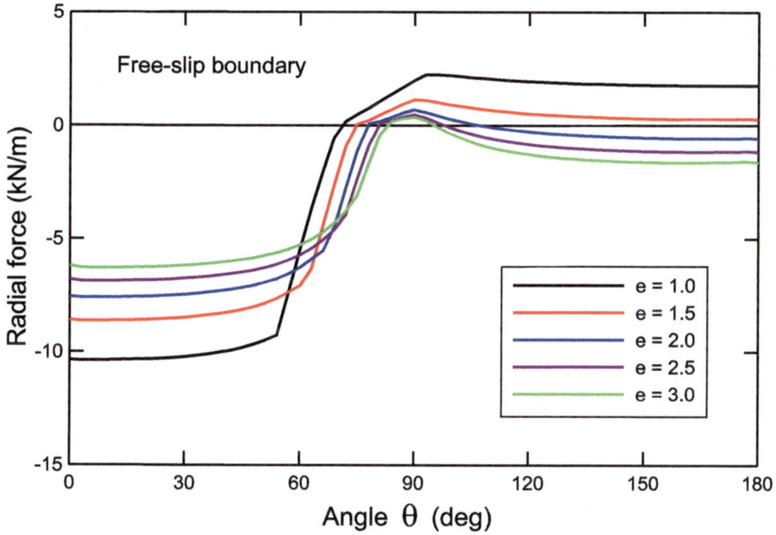

Fig. 4.31 Distribution of the radial forces N_r along the cylinder wall for free-slip boundary conditions and different values of the rheological parameter e

angles θ, some rapid changes in the magnitudes of N_r are predicted by the viscous-plastic rheological model in the case of the free-slip boundary. These changes occur within the range of the angles $60° \lesssim \theta \lesssim 90°$, where the normal forces switch from compressive to tensile ones, with the maximum tensile forces occurring at $\theta \sim 90°$. It turns out that the above dramatic changes in the contact forces within the range $60° \lesssim \theta \lesssim 90°$ are associated with the change in the creep behaviour of ice, which is in viscous flow for the latter range of θ, in contrast to the rest of the wall, where it is in plastic yield (Staroszczyk 2006). As concerns the case of the no-slip boundary conditions illustrated in Fig. 4.29, the ice is in plastic flow for all e and θ, except for the case of $e = 1$ (the solid line) when, for $\theta \lesssim 30°$, viscous deformation of ice takes place.

More results, regarding the ice horizontal deformation-rates and transverse plate deflection variations in time and space in the vicinity of a cylindrical structure interacting with sea ice, can be found in the papers by Staroszczyk (2005, 2006).

4.3 Ice Floe Impact on an Engineering Structure

In the previous part of this section, ice–structure interaction problems are considered in which, due to the stress, deformation or deformation-rate levels involved, the behaviour of ice can be sufficiently well approximated by that of a continuous slab of ice floating on the surface of water. Hence, the ice can be treated as an either elastic or creeping material which deforms in a continuous (ductile) manner, and it has been assumed that the ice remains in perfect contact with an engineering structure walls throughout an interaction event.

A different situation arises when strains, strain-rates and stresses in ice reach the magnitudes at which cracks start to develop in the material, giving rise to the brittle fracture of ice. Typically, sea ice undergoes a transition from ductile to brittle behaviour when stresses exceed the fracture strength of the material (that is, about 5 MPa in compression and about 1 MPa in tension), or strains exceed the value of about 0.01, or strain-rates reach the level of about 10^{-4} to 10^{-3} s^{-1} (Hawkes and Mellor 1972; Sanderson 1988; Schulson and Duval 2009) . During this creep-to-brittle transition phase (see Sect. 3.4), the loads exerted by floating ice on a structure attain their maximum or near-maximum values, and further increase in the ice deformation and its rate, usually associated with the fast-progressing process of crack formation and their subsequent growth, does not increase the forces in ice. Typically, the brittle failure of ice takes place, at a given time instant, only at a number of relatively small regions of the ice–structure interface, therefore the total loads exerted by ice on the object show a highly irregular variation in time, with characteristic sharp spikes appearing at irregular time intervals.

In order to investigate the main features of the mechanism of brittle failure of floating ice during its interaction with an engineering object, a problem is considered in which an ice floe impacts dynamically on a rigid cylindrical structure. It is believed that, in spite of a number of simplifying assumptions adopted in the course of the analysis, the results obtained will realistically describe the complex nature of the dynamic ice–structure interaction phenomenon, and will be in reasonable agreement with the behaviour of ice observed in Arctic seas (Sanderson 1988; Jordaan 2001).

As already noted in Sect. 3.4, the modelling of the ice fracture mechanism is difficult, and requires the knowledge of advanced methods of mechanics (Ashby and Hallam 1986; Sjölind 1987; Nixon 1996; Pralong et al. 2006). The derived theoretical solutions seem to be by far too complicated to be effectively implemented into realistic, engineering applications. Therefore, a simple approach (Staroszczyk 2007), extending the method suggested by Sanderson (1988), is applied to construct a model that enables the estimation of forces exerted by brittle-failing ice on a vertically-walled rigid structure. In this approach, the mechanism of ice fracture is described, essentially, by only three physical parameters: (1) axial compressive fracture stress, (2) an associated axial strain at which the fracture occurs, and (3) ice clearing axial stress, which is a stress occurring in already fractured blocks of ice. There is no doubt that many interesting small-scale processes occurring during the fracture of ice are disregarded in this way. However, such small-scale effects are deemed unimportant for the purpose of this analysis, the objective of which is to evaluate total net forces sustained by the structure during its dynamic impact by a large ice floe.

An ice floe that hits the rigid structure is treated as a compact plate of uniform thickness. The interaction between the moving floe and the structure vertical wall is assumed to occur, at any time instant, at a number of small zones, with local fracture events taking place non-simultaneously at different points of the ice-structure contact interface. The local failure of ice at each small zone is supposed to occur independently of the other zones, and all these independent local fracture events are treated as separate random processes. The total force sustained by the structure is then determined as a statistical sum of individual loads occurring at all the contact zones. By simulating numerically a large number of separate floe–structure collision events, with randomly varied input parameters, the probability distributions of the total contact force magnitudes during an ice–structure impact phenomenon are calculated. In particular, the probability distributions illustrating the dependence of peak interaction forces on the floe thickness, its size and its initial velocity are presented. These distributions can be used by an engineer to perform a risk assessment analysis for an off-shore structure during the stage of its designing.

4.3.1 Fracture of Ice at a Structure Wall

The problem under consideration is sketched in Fig. 4.32. An ice floe, being initially at some distance from a rigid structure, is driven by wind and/or water current drag forces towards the object at a horizontal free-drift velocity V_0. After arriving at the structure at time t_0 and establishing first contact with its walls, the impacting ice floe starts to break at points at which the magnitudes of local contact stresses exceed the brittle fracture strength of ice. As the ice undergoes crushing, its chunks pile up or sink near the structure walls, and the initial kinetic energy of the floe is gradually dissipated. The floe velocity, $v(t)$, with t measuring the time elapsed from the instant t_0, steadily decreases, until the floating ice slab eventually comes to rest.

In real field conditions, an ice floe leading edge is commonly irregular in shape. Thus, the contact between the floe and the structure wall is unlikely to take place

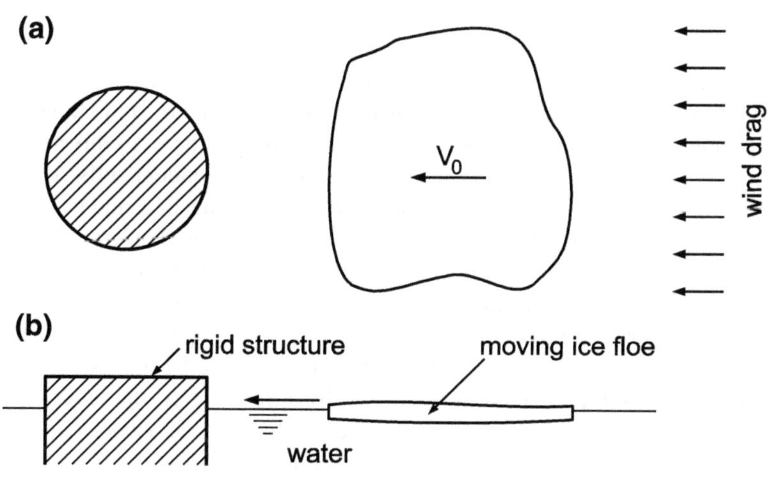

Fig. 4.32 Definition of the ice floe impact problem: **a** planar and **b** cross-sectional views

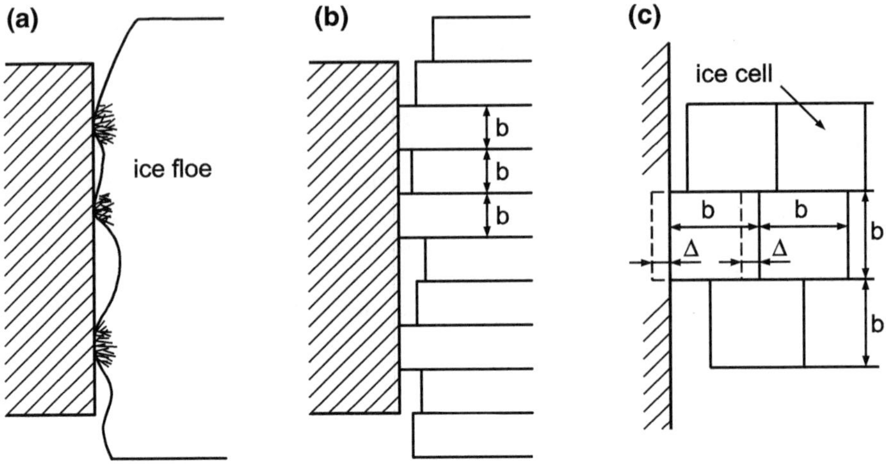

Fig. 4.33 **a** Imperfect contact between an ice floe and a structure wall, **b** problem idealization, **c** detailed view and definitions

over the entire possible interface between the ice and the object. Instead, as illustrated in Fig. 4.33a, the ice interacts with the structure walls at a number of locations of small areas, and these contact locations change all the time as the ice floe advances. Once the ice–structure interaction has been initiated and the ice starts to fail, broken blocks of ice of various size and shape are formed in a chaotic manner at each local interaction zone. These local ice fracture events occur non-simultaneously, as different ice fragments fail at different times at different places at the interface surface. At any one small contact zone, the ice fragments are supposed to arrive and fail one by one: as one fragment fails and the debris is cleared by the process of ice piling up or sinking, another ice fragment arrives immediately to start its interaction with the wall and to fail after some time, etc.

In order to model such a complex interaction phenomenon as depicted above, a method that refines the approach originally proposed by Ashby and Hallam (1986) and subsequently followed by Sanderson (1988) is applied here. In this method, the floe is treated as a collection of regular in shape and independent cells, as shown in Fig. 4.33b, c. Each discrete cell is assumed to have the same size, and to be a square of dimensions $b \times b$ in the horizontal plane. As observed by Sanderson (1988), the characteristic dimension b of the fractured ice blocks is similar to the ice thickness h; therefore, it is assumed in this analysis that $b \sim h$. As a particular cell starts to interact with the structure (see Fig. 4.33c), it is supposed that it fails (that is, the axial stress component normal to the contact surface reaches the fracture strength of ice) when the whole discrete element is advanced by a distance Δ; the latter parameter represents a critical displacement at which the crushing of ice occurs.

The history of loading experienced by a discrete zone (of width b) at the contact interface, as successive discrete ice blocks arrive and fail there, is idealized in a manner illustrated in Fig. 4.34, adapted from Sanderson (1988). When a given ice

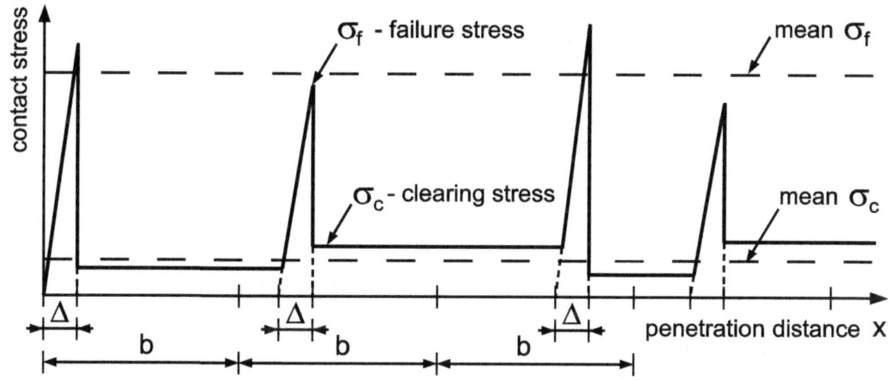

Fig. 4.34 Contact stress history for a single discrete zone

block comes to the rigid wall and then moves by a distance Δ, the contact stress is assumed to grow linearly from zero to its peak value, equal to the ice fracture strength, σ_f. Next, immediately after the failure of ice, the contact stress falls sharply to a much lower level, σ_c, which is a stress in ice caused by forces that are needed to clear the debris formed during the failure (that is, to move the fractured ice fragments up or down, since the debris cannot be cleared by pushing it aside, in the direction lateral to the impact direction). This clearing stress is supposed to remain constant until the time when the next ice block arrives at the wall and starts to fail, rising the stress gradually to the σ_f level again, etc. The failure and clearing stresses, σ_f and σ_c, are assumed to have different magnitudes for different ice cells, in order to reflect both a stochastic character of the fracture mechanism and an associated statistical scatter in available empirical data. The respective mean values of σ_f and σ_c are illustrated in the figure by the two horizontal dashed lines. For simplicity, the variations of σ_f and σ_c about their mean values are supposed to follow the normal Gauss distribution (though it is possible that the Weibull distribution might be more appropriate). Moreover, the distance between consecutive failure stress peaks is not uniform, but also varies in a stochastic manner; in such a way the randomness of an individual ice block size is accounted for. It is assumed in the model that the average distance separating two successive failure stress peaks is equal to b—the average size of a fractured block. Further, it is supposed that there is an equal probability of a stress peak to lie anywhere within a given stretch of length b (that is, a uniform probability distribution function is used for this purpose).

The model for the mechanism of the ice–structure brittle interaction developed on the basis of the above-described ideas involves three main parameters: the stress magnitudes σ_f and σ_c (their mean values and statistical variation) and the critical displacement Δ at which an ice block of length b fails. The latter parameter will be expressed by means of a critical axial strain, ϵ_f, being the strain at which brittle fracture occurs.

Of the above three parameters, the failure stress level σ_f, equal to the brittle fracture strength of ice, is the most significant. As already noted in Sect. 3.4, the ice fracture mechanism exhibits a pronounced scale-dependence, illustrated by the pressure–area curve plotted in Fig. 3.11 on p. 44. The weakening of ice strength with increasing contact area can be roughly approximated by a functional relationship expressed by

$$p \propto A^{-\beta}, \quad \beta > 0, \tag{4.108}$$

where p is the pressure in ice at failure, A denotes the contact area, and the symbol '$\propto$' means 'proportional to'. There is some discussion in the literature, concerning the most appropriate value of the parameter β in the above *pressure–area relationship*. Generally, it is accepted that β takes a value from the range $1/4$ to $1/2$, with the lower limit value appropriate for smaller scales ($A \lesssim 0.1\,\mathrm{m}^2$), and the upper limit value relevant for larger scales ($A \gtrsim 10^3\,\mathrm{m}^2$). In the model proposed here, the value $\beta = 1/4$ is adopted, which has been derived theoretically by Palmer and Sanderson (1991) and Xu et al. (2004) on the basis of a fractal analysis of the size distribution of fragmented sea ice. Hence, the ice size effect on the failure pressure, or the compressive fracture strength σ_f, is expressed in the following, normalized, form as

$$\sigma_f = \sigma_f^* \left(\frac{A}{A_0} \right)^{-1/4}, \tag{4.109}$$

where A_0 is a reference contact area, assumed here to be equal to $1\,\mathrm{m}^2$, and σ_f^* is a normalized ice failure strength (that is, that corresponding to A_0). Xu et al. (2004) recommend a value of $\sigma_f^* = 1.66\,\mathrm{MPa}$, as the one providing the best fit to empirical data. The latter value is particularly suitable for the case of contact areas of the order of $1\,\mathrm{m}^2$; that is, those occurring in typical engineering problems.

The ice failure strength defined by Eq. (4.109) represents its mean value. The experimental data for sea ice, however, show a significant statistical scatter. Sanderson (1988) carried out some detailed statistical calculations for the Arctic sea ice and found that the variation coefficient (the ratio of the standard deviation to the mean value) of the data for first-year ice is as high as about 45%, and for multi-year ice it is about 65%. As in our ice–structure interaction problem we are concerned mostly with first-year ice, the variation coefficient equal to 50% has been chosen to describe the scatter in possible values of σ_f^*. Moreover, on the basis of statistical analysis, Sanderson (1988) observed that the probability distribution of experimental data for σ_f^* is approximately normal, and, therefore, such a type of distribution will be used in the numerical simulations presented further in this section.

While the failure strength of ice, σ_f, can be relatively easily determined by small-scale indentation tests or large-scale observations, the magnitude of the ice clearing stress, σ_c, the second free parameter in the proposed model, is more difficult to identify, since, to the author's knowledge, no in situ measurements of this quantity have been conducted yet during real ice floe impact events. For this reason, several authors have made attempts to estimate the ice clearing stress levels indirectly, by theoretical arguments. For instance, Sanderson (1988) analysed the work done against gravity which is required to rise or sink ice fragments after their failure, and compared it to the work done by the forces acting in impacting ice. On this basis it has been inferred that the typical clearing stress is appreciably smaller than the failure stress, and the value of $\sigma_c = 0.05\,\text{MPa}$ has been proposed for multi-year ice. It seems, however, that for thinner, first-year ice considered here the clearing stresses are even smaller. Therefore, the value of 0.02 MPa has been adopted as a mean magnitude of σ_c, together with a 50% variation coefficient accounting for its possible scatter, which is the same value of the variation coefficient as that assumed for the failure stress scatter.

Finally, the third free parameter in the proposed ice floe impact model is the critical strain ϵ_f, developing during the process of brittle crushing of ice, and determining the critical axial displacement Δ (see Fig. 4.34) through the relation $\Delta = \epsilon_f b$. There are some experimental data regarding the strain magnitudes at which ice fractures, but these are limited to small-scale laboratory tests on fresh-water ice samples (Schulson and Gratz 1999; Iliescu and Schulson 2002), and thus have little relevance to large-scale field conditions. Therefore, as in the above case of σ_c, the value of the critical strain ϵ_f has been inferred by theoretical considerations (Sanderson 1988), and the values ranging from $\epsilon_f = 0.02$ to $\epsilon_f = 0.05$ have been obtained for multi-year ice. In the present analysis the lower value is adopted, that is $\epsilon_f = 0.02$, in belief that pre-failure strains that develop in young ice are smaller than those occurring in thick, multi-year ice.

4.3.2 Numerical Method

The ice floe impact model, based on the assumptions and simplifications discussed above, has been applied to simulate dynamic ice–structure interaction events. In these simulations, the direction of the floe movement is defined by an axis x, with the origin $x = 0$ on the structure wall and its increasing coordinate measuring the ice penetration distance, shown in Fig. 4.34 (see also Fig. 4.35 on p. 123). Further, it is assumed that the first contact between the floe and the structure takes place at time $t = 0$ and the floe velocity is then $v = V_0$.

The computations proceed in the following steps (Staroszczyk 2007):

1. Given the initial geometry of an ice floe, its in-plane dimensions and a mean thickness h, the ice sheet is discretized in the way shown in Fig. 4.33, by choosing the ice cell size b of a magnitude close to h. Also the floe mass, m, and its initial kinetic energy are evaluated.
2. For each discrete contact zone at the ice–structure interface, a separate stochastic realization of loading, as illustrated in Fig. 4.34, is prescribed, with the values of the failure and clearing stress as well as the distance between consecutive failure stress peaks randomized about their mean values. For this purpose, standard

random number generators for the uniform and normal probability distributions are used.

3. At each calculation step, for $t > 0$, the floe is advanced by a small increment δx, chosen to be a fraction of the critical displacement $\Delta = \epsilon_f b$. For the current value of x, a local contact stress, determined from the respective realization of loading, is calculated for each discrete zone, and all these local stresses, multiplied by the respective local contact areas, are summed up to yield a total impact force, F, at current x and t.
4. Assuming that during a given displacement step k ($k = 1, 2, 3, \ldots$) the interaction force, F_k, is constant, the total work done by this force over the distance δx is determined as $F_k \delta x$. Equating that work with the amount of the total kinetic energy of the floe lost due to the decrease in its velocity from the value of v_{k-1} to v_k, the current velocity v_k can be evaluated from the relation

$$v_k^2 = v_{k-1}^2 - \frac{2F_k}{m}\,\delta x, \quad k = 1, 2, 3, \ldots, \quad v_0 = v(t = 0) = V_0. \tag{4.110}$$

5. Assuming a linear variation of the floe velocity at each step, the time that elapsed during the advance of the floe at the k-th step, denoted by $(\delta t)_k$, is calculated from the formula

$$(\delta t)_k = \frac{2\,\delta x}{v_{k-1} + v_k}\,. \tag{4.111}$$

All the time increments, added up over all preceding displacement steps, determine the current value of time t elapsed since the beginning of the interaction event.

The procedure outlined above yields time histories of the total interaction force F, the floe velocity v, and the penetration distance x for one particular stochastic realization of an impact event. For each such a realization, a magnitude of the maximum force F occurring during an entire interaction event is found, and then, by simulating a large number of random realizations, probability distributions for the peak ice impact loads F are calculated.

4.3.3 Simulations of Forces Exerted on a Structure

The proposed model has been used to simulate a series of dynamic impact events of the geometries shown in Fig. 4.35. The basic configuration investigated in the simulations is that depicted in Fig. 4.35a, showing a circular cylindrical structure of radius r_0 interacting with a circular ice floe of radius R_0 and an average thickness h. In order to examine the effect of the shape of the impacting floe edge on the forces sustained by the structure, also the configuration presented in Fig. 4.35b has been considered, in which the projection of the leading edge on the horizontal plane is a straight line normal to the direction of the floe movement.

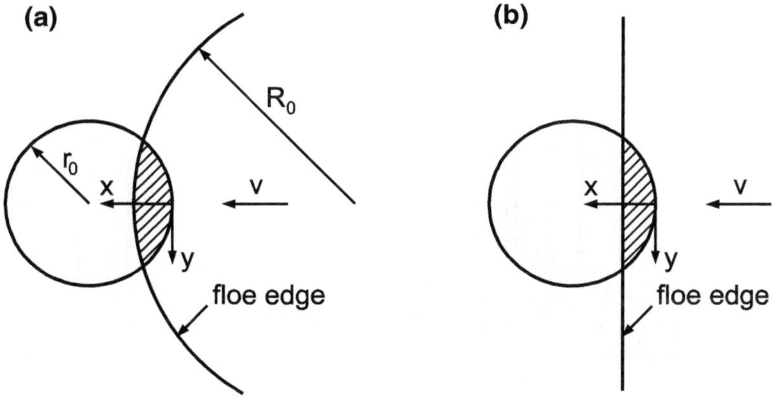

Fig. 4.35 Cylindrical structure of radius r_0 in contact **a** with a floe of radius R_0, **b** with a straight-edged floe. The hatched areas show the regions of ice penetrated by the structure

The numerical computations have been carried out for a vertically-walled circular cylinder of radius $r_0 = 10\,\mathrm{m}$. As discussed earlier in Sect. 4.3.1, the three basic parameters of the model have the values: $\sigma_f^* = 1.66\,\mathrm{MPa}$, $\sigma_c = 0.02\,\mathrm{MPa}$ and $\epsilon_f = 0.02$. The ice density (needed to determine the total mass of the floe) was taken as $900\,\mathrm{kg\,m^{-3}}$. All probability distributions presented below have been obtained by running the model repeatedly for 10,000 times.

Typical time histories of (a) the total force, $F_x(t)$, acting on the structure in the direction of the floe advance, (b) the floe velocity, $v(t)$, and (c) the ice penetration distance, $x(t)$, are shown in Fig. 4.36. These results have been obtained for a floe of radius 100 m and thickness 0.5 m, moving towards the structure at the velocity of $0.5\,\mathrm{m\,s^{-1}}$. It can be seen that, for the particular realization illustrated, the impact event lasts nearly 25 s, the ice floe moves a distance of about 6.8 m before it comes to rest, and the peak forces exerted on the structure have magnitudes close to 1.6 MN. As anticipated, the time-variation of the impact forces is very irregular. On the contrary, the floe velocity v and its position x vary in a relatively smooth way. A characteristic feature is a gradual increase of peak forces as the collision progresses—this is because the total ice–cylinder contact area increases with the interaction time, so that the number of local zones at which ice fails increases.

The results plotted in Fig. 4.36 illustrate a single impact event, which due to the intrinsic randomness of the process may not be representative for the dynamic phenomenon under consideration. More general information can be obtained by a statistical analysis of a series (in our case 10,000) of randomized runs. It turns outs that, for the same input data as above, the mean values of the peak load x- and y-components (see Fig. 4.35) are $\bar{F}_x = 1.58\,\mathrm{MN}$ and $\bar{F}_y = 0.59\,\mathrm{MN}$, with respective standard deviations 0.18 and 0.08 MN. Similar ratios of the F_y to F_x components have been obtained for other combinations of the floe parameters (Staroszczyk 2007). This means that the average lateral peak force equals nearly $2/5$ of the average longitudinal component, indicating thus that the ice–structure interaction loads are

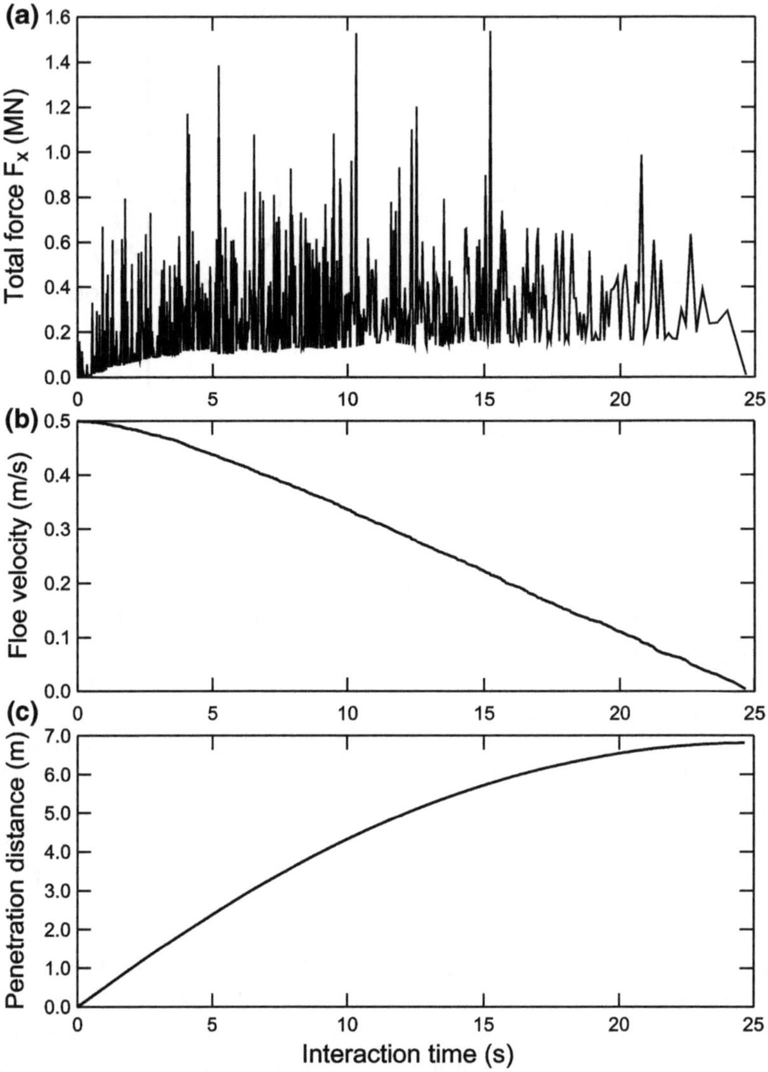

Fig. 4.36 **a** Typical history of the total force F_x, **b** the floe velocity and **c** the ice penetration distance during an impact event, for the floe radius $R_0 = 100\,\mathrm{m}$ and thickness $h = 0.5\,\mathrm{m}$, and its initial velocity $V_0 = 0.5\,\mathrm{m\,s^{-1}}$

far from symmetric with respect to the x-axis direction (the direction of the floe advancement).

The following three diagrams show density probability distributions of the peak ice–structure interaction forces. These figures illustrate the influence of the floe velocity and its thickness and planar size on the magnitude of the total longitudinal contact force F_x and the statistics of its occurrence. In the plots, for each value of

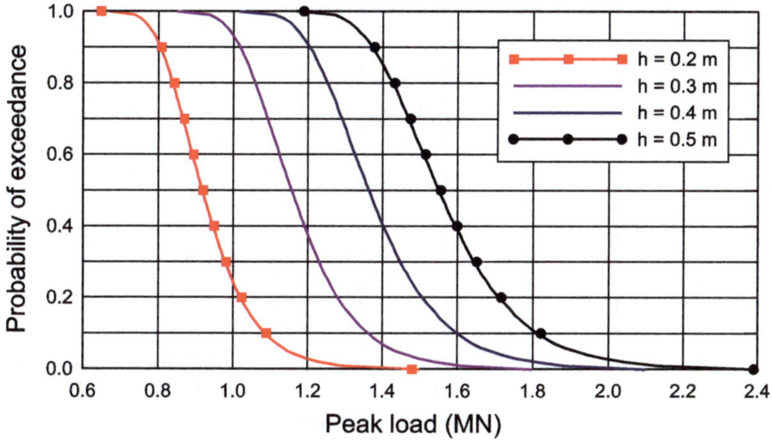

Fig. 4.37 Exceedance probability distributions of peak impact loads exerted on a structure for different floe thicknesses h ($R_0 = 100\,\mathrm{m}$, $V_0 = 0.5\,\mathrm{m\,s^{-1}}$). Compared are the results for circular (lines) and straight-edged (symbols) floes

the load F_x obtained from the simulations, the probability that this particular value will be exceeded is shown. Fig. 4.37 demonstrates the effect of the floe thickness h on the exceedance probabilities of the total impact loads, for the floe of radius $R_0 = 100\,\mathrm{m}$ and its initial velocity $V_0 = 0.5\,\mathrm{m\,s^{-1}}$. It is seen that the influence of the floe edge geometry on the load probability distributions is negligibly small—the maximum relative discrepancies are of order 1%. On the other hand, the influence of the ice thickness on the impact load magnitudes is, obviously, significant. However, the total loads sustained by the cylinder are not roughly proportional to the ice floe thickness, as could be expected at first sight, which is due to the scale-effects (the thinner ice has larger fracture strength σ_f^* than the thicker one).

The plots in Fig. 4.38 display the exceedance probability curves for the peak loads F_x as a function of the initial velocity of ice, V_0, with the ice floe thickness and radius kept constant. Hence, the probability distributions for F_x are plotted for the values of V_0 ranging from 0.2 to $0.5\,\mathrm{m\,s^{-1}}$. It can be noted that, despite an increase in the floe kinetic energy by a factor of 6.25 when the velocity changes from 0.2 to $0.5\,\mathrm{m\,s^{-1}}$, the magnitudes of the impact forces, at the same exceedance probability, increase only by a factor of about 1.25 to 1.4.

In Fig. 4.39 the floe size effect on the exceedance probabilities of the peak impact forces is illustrated, by presenting the results of simulations carried out for circular floes of radii R_0 varying between 50 and 200 m. In a way, the character of the exceedance curves resembles that in the previous diagram. Although the total kinetic energy of the moving floe increases significantly with increasing floe radius (by a factor of 16 between the smallest and the largest floes considered), the corresponding peak load magnitudes vary merely by a factor of about 1.5 at the same exceedance probability. This demonstrates once again the complexity of the interaction mechanism in which extensive brittle fracture of ice takes place.

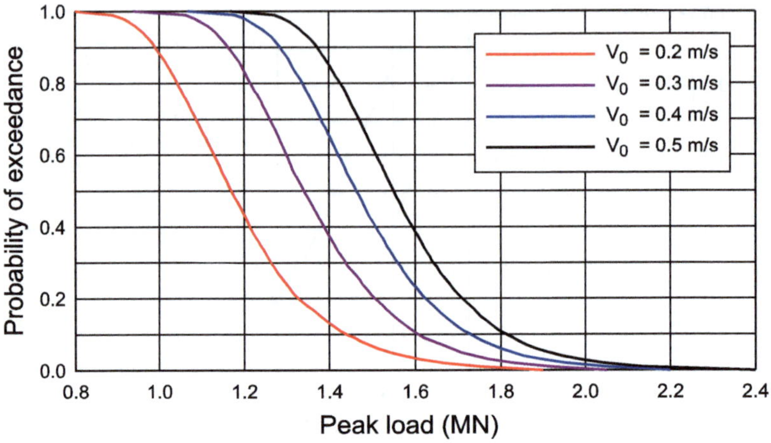

Fig. 4.38 Exceedance probability distributions of peak impact loads exerted on a structure for different initial floe velocities V_0 ($h = 0.5\,\mathrm{m}$, $R_0 = 100\,\mathrm{m}$)

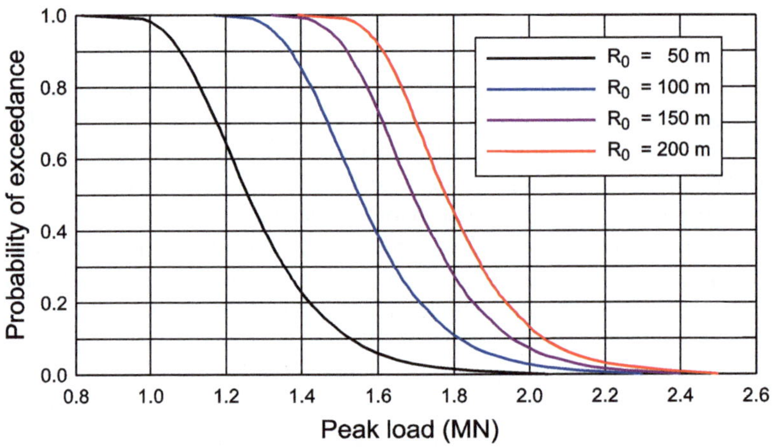

Fig. 4.39 Exceedance probability distributions of peak impact loads exerted on a structure for different floe radii R_0 ($h = 0.5\,\mathrm{m}$, $V_0 = 0.5\,\mathrm{m\,s^{-1}}$)

There is no doubt that the above-presented model considerably simplifies the real phenomenon of the ice floe impact on an engineering structure. However, its main purpose has been to provide an engineer with the estimations of the magnitudes of the total forces exerted by ice on a structure in order to carry out a risk assessment analysis, without a detailed consideration of the local mechanisms occurring in the immediate vicinity of the structure walls and the processes taking place in the ice itself. Such a more insightful analysis of the brittle behaviour of ice impacting a structure has become possible in the past decade with the fast development of the discrete-element method (DEM) and its application to sea ice problems. Some examples of the application of this still relatively new discrete method to the ice–structure interaction problems can be found in the papers by Polojärvi and Tuhkuri (2009) and Polojärvi et al. (2015). In recent papers by Herman (2016, 2017), the DEM has been employed for the numerical analysis of the problem of surface wave-induced breaking of floating ice. It seems that an extension of the latter model by accounting for the ice–structure and wave–structure interaction mechanisms would make possible a very realistic dynamic analysis of the coupled wave–ice–structure system (though, certainly, such an analysis would involve high computational costs typical of the discrete-element method applications).

References

Ashby MF, Hallam SD (1986) The failure of brittle solids containing small cracks under compressive stress-states. Acta Metall 34(3):497–510

Chadwick P (1999) Continuum mechanics: concise theory and problems, 2nd edn. Dover, Mineola, New York

Flato GM, Hibler WD (1992) Modeling pack ice as a cavitating fluid. J Phys Oceanogr 22(6):626–651

Gray JMNT, Morland LW (1994) A two-dimensional model for the dynamics of sea ice. Philos Trans R Soc Lond A 347(1682):219–290. https://doi.org/10.1098/rsta.1994.0045

Hawkes I, Mellor M (1972) Deformation and fracture of ice under uniaxial stress. J Glaciol 11(61):103–131

Herman A (2016) Discrete-element bonded-particle Sea Ice model DESIgn, version 1.3a—model description and implementation. Geosci Model Dev 9(3):1219–1241. https://doi.org/10.5194/gmd-9-1219-2016

Herman A (2017) Wave-induced stress and breaking of sea ice in a coupled hydrodynamic discrete-element wave-ice model. Cryosphere 11(6):2711–2725. https://doi.org/10.5194/tc-11-2711-2017

Hibler WD (1977) A viscous sea ice law as a stochastic average of plasticity. J Geophys Res 82(27):3932–3938

Hibler WD (1979) A dynamic thermodynamic sea ice model. J Phys Oceanogr 9(4):815–846

Hibler WD (2001) Sea ice fracturing on the large scale. Eng Fract Mech 68(17–18):2013–2043. https://doi.org/10.1016/S0013-7944(01)00035-2

Hibler WD, Ip CF (1995) The effect of sea ice rheology on Arctic buoy drift. ASME AMD 207:255–263

Hunke EC, Dukowicz JK (1997) An elastic-viscous-plastic model for sea ice dynamics. J Phys Oceanogr 27(9):1849–1867

Hutter K (1983) Theoretical glaciology. Material science of ice and the mechanics of glaciers and ice sheets. Reidel, Dordrecht

Iliescu D, Schulson EM (2002) Brittle compressive failure of ice: monotonic versus cyclic loading. Acta Mater 50(8):2163–2172

Ip CF, Hibler WD, Flato GM (1991) On the effect of rheology on seasonal sea-ice simulations. Ann Glaciol 15:17–25

Jordaan IJ (2001) Mechanics of ice–structure interaction. Eng Fract Mech 68(17–18):1923–1960

Kara AB, Wallcraft AJ, Metzger EJ, Hurlburt HE, Fairall CW (2007) Wind stress drag coefficient over the global ocean. J Clim 20(23):5856–5864. https://doi.org/10.1175/2007JCLI1825.1

Kerr AD (1978) On the determination of horizontal forces a floating ice plate exerts on a structure. J Glaciol 20(82):123–134

Kerr AD, Palmer WT (1972) The deformation and stresses in floating ice plates. Acta Mech 15(1–2):57–72. https://doi.org/10.1007/BF01177286
Lu P, Li Z, Cheng B, Leppäranta M (2011) A parameterization of the ice-ocean drag coefficient. J Geophys Res 116(C07):C07019. https://doi.org/10.1029/2010JC006878
Mellor M (1980) Mechanical properties of polycrystalline ice. In: Tryde P (ed) Proceedings of IUTAM symposium on physics and mechanics of ice, Copenhagen 1979. Springer, Berlin, pp 217–245
Morland LW (1993) The flow of ice sheets and ice shelves. In: Hutter K (ed) Continuum mechanics in environmental sciences and geophysics. Springer, Wien, pp 403–466
Morland LW (2001) Influence of bed topography on steady plane ice sheet flow. In: Straughan B, Greve R, Ehrentraut H, Wang Y (eds) Continuum mechanics and applications in geophysics and the environment. Springer, Berlin, pp 276–304
Morland LW, Staroszczyk R (1998) A material coordinate treatment of the sea-ice dynamics equations. Proc R Soc Lond A 454(1979):2819–2857. https://doi.org/10.1098/rspa.1998.0283
Nevel DE (1980) Bending and buckling of a wedge on an elastic foundation. In: Tryde P (ed) Proceedings of IUTAM symposium on physics and mechanics of ice, Copenhagen 1979. Springer, Berlin, pp 278–288
Nixon WA (1996) Wing crack models of the brittle compressive failure of ice. Cold Reg Sci Technol 24(1):41–55
Overland JE, Pease CH (1988) Modeling ice dynamics of coastal seas. J Geophys Res 93(C12):15 619–15 637. https://doi.org/10.1029/JC093iC12p15619
Palmer AC, Sanderson TJO (1991) Fractal crushing of ice and brittle solids. Proc R Soc Lond A 433:469–477
Polojärvi A, Tuhkuri J (2009) 3D discrete numerical modelling of ridge keel punch through tests. Cold Reg Sci Technol 56(1):18–29. https://doi.org/10.1016/j.coldregions.2008.09.008
Polojärvi A, Tuhkuri J, Pustogvar A (2015) DEM simulations of direct shear box experiments of ice rubble: force chains and peak loads. Cold Reg. Sci. Technol. 116:12–23. https://doi.org/10.1016/j.coldregions.2015.03.011
Pralong A, Hutter K, Funk M (2006) Anisotropic damage mechanics for viscoelastic ice. Continuum Mech Thermodyn 17(5):387–408
Rothrock DA (1975) The energetics of the plastic deformation of pack ice by ridging. J Geophys Res 80(33):4514–4519. https://doi.org/10.1029/JC080i033p04514
Sanderson TJO (1988) Ice mechanics. Risks to offshore structures. Graham and Trotman, London
Schulkes RMSM, Morland LW, Staroszczyk R (1998) A finite-element treatment of sea ice dynamics for different ice rheologies. Int J Numer Anal Methods Geomech 22(3):153–174
Schulson EM, Duval P (2009) Creep and fracture of ice. Cambridge University Press, Cambridge
Schulson EM, Gratz ET (1999) The brittle compressive failure of orthotropic ice under triaxial loading. Acta Mater 47(3):745–755
Sjölind SG (1985) Viscoelastic buckling analysis of floating ice sheets. Cold Reg Sci Technol 11(3):241–246
Sjölind SG (1987) A constitutive model for ice as a damaging visco-elastic material. Cold Reg Sci Technol 14(3):247–262
Smith GD, Morland LW (1981) Viscous relations for the steady creep of polycrystalline ice. Cold Reg Sci Technol 5(2):141–150
Smith RB (1983) A note on the constitutive law for sea ice. J Glaciol 29(101):191–195
Sodhi DS, Haynes FD, Kato K, Hirayama K (1983) Experimental determination of the buckling loads of floating ice sheets. Ann Glaciol 4:260–265
Specht B (1988) Modified shape functions for the three-node plate bending element passing the patch test. Int J Numer Methods Eng 26(3):705–715. https://doi.org/10.1002/nme.1620260313
Staroszczyk R (2002) On the maximum horizontal forces exerted by floating ice on engineering structures. Arch Hydro-Eng Environ Mech 49(4):17–35
Staroszczyk R (2003) Finite element simulations of floating ice–engineering structure interactions. Arch Hydro-Eng Environ Mech 50(3):251–268

Staroszczyk R (2005) Loads exerted by floating ice on a cylindrical structure. Arch Hydro-Eng Environ Mech 52(1):39–58

Staroszczyk R (2006) Loads exerted on a cylindrical structure by floating ice modelled as a viscous-plastic material. Arch Hydro-Eng Environ Mech 53(2):105–126

Staroszczyk R (2007) Loads on an off-shore structure due to an ice floe impact. Arch Hydro-Eng Environ Mech 54(2):77–94

Staroszczyk R (2018) Floating ice plate failure due to its thermal expansion at the surface. Ocean Eng 158:331–337. https://doi.org/10.1016/j.oceaneng.2018.03.072

Staroszczyk R, Hedzielski B (2004) Creep buckling of a wedge-shaped floating ice plate. Eng Trans 52(1–2):111–130

Timco GW, O'Brien S (1994) Flexural strength equation for sea ice. Cold Reg Sci Technol 22(3):285–298. https://doi.org/10.1016/0165-232X(94)90006-X

Timoshenko S, Woinowsky-Krieger S (1959) Theory of plates and shells, 2nd edn. McGraw-Hill, New York

Tremblay LB, Mysak LA (1997) Modeling sea ice as a granular material, including the dilatancy effect. J Phys Oceanogr 27(11):2342–2360

Wang YS, Ralston TD (1983) Elastic-plastic stress and strain distributions in an ice sheet moving against a circular structure. In: Proceedings of seventh international conference on port and ocean engineering under arctic conditions, Helsinki 1983, pp 940–951

Xu Y, Xu J, Wang J (2004) Fractal model for size effect on ice failure strength. Cold Reg Sci Technol 40(1–2):135–144

Zienkiewicz OC, Taylor RL (2005a) The finite element method for solid and structural mechanics, 6th edn. Elsevier Butterworth-Heinemann, Amsterdam

Zienkiewicz OC, Taylor RL, Zhu JZ (2005b) The finite element method: its basis and fundamentals, 6th edn. Elsevier Butterworth-Heinemann, Amsterdam

Chapter 5
Sea Ice in Geophysical Applications

In Chap. 4, a number of problems in which an interaction of sea ice with an engineering structure occurs is considered. In these problems the ice can be treated as a coherent sheet floating on the free surface of water, and the characteristic lengths involved in this kind of ice–structure interaction phenomena are measured in tens to hundreds of metres (at most to a few kilometres in the case of undisturbed land-fast ice, but such conditions are very rare).

This chapter is devoted to the description of the sea ice behaviour on geophysical length scales; that is, the scales ranging from at least tens of kilometres up to the scales of thousands of kilometres. These are the scales which are of interest to those dealing with the weather forecasting, or those involved in the climate modelling. On such length scales, the sea ice cover cannot be regarded as a continuous layer covering the surface of ocean water. Instead, it must be treated as a collection of a large number of ice floes of vastly differing sizes (varying from metres to kilometres), with ocean water interspersed between the floes. Depending on the wind and water current conditions, and the thermal interaction of ice with other components of the climatic system, the large-scale behaviour of ice can be very complex and difficult to predict. Also, in contrast to typical civil engineering problems analysed in Chap. 4, the time scales encountered in the problems considered in this section are much longer, since they can range from weeks up to several years.

In the first section of this chapter key mechanisms occurring in sea ice are briefly described. Next a system of equations governing the large-scale dynamic behaviour of a sea-ice pack is formulated, first in the spatial, and then in the material coordinates. Based on these equations, two discrete models for sea ice dynamics are developed. The first one is constructed by following the classical finite-element method, while the second model is based on a newer, mesh-free approach, using the method of smoothed particle hydrodynamics (SPH). Both numerical models have been used to simulate the evolution of a large ice pack under the action of wind, and the results of calculations illustrating the behaviour of sea ice are presented.

R. Staroszczyk, *Ice Mechanics for Geophysical and Civil Engineering Applications*, GeoPlanet: Earth and Planetary Sciences, https://doi.org/10.1007/978-3-030-03038-4_5

5.1 Important Physical Mechanisms in Sea Ice

The process of formation of sea ice is described in detail in Sect. 2.1. Apart from thermodynamic phenomena which dominate the formation process during its early stages, the dynamics of a fully developed ice pack is primarily driven by the action of wind and water currents, which make the ice move relative to the sea, with a typical velocity magnitude of about $10\,\mathrm{cm\,s^{-1}}$ (8.64 km per day). On large geophysical scales, also the Coriolis effects play a role, and therefore they must be accounted for in analytical descriptions of the sea-ice dynamics.

As a multi-floe ice pack drifts as a whole, the pattern of individual floes (their mutual layout) continuously change in time due to non-uniform wind and water drag stresses acting on each floe because of their different shapes and sizes, and also because of the collisions of adjacent floes. As already emphasized in Chap. 4, a very important feature is that regions of either diverging or converging flow develop in an ice pack, with very distinct mechanical behaviour of ice in these regions. When the ice is in diverging flow, then there is practically no dynamic interactions between floes, since collisions between individual floes are rare and, therefore, they mostly move away from each other in a free drift controlled by the wind and water current drag forces. On the contrary, when the ice is in converging flow, then frequent collisions of adjacent floes take place and the 'macroscopic' behaviour of the ice cover is very complex. As a result of dynamic contact forces that arise between colliding floes, the latter first start to crush along their edges, then they bend and fracture, and due to the wind and water wave action they subsequently over- or under-ride each other, or raft on each other. In the outcome of this series of events, ice ridges form in the pack. This important mechanism is illustrated in Fig. 5.1.

The ridges include *compression ridges* forming as ice floes are driven towards each other, or *shear ridges*, formed by lateral movements of interacting floes. Compression ridges are usually very irregular in shape and height, with their surface sails reaching 4 m and underwater keels up to 10 m below sea level (Sanderson 1988). Shear ridges, which develop much rarer than compression ridges, are generally very regular in shape and height, and can have a length of a few kilometres. Essentially, the ice ridging process can be regarded as irreversible. Once an ice floe has increased its thickness as a result of ridging, then this thickness either remains constant, or can increase again, as a result of another series of inter-floe collision events, provided that converging flow is maintained and, of course, no ice melting is involved. When, subsequently, ice flow becomes diverging, then the floes with ridges move away from each other, with their mean thicknesses remaining unchanged.

The above-said concerns the first-year ice before its first summer season. As summer temperatures increase, the process of ice deterioration begins due to its melting. The thickness of ice decreases, its surface becomes smoother, and the initially coherent winter ice cover gradually disintegrates into a system of smaller floes, before the next winter season starts. As the cycles of winter ice formation and its summer deterioration repeat over several years, the multi-year ice in the Arctic waters gradually grows thicker, with the mean thickness of about 2–6 m for the ice which is 5–10 years old. Ridges on such multi-year floes can have sails 5 m high, and keels up to 20 m deep.

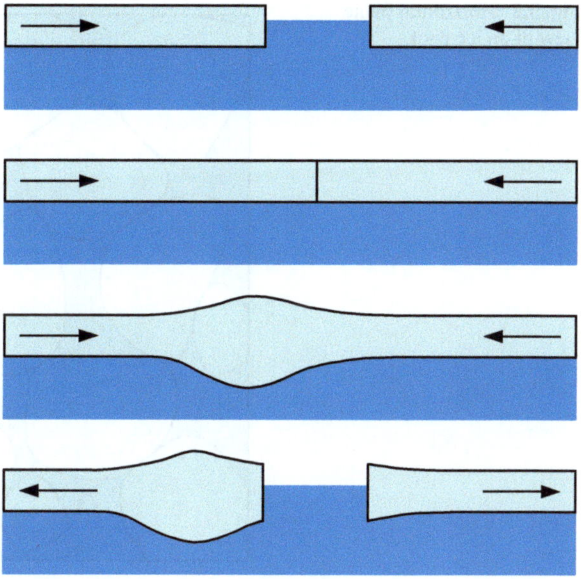

Fig. 5.1 Idealization of ice ridging mechanism

5.2 Sea-Ice Dynamics Equations

The first attempts to describe analytically the mechanics of sea ice were undertaken at the beginning of the 1900s, when a number of scientific expeditions to the Arctic Basin and Antarctica were carried out. In these early descriptions, sea ice was treated as a collection of rigid floating bodies representing ice floes, which are driven by wind and ocean current drag forces, but no interactions between these bodies were considered. Only around the 1930s, first theoretical models appeared, in which the interaction between adjacent floes was described in a simplified way by introducing friction forces acting along floe boundaries. A more rigourous research, by the modern standards, was initiated only in the middle of the previous century, when first theories were formulated in which large-scale behaviour of sea ice was described by the methods of continuum mechanics. This research was intensified in the 1970s, especially when the AIDJEX (*Arctic Ice Dynamics Joint Experiment*) project was run in the United States between 1970 and 1978. This was a very substantial undertaking, which combined empirical observations, theoretical investigations and numerical modelling within the framework of one research programme. The interest in the project's results was additionally increased by the outbreak of the world oil crisis in the mid 1970s, in response to which the exploration of the Arctic off-shore waters for oil resources was started.

When viewed on the horizontal plane, see Fig. 5.2, a sea-ice pack visually resembles a cross-section through a granular material, like a soil. This observation is the explanation why the first theories describing the large-scale behaviour of sea ice, and first of all its rheology, were the adaptations of models that had been already

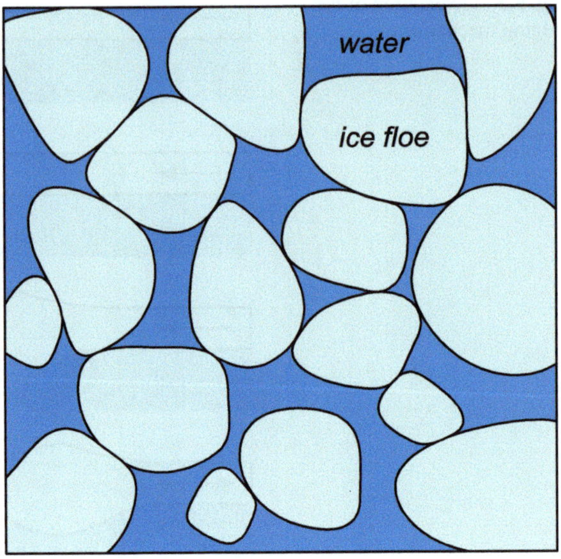

Fig. 5.2 Horizontal plane view of an ice pack

well established in soil mechanics, such as elastic-plastic (Pritchard 1975) or plastic (Hibler 1977, 1979) models used for cohesionless soils, with an idea to capture in this way irreversible processes occurring in ice. However, inclusion of the ice ridging process into the models made them more complex than the corresponding models in soil mechanics, since individual floes can change significantly their planar size during the deformation of an ice pack, whereas in soil mechanics it is commonly assumed that single grains (counterparts of ice floes) do not deform (deforms only the soil skeleton treated as a whole).

Another approach to sea ice modelling, prompted by satellite observations of sea ice fields in which large-scale deformations resemble the behaviour of two-constituent mixtures, is based on the application of a continuum mixture theory. In this way, all the equations governing the two-dimensional flow of sea ice can be derived in a consistent manner from the fundamental three-dimensional equations expressing the conservation laws of thermodynamics. Such a rigorously derived theoretical model was formulated by Gray and Morland (1994), in which the two-dimensional flow equations are obtained by integrating full three-dimensional mass and momentum balances through the ice thickness. By following this method, the introduction of some non-physical terms, adopted in an ad hoc manner in the equations appearing in the Hibler (1979) approach, is avoided. Also, a new non-linearly viscous fluid constitutive model with a stress bounding envelope was proposed by Gray and Morland (1994), in an attempt to circumvent numerical difficulties associated with the application of a viscous-plastic rheological model with a yield surface.

In this section, the equations that describe the large-scale dynamics of sea ice are presented. Hence, mass conservation and linear momentum equations are formulated for the two-dimensional space, supplemented by thermal balance equations describing the phase changes (freezing and melting) in ice and heat fluxes between the ice and the air and ocean water. Finally, constitutive models which are applicable to the large-scale sea ice deformations are discussed; these models essentially extend those already formulated in Chap. 4 for the description of the behaviour of sea ice on civil engineering scales.

5.2.1 Mass Conservation Equations

Sea-ice pack behaviour is analysed in a rectangular Cartesian coordinate system, with origin O and the two coordinate axes, x_1 and x_2, placed on the horizontal plane defined by the mean sea level; thus, for simplicity, Earth's surface curvature is ignored in the analysis. Let the current position of an ice particle on the horizontal plane be defined by the position vector $\boldsymbol{x}(t)$, with components $x_i(t)$, $i = 1, 2$. The motion of the ice pack in the plane Ox_1x_2 is described by the horizontal velocity vector $\boldsymbol{v}(t)$, with components v_1 and v_2. It is assumed that the lead water (the water in open spaces between ice floes) moves horizontally with the velocity of the local ice field. The pack layer has a variable thickness, defined by the function $h(\boldsymbol{x}, t)$. The latter is supposed to be a smooth function of $\boldsymbol{x}$, which is achieved by continuous extension of the surfaces of adjacent ice floes in places where there is lead water, as shown in Fig. 5.3. On the horizontal plane Ox_1x_2, ice floes occupy, in general, only some fraction $A(\boldsymbol{x}, t)$ of the total surface of the ice–water system. This area fraction of ice, also referred to as the ice concentration, is defined as the area of the sea surface covered by the floating ice per the unit area of the sea; hence, it has the property $0 \leq A \leq 1$.

In order to formulate the equations describing the principle of mass conservation of ice let us neglect, at this point, all thermodynamic effects taking place in the ice–water–atmosphere system, such as phase changes (melting and freezing) that can change the mass of ice. Further, also suppose that there is no ice accumulation (due to the atmospheric precipitation) on the top surface of ice, so that the total mass of ice in the system considered remains constant. Under such simplifications, the mass conservation balance for the ice pack is expressed in terms of two equations, describing the evolution of the local ice concentration A and the local ice thickness

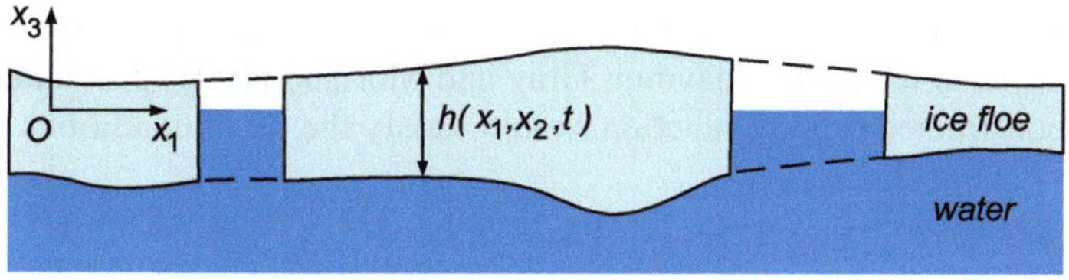

Fig. 5.3 Vertical cross-section through an ice pack layer

h. These two equations are adopted in the following forms (Gray and Morland 1994; Morland and Staroszczyk 1998):

$$\frac{\mathrm{D}A}{\mathrm{D}t} + A\eta\,[1 - \alpha(A)H(-\eta)] = 0, \tag{5.1}$$

$$\frac{\mathrm{D}h}{\mathrm{D}t} + h\eta\alpha(A)H(-\eta) = 0. \tag{5.2}$$

In the above equations, $\mathrm{D}/\mathrm{D}t$ denotes the material (convected with the velocity field $\boldsymbol{v}$) time derivative, and $H(-\eta)$ is the Heaviside step function defined by relations (4.35) on p. 76. Recall that $\eta(\boldsymbol{x}, t) = \operatorname{div} \boldsymbol{v}$ is the horizontal dilatation-rate which is positive in diverging flow and negative in converging flow. The function $\alpha(A)$ will be defined shortly.

The Heaviside function is used to describe the distinct ice pack behaviour in diverging and converging flows. In the case of diverging flow ($\eta > 0$), on account of the definitions (4.35), the mass balance equations (5.1) and (5.2) reduce to the forms:

$$\frac{\mathrm{D}A}{\mathrm{D}t} = -A\eta < 0, \quad \frac{\mathrm{D}h}{\mathrm{D}t} = 0. \tag{5.3}$$

These relations imply that in diverging flow the ice concentration A decreases with time, whereas the ice thickness h (understood as shown in Fig. 5.3) remains unchanged. This is in contrast to the case of converging flow ($\eta < 0$), for which the full forms of equations (5.1) and (5.2) apply, leading to the expressions

$$\frac{\mathrm{D}A}{\mathrm{D}t} = -A\eta\,[1 - \alpha(A)] \geq 0, \quad \frac{\mathrm{D}h}{\mathrm{D}t} = -h\eta\alpha(A) \geq 0, \tag{5.4}$$

provided that the condition $0 \leq \alpha(A) \leq 1$ holds. It will be seen that, by construction of the function $\alpha(A)$, this condition is satisfied indeed. Relations (5.4) show that in converging flow both the ice concentration and ice thickness increase in time.

It is seen from (5.3) and (5.4) that the ice thickness h can only either grow or remain unchanged in time, depending on the sign of η, reflecting thus the irreversible character of the ice ridging mechanism. In order to incorporate this mechanism into the theoretical description of the ice-pack dynamics, a so-called ice ridging function, denoted here by $\alpha(A)$, is introduced. In the first sea-ice dynamics model, formulated by Hibler (1979), this function was proposed in a form resulting in the ice concentrations A exceeding unity during sustained converging flows. To avoid this non-physical ice pack behaviour, Gray and Morland (1994) postulated that the properly constructed ridging function should satisfy the set of conditions

$$\eta \geq 0: \quad \alpha = 0; \quad \eta < 0: \quad 0 \leq \alpha \leq 1; \quad \alpha \to 1 \quad \text{as} \quad A \to 1. \tag{5.5}$$

Accordingly, the following form of the ridging function $\alpha(A)$ is adopted

$$\alpha(A) = \begin{cases} \frac{A-A_f}{1-A_f}, & \text{for } 0 < A_f < A \leq 1, \\ 0, & \text{for } 0 \leq A \leq A_f. \end{cases} \tag{5.6}$$

In (5.6), A_f is a critical ice concentration level, below which no ice ridging occurs (in spite of the converging flow regime), and above which $\alpha(A)$ increases continuously to approach the unit limit as A approaches unity. Two particular values of $A_f = 0.5$ and $A_f = 0.75$ were used in numerical simulations carried out by Gray and Morland (1994).

5.2.2 Linear Momentum Equation

As stated earlier, the deformation of the sea-ice pack is assumed to be two-dimensional, restricted to the horizontal plane Ox_1x_2; that is, the possible motion of ice in the vertical x_3-direction is neglected. The horizontal linear momentum equation governing the motion of ice can be derived (Gray and Morland 1994) by integrating the full three-dimensional balances through the ice thickness, to yield the expression:

$$\varrho h \frac{\mathrm{D}\boldsymbol{v}}{\mathrm{D}t} = \operatorname{div} \boldsymbol{N} + \boldsymbol{f}_a + \boldsymbol{f}_w + \boldsymbol{f}_c\,. \tag{5.7}$$

In (5.7), ϱ is the (constant) intrinsic density of ice, with the value of $\varrho = 917$ kg m^{-3}, $\boldsymbol{N}$, with components N_{11}, N_{12} and N_{22}, is the depth-integrated horizontal stress tensor (with physical dimension Pa m), $\boldsymbol{f}_a$ and $\boldsymbol{f}_w$ denote external tractions exerted on the top and the bottom surfaces of the ice cover by wind stress and water drag, respectively, and the term $\boldsymbol{f}_c$ represents the Coriolis force effect.

The stress $\boldsymbol{N}$ is a mean stress acting on a unit area of the aggregate consisting of ice floes and lead water. Hence, by analogy to the theory of two-phase media, it can be interpreted as so-called partial stress, which is the product of the intrinsic stress times the ice area fraction A. Hence, it can be related to the actual stress in ice, denoted by $\boldsymbol{\sigma}$, by the formula

$$\boldsymbol{N} = A r(h) f(A)\, \boldsymbol{\sigma}. \tag{5.8}$$

The stress $\boldsymbol{\sigma}$ describes a mean value of stress per unit thickness of ice, and it acts when there is full contact between adjacent floes along their edges, corresponding to unit ice concentration A. This stress is prescribed by constitutive relations in terms of the horizontal deformation-rates of the ice pack, see Sect. 5.2.4.

The proportionality factors $r(h)$ and $f(A)$ in (5.8) measure the effects of increasing ice thickness h, and of increasing contact length along floe boundaries with increasing ice concentration A, on the depth-integrated stress. Two simple forms of the factor $r(h)$ are

$$r(h) = h, \quad \text{or} \quad r(h) = \frac{h^2}{h^*}. \tag{5.9}$$

The first, linear, dependence on h supposes a uniform contact stress variation across the ice thickness, whereas the second, quadratic form, supposes a linear variation with depth, with h^* denoting a typical ice thickness magnitude. In this work only the first, linear form $r(h) = h$, will be adopted for computations. The dimensionless contact length function $f(A)$, necessarily with the properties $f(0) = 0$ and $f(1) = 1$, has been adopted in the form proposed by Gray and Morland (1994):

$$f(A) = \frac{\exp[-K(1-A)] - \exp(-K)}{1 - \exp(-K)}, \quad K \gg 1, \tag{5.10}$$

which is a modification of an earlier (improper) formula by Hibler (1979), introduced here to ensure the condition $f(0) = 0$. According to Hibler, the empirical parameter K should have a value of about 20.

The terms $\boldsymbol{f}_a$ and $\boldsymbol{f}_w$ entering the motion equation (5.7) represent the tangential tractions acting at the top surface and the base of the ice cover. Like the stress $\boldsymbol{N}, \boldsymbol{f}_a$ and $\boldsymbol{f}_w$ have also the meaning of partial quantities. Accordingly, they are expressed in terms of the intrinsic tractions $\boldsymbol{\tau}_a$ and $\boldsymbol{\tau}_b$ exerted by wind and water currents, respectively, as follows

$$\boldsymbol{f}_a = A\boldsymbol{\tau}_a, \quad \boldsymbol{f}_w = A\boldsymbol{\tau}_w. \tag{5.11}$$

The intrinsic surface tractions $\boldsymbol{\tau}_a$ and $\boldsymbol{\tau}_b$ are adopted in the forms given by the quadratic relations (4.77) on p. 98, which define the tractions in terms of dimensionless wind stress and water drag coefficients C_a and C_w. A detailed discussion on the magnitudes of these coefficients can be found in the papers by Kara et al. (2007) and Lu et al. (2011). In the first paper, devoted to the wind stress drag analysis, detailed maps of the global ocean are presented, showing typical monthly-averaged magnitudes of C_a during the 7-year-long period 1998–2004. In their conclusion, the authors suggest an annual mean value of $C_a \approx 1.5 \times 10^{-3}$ for high-latitude regions of both hemispheres. In the second paper, investigating the ice–ocean drag coefficient magnitude, specific values for C_w are compiled (in Table 1 of that paper) for various sites (seas and gulfs) in polar regions. It turns out that these values differ significantly, depending on a particular site, type of ice (smooth or rough), typical floe size, etc. For instance, $C_w = 5.3 \times 10^{-3}$ to 7.8×10^{-3} for smooth ice in the Bering Sea, $C_w = 7.1 \times 10^{-3}$ to 8.3×10^{-3} in the Greenland Sea, and $C_w = 1.3 \times 10^{-3}$ to 1.6×10^{-3} in the Weddell Sea. For moderately-ridged ice in the Baltic Sea, a value of $C_w = 3.5 \times 10^{-3}$ is given. In the present work, the values $C_a = 2 \times 10^{-3}$ and $C_w = 4 \times 10^{-3}$ will be applied in numerical simulations, which are the same values as those used in Chap. 4 in the analysis of civil engineering problems.

The term f_c describing the Coriolis force effect in the momentum equation (5.7) is given by

$$\boldsymbol{f}_c = \varrho h \boldsymbol{a}_c , \tag{5.12}$$

where $\boldsymbol{a}_c$ is the Coriolis acceleration vector defined by

$$\boldsymbol{a}_c = 2\boldsymbol{v} \times \boldsymbol{\Omega} . \tag{5.13}$$

In this formula, $\boldsymbol{\Omega}$ is the vector of the angular velocity of Earth in its rotation about the north–south axis, with Earth's rotation-rate equal to $\Omega = 7.292 \times 10^{-5}$ rad s^{-1}.

On account of the definitions (5.8), (5.9)$_1$, (5.11), (5.12) and (5.13), the linear momentum equation (5.7) becomes

$$\varrho h \frac{\mathrm{D}\boldsymbol{v}}{\mathrm{D}t} = Ahf(A)\operatorname{div}\boldsymbol{\sigma} + A\,(\boldsymbol{\tau}_a + \boldsymbol{\tau}_w) + 2\varrho h \boldsymbol{v} \times \boldsymbol{\Omega} , \tag{5.14}$$

which in components is expressed as

$$\varrho h \frac{\mathrm{D}v_i}{\mathrm{D}t} = Ahf(A)\frac{\partial \sigma_{ij}}{\partial x_j} + A\,(\tau_{ai} + \tau_{wi}) + \varrho h f_c\, \epsilon_{ij3}\, v_j \quad (i, j = 1, 2). \tag{5.15}$$

In the latter equation, ϵ_{ij3} is the permutation symbol, with the values $\epsilon_{123} = +1$, $\epsilon_{213} = -1$ and zero otherwise, and the parameter f_c is the Coriolis coefficient (or the Coriolis frequency) defined by

$$f_c = 2\Omega \sin\phi , \tag{5.16}$$

where ϕ is the geographic latitude angle. The typical values for f_c are about 10^{-4} rad s^{-1}.

5.2.3 *Thermal Balance Equation*

Assume that the temperature field in an ice pack is described by a function $T(x_1, x_2, x_3, t)$. Note that, contrary to the preceding analysis, in which all fields were treated as two-dimensional, the temperature is a three-dimensional field, with the vertical variation of T across the ice pack depth. This is done to account for the fact that the vertical gradient of temperature in ice can be significant, and is much larger than the respective gradients in the horizontal plane. The heat conduction through the medium is assumed to obey the standard Fourier law, with constant thermal conductivity κ (equal to 2.22 W m^{-1} K^{-1} at T = 272 K). Further, it is assumed that heat production due to mechanical working can be neglected due to the low shear stress and strain-rate levels occurring in ice. Under these simplifications, the equation of energy conservation is given by the thermal balance (Gray and Morland 1994) expressed in the form

$$\varrho AC\frac{\mathrm{D}T}{\mathrm{D}t} = \kappa \mathrm{div}\,(A \mathrm{grad}\, T) + Ar + \frac{1}{2}kL. \tag{5.17}$$

In the above equation, C denotes the specific heat capacity of ice (which at the melting point equals 2.04×10^3 J kg^{-1} K^{-1}), r is the energy deposit per unit volume per unit time, k is the mass transfer per unit pack volume per unit time, and $L = 3.34 \times 10^5$ J kg^{-1} is the specific latent heat of ice melting.

By taking into account that the temperature gradient components in the x_1 and x_2 directions are by several orders of magnitude smaller than those in the vertical x_3 direction, hence the former can be ignored in the energy equation (5.17), the thermal balance can be expressed in the following reduced form

$$\frac{\mathrm{D}T}{\mathrm{D}t} = \frac{\kappa}{\varrho C}\frac{\partial^2 T}{\partial x_3^2} + \frac{kL}{2\varrho AC} + \frac{r}{\varrho C}, \tag{5.18}$$

The latter equation is subject to the conditions $T = T_m$ at the ice base and $T = T_s$ at the ice top surface, where T_m denotes the ice melting point and T_s is the prescribed temperature at the ice surface. In polar regions, the difference between the ice pack basal and top surface temperatures can be as high as 20–30 K.

It should be noted in passing that when the thermal balance equation (5.18) is solved to incorporate thermodynamic effects in a sea-ice pack model, then the mass conservation equations (5.1) and (5.2) for the ice concentration A and thickness h should be supplemented with additional terms describing the mass fluxes into the system due to phase changes taking place in ice.

5.2.4 Constitutive Models

It is a general belief that the large-scale behaviour of a sea-ice pack, a complex system of ice floes of widely differing sizes, either interacting with each other or separated by lead water, resembles the behaviour of a viscous-plastic material. Therefore, the viscous-plastic rheological models are in common use to describe and numerically model this type of the material behaviour. Most of the large-scale viscous-plastic models stem from the formulation originated by Hibler (1979) and subsequently modified and extended by him and his co-workers and followers (Leppäranta and Hibler 1985; Ip et al. 1991; Hibler and Ip 1995; Hunke and Dukowicz 1997). Hence, the rheology of sea ice is described by the viscous-plastic flow law given by Eq. (4.49) on p. 81, but without the term P_2 defining the tensile strength of ice (or in other words, with $P_2 = 0$). A simpler variant of this rheological model, known as the cavitating fluid rheology, in which shear stress in ice is zero, was also tried (Parkinson and Washington 1979; Flato and Hibler 1992), but without much success.

The viscous-plastic model with the elliptic yield curve in principal stress axes (Hibler 1979) requires distinct responses during yield (plastic flow) and during the flow when the yield condition is not applying (viscous flow), which adds numerical

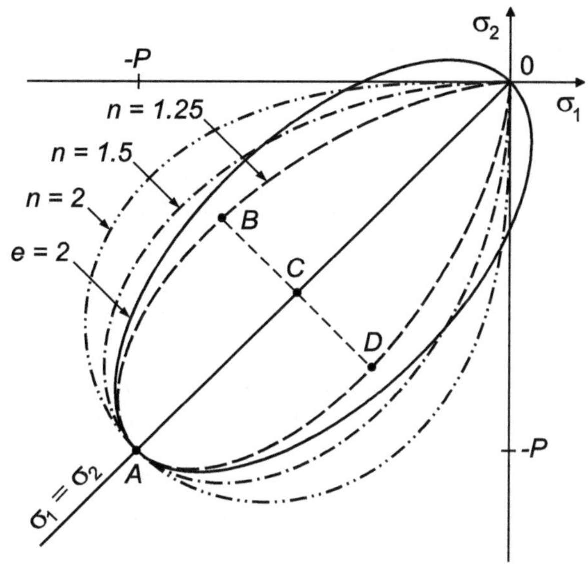

Fig. 5.4 Different bounding envelopes: elliptic yield curve for $e = 2$ (Hibler 1979), and tear-drop curves for $n = 1.25$, $n = 1.5$ and $n = 2$ (Morland and Staroszczyk 1998)

complexity to the model. The elliptic yield curve also allows tensile stresses in ice, since parts of the curve lie outside the negative principal stress quadrant, as seen in Fig. 5.4. Gray and Killworth (1995) and Gray (1999) demonstrated that the above two features of the model (disjoint stress relations in different flow regimes and the possibility of tensile stress occurrence) can give rise to numerical instabilities and unsatisfactory computational performance of the model. The issue of the ill-posedness of the viscous-plastic model was also analysed more recently by Guba et al. (2013). In order to improve the stability of the numerical algorithm, artificial dumping terms can be introduced into the model (Hibler 1979), which has become a common practice in sea-ice modelling, but these additional terms can give rise to fictitious responses of ice.

While the motivation for the application of a plasticity theory is the large-scale response of sea ice being similar to that observed in granular materials, the assumption that stresses should lie on a precise yield curve is not obvious, and the associated flow rule (4.42) which gives a strain-rate vector that is normal to the yield curve is less firm for granular materials than for metals. Gray and Morland (1994) showed that a non-linearly viscous flow law, continuous for all strain-rates, can restrict the stress states to lie within an envelope bounded by two close curves at high convergence rates. Based on the concept by Gray and Morland (1994), a viscous flow law was developed for sea ice by Morland and Staroszczyk (1998), in which all possible stress states lie entirely in the negative principal stress quadrant. For this purpose, a class of bounding envelopes, which asymptote to the principal stress axes $O\sigma_1\sigma_2$ as the principal stresses approach zero, was constructed to be used in combination with the viscous fluid constitutive relation (4.39) on p. 77. Several such bounding envelopes, the shapes of which resemble tear drops, are sketched in Fig. 5.4.

Accordingly, let us consider the class of closed curves in the negative principal stress quadrant defined by

$$\left[(\sigma_1)^2 + (\sigma_2)^2\right]^n = 2^n\, P^{\,2n-2}\, \sigma_1\, \sigma_2\,, \quad (n > 1), \tag{5.19}$$

where P denotes the compressive strength of ice, and n is a free parameter greater than unity. Each curve passes through the stress origin and is symmetric about the axis $\sigma_1 = \sigma_2$, and

$$\frac{\mathrm{d}\,\sigma_2}{\mathrm{d}\,\sigma_1} \to 0 \;\; \text{as} \;\; \sigma_1 \to 0, \quad \frac{\mathrm{d}\,\sigma_1}{\mathrm{d}\,\sigma_2} \to 0 \;\; \text{as} \;\; \sigma_2 \to 0, \tag{5.20}$$

confirming the asymptotic behaviour as stress approaches zero. Let C be the point on the major axis OA at which the width BD is a maximum, then

$$OA = 2^{1/2}\, P, \quad BD = 2\left[(n-1)\left(\frac{1}{n}\right)^{n/(n-1)}\right]^{1/2} P\,, \tag{5.21}$$

and the length to width ratio, equivalent to the ratio e for the Hibler ellipse, is

$$\frac{OA}{BD} = \left[\frac{n^{n/(n-1)}}{2\,(n-1)}\right]^{1/2}. \tag{5.22}$$

Corresponding to $e = 2$ used by Hibler (1979), this ratio is 2.000 if $n = 1.407$, and we use in the further calculations the rounded value $n = 1.5$, giving a ratio 2.015.

In order to ensure that the stress states (σ_1, σ_2) described by the viscous flow law (4.39) always lie inside an adopted bounding envelope, the ice response functions $\phi_1(\eta, \gamma)$ and $\phi_2(\eta, \gamma)$ need to be constructed in special forms. The details concerning the derivation of these forms can be found in the paper by Morland and Staroszczyk (1998). Here it is only noted that the formulae defining the shear and bulk viscosities of ice, μ and ζ respectively, are much more involved than those given by relations (4.46) and (4.47) on p. 80 in the case of the viscous-plastic rheology. Further, the ratio of the shear to bulk viscosities depends on the strain-rate invariants η and γ, unlike the viscous-plastic model in which this ratio is constant and equal to e^{-2}.

In Sects. 5.4 and 5.6, discrete (finite-element and smoothed particle hydrodynamics) models for sea-ice dynamics will be formulated, with the aim to use them for numerical simulations of the large-scale behaviour of an ice pack driven by wind stresses. It turns out that for the purposes of the numerical modelling, it is convenient to re-write the general non-linearly viscous fluid law (4.39) in a slightly different form, namely

$$\boldsymbol{\sigma} = [\phi_1(\eta, \gamma)\boldsymbol{I} + \phi_2(\eta, \gamma)\boldsymbol{D} + \phi_3(\eta, \gamma)\eta\boldsymbol{I}]\, H(-\eta), \tag{5.23}$$

since this form also incorporates other rheological models, such as the viscous-plastic (Hibler 1979), cavitating fluid (Flato and Hibler 1992) or viscous fluid model by Overland and Pease (1988). Thus, a range of different sea-ice rheologies can be easily incorporated in numerical codes, provided that specific forms of the ice response functions ϕ_1, ϕ_2 and ϕ_3 expressed in terms of the strain-rate invariants are implemented.

Although only non-linearly viscous and viscous-plastic rheological models for large-scale behaviour of sea ice are considered here and then implemented in numerical models described further in this chapter, one has to mention alternative approaches which have been developed since the mid 2000s. One such an approach is an elastic-decohesive model, proposed by Schreyer et al. (2006). This model is based on methods known from fracture mechanics and, compared to the viscous-plastic theories, is much more complex in terms of the formal description. Some examples of the application of the elastic-decohesive model to the sea-ice dynamics can be found in the paper by Sulsky et al. (2007). The model was later extended by Sulsky and Peterson (2011) and used to numerically simulate the motion of the Arctic sea ice.

Usually, on geophysical scales of thousands of kilometres, sea ice is treated by modellers as an isotropic material. However, observations from satellite imagery indicate the presence of large-scale oriented features in sea ice caused by the formation and propagation of wide cracks, the length of which can be in excess of tens of kilometres. There have already been a few attempts to capture such anisotropic properties of the sea-ice cover. The most significant contributions to this, still a relatively new, subject are due to Wilchinsky and Feltham (2004, 2011). The results of the application of their anisotropic model to the Arctic sea ice behaviour, and comparisons with the predictions of the standard isotropic elastic-viscous-plastic model by Hunke and Dukowicz (1997), can be found in the paper by Tsamados et al. (2013).

Some aspects of the sea ice rheology modelling, including the large-scale anisotropy of ice, are reviewed in the paper by Feltham (2008).

5.3 Material Formulation of Sea-Ice Equations

Numerical simulations of sea-ice pack flows are usually based on the spatial (Eulerian) formulation of the governing equations. This well-established approach has been successful in solving a variety of problems, but has its limitations. In particular, it is difficult to deal with problems in which ice pack configuration undergoes large changes, and the ice pack–open ocean boundaries move over large distances and substantially change their geometry. These difficulties arise due to problems associated with the sufficiently accurate tracing of moving boundaries, which is required for maintaining the stability of numerical schemes applied. For this reason, a natural way of solving sea-ice flow problems in which large deformations of the initial ice pack configuration occur is to apply the material (Lagrangian) description, in which individual material points of the pack are followed in time during an analysis. Compared to the spatial description, the Lagrangian method has a potential of increased numerical stability due to the lack of convective terms appearing in the momentum equations set in the Eulerian framework, thus avoiding spatial interpolations between discrete nodes of a numerical model. Additionally, the treatment of boundary conditions is much easier in the Lagrangian formulation, since these are imposed on material surfaces, that is, on the surfaces which do not move in the fixed material (referential) coordinates.

In order to trace the motion of individual material elements of a sea-ice pack on the horizontal plane, a fixed in space reference coordinate frame is introduced, which defines the positions of material points at a reference time, say at $t = 0$. Hence, a fixed rectangular Cartesian coordinate system with the origin O is adopted. Let $\boldsymbol{X}$ denote the position vector, with components X_i $(i = 1, 2)$, describing an ice particle position relative to O at the reference time; that is, $\boldsymbol{x} = \boldsymbol{X}$ at $t = 0$. As the ice pack flow proceeds, at a subsequent time $t > 0$ a given ice particle moves to a new position $\boldsymbol{x} = \boldsymbol{x}(\boldsymbol{X}, t)$. The motion of this particle relative to its neighbourhood, measured in the material coordinates X_i, is described by the deformation gradient tensor $\boldsymbol{F}$. The components of this tensor, together with the components of the velocity vector $\boldsymbol{v}$, are given by

$$F_{ij} = \frac{\partial x_i}{\partial X_j}, \quad v_i = \frac{\partial x_i}{\partial t}, \quad (i, j = 1, 2). \tag{5.24}$$

Now consider all variables as functions of $(\boldsymbol{X}, t)$, but without introducing new function notation, then the material time derivative D/Dt becomes the partial time derivative $\partial/\partial t$ at fixed $\boldsymbol{X}$. The initial, or reference, distributions of ice thickness, ice area fraction and ice velocity will be denoted by $h_0(\boldsymbol{X})$, $A_0(\boldsymbol{X})$ and $\boldsymbol{v}_0(\boldsymbol{X})$. The required transformation relations are

$$\frac{\partial F_{ij}}{\partial t} = \frac{\partial v_i}{\partial X_j}, \quad \eta = G_{ij}\frac{\partial v_i}{\partial X_j}, \quad \text{where} \quad G_{ij} = \frac{\partial X_j}{\partial x_i}, \tag{5.25}$$

from which $\boldsymbol{F}$ is calculated by time integration of the material velocity gradient, and the horizontal dilatation-rate η is expressed in terms of the material velocity gradient and the components of the tensor $\boldsymbol{G} = (\boldsymbol{F}^T)^{-1}$. Explicit component expressions for the tensors $\boldsymbol{F}$ and $\boldsymbol{G}$ are given by

$$\boldsymbol{F} = \begin{pmatrix} F_{11} & F_{12} \\ F_{21} & F_{22} \end{pmatrix}, \quad \boldsymbol{G} = J^{-1}\begin{pmatrix} F_{22} & -F_{21} \\ -F_{12} & F_{11} \end{pmatrix}, \tag{5.26}$$

where

$$J = \det \boldsymbol{F} = (\det \boldsymbol{G})^{-1} = F_{11}F_{22} - F_{12}F_{21}, \tag{5.27}$$

with $J = 1$ in the initial, undeformed configuration, in which $\boldsymbol{F} = \boldsymbol{G} = \boldsymbol{I}$. By means of the components of the tensor $\boldsymbol{G}$, the spatial derivatives of a scalar quantity ϕ are expressed in terms of the material derivatives by

$$\frac{\partial \phi}{\partial x_i} = G_{ij} \frac{\partial \phi}{\partial X_j}, \quad (i, j = 1, 2), \tag{5.28}$$

where the summation convention for a repeated subscript applies.

Let the unit vectors which are normal and tangent to a fixed contour in the reference configuration be $\bar{\boldsymbol{n}}$ and $\bar{\boldsymbol{s}}$, respectively. Then their spatial counterparts, $\boldsymbol{n}$ and $\boldsymbol{s}$, are given by

$$\boldsymbol{n} = \frac{\boldsymbol{G}\bar{\boldsymbol{n}}}{|\boldsymbol{G}\bar{\boldsymbol{n}}|}, \quad \boldsymbol{s} = \frac{\boldsymbol{G}\bar{\boldsymbol{s}}}{|\boldsymbol{G}\bar{\boldsymbol{s}}|}. \tag{5.29}$$

The latter relations are needed for proper expression of boundary conditions at the ice pack edges.

With the above transformation rule (5.28), the ice area fraction and thickness equations (5.1) and (5.2) on p. 136 now become

$$\frac{\partial A}{\partial t} + A\eta \left[1 - \alpha(A) H(-\eta)\right] = 0, \tag{5.30}$$

$$\frac{\partial h}{\partial t} + h\eta\alpha(A) H(-\eta) = 0 \tag{5.31}$$

the linear momentum equation (5.15) transforms to

$$\varrho h \frac{\partial v_i}{\partial t} = A h f(A) G_{jk} \frac{\partial \sigma_{ij}}{\partial X_k} + A\left(\tau_{ai} + \tau_{wi}\right) + \varrho h f_c \, \epsilon_{ij3} \, v_j \,, \tag{5.32}$$

and the energy equation (5.18) takes the form

$$\frac{\partial T}{\partial t} = \frac{\kappa}{\varrho C} \frac{\partial^2 T}{\partial x_3^2} + \frac{kL}{2\varrho A C} + \frac{r}{\varrho C}. \tag{5.33}$$

Relations (5.30)–(5.33) form a system of five differential equations for A, h, v_1, v_2 and T, once the stress components σ_{ij} are determined from the constitutive law (5.23). Recall that the dilatation-rate η is expressed in terms of the components of $\boldsymbol{G}$ by relation $(5.25)_2$.

5.4 Finite-Element Model

The system of sea-ice dynamics equations, formulated in the material coordinates in Sect. 5.3, is solved by a finite-element method. The attention is restricted here to the mechanical behaviour of a sea-ice pack; hence, the thermal energy balance equation (5.33) describing the heat conduction and the ice phase changes is excluded from the analysis. Accordingly, the following system of four first-order partial differential equations is solved:

$$\frac{\partial A}{\partial t} + A\eta\,[1 - \alpha(A)H(-\eta)] = 0, \tag{5.34}$$

$$\frac{\partial h}{\partial t} + h\eta\alpha(A)H(-\eta) = 0, \tag{5.35}$$

$$\varrho h \frac{\partial v_i}{\partial t} - Ahf(A)G_{jk}\frac{\partial \sigma_{ij}}{\partial X_k} - \varrho h f_c\, \epsilon_{ij3}\, v_j = A\,(\tau_{ai} + \tau_{wi})\,, \tag{5.36}$$

with the right-hand side of (5.36) representing ice driving forces due to the wind stress and water drag, respectively.

The system of equations (5.34)–(5.36), coupled by non-linear terms, is solved by a weighted residual method (Zienkiewicz et al. 2005), which transforms the problem to the solution of the system of equations

$$\boldsymbol{M}\dot{\boldsymbol{w}} + \boldsymbol{K}\boldsymbol{w} = \boldsymbol{f}, \tag{5.37}$$

where the vector $\boldsymbol{w}$ specifies the values of the ice area fraction A, the ice thickness h and of the two components of the ice velocity $\boldsymbol{v}$ at all N discrete points of the system under consideration. This vector is composed of the component vectors $\boldsymbol{w}_k$, $k = 1, \ldots, N$:

$$\boldsymbol{w} = (\boldsymbol{w}_1, \ldots, \boldsymbol{w}_k, \ldots, \boldsymbol{w}_N)^T, \tag{5.38}$$

with $\boldsymbol{w}_k$ being a four-element vector containing the values of A, h and $\boldsymbol{v}$ at k-th node of the discrete system; that is

$$\boldsymbol{w}_k = (A_k\,, h_k\,, v_{1k}\,, v_{2k})^T. \tag{5.39}$$

The matrices $\boldsymbol{M}$ and $\boldsymbol{K}$, and the forcing vector $\boldsymbol{f}$, entering (5.37), are aggregated from the respective element matrices $\boldsymbol{M}^e$, $\boldsymbol{K}^e$ and element vectors $\boldsymbol{f}^e$ in a way typical of the finite-element method. Accordingly, the element matrices $\boldsymbol{M}^e$ and $\boldsymbol{K}^e$ are composed of $N_e \times N_e$ submatrices $\boldsymbol{m}^e_{ij}$ and $\boldsymbol{k}^e_{ij}$ $(i, j = 1, \ldots, N_e)$, respectively, each of dimension 4×4, with N_e being the number of nodes in a finite element. The element matrix $\boldsymbol{M}^e$ has the structure

$$\boldsymbol{M}^e = \begin{bmatrix} \boldsymbol{m}^e_{11} & \ldots & \boldsymbol{m}^e_{1N_e} \\ \ldots & \boldsymbol{m}^e_{ij} & \ldots \\ \boldsymbol{m}^e_{N_e1} & \ldots & \boldsymbol{m}^e_{N_eN_e} \end{bmatrix}, \tag{5.40}$$

where $\boldsymbol{m}^e_{ij}$ is the diagonal matrix

$$\boldsymbol{m}^e_{ij} = \begin{bmatrix} m^{11}_{ij} & 0 & 0 & 0 \\ 0 & m^{22}_{ij} & 0 & 0 \\ 0 & 0 & m^{33}_{ij} & 0 \\ 0 & 0 & 0 & m^{44}_{ij} \end{bmatrix}. \tag{5.41}$$

The diagonal entries in $\boldsymbol{m}^e_{ij}$ are given by

$$m^{11}_{ij} = m^{22}_{ij} = \int\limits_{\Omega_e} \Phi_i \Phi_j \,\mathrm{d}\,\Omega_e\,, \quad m^{33}_{ij} = m^{44}_{ij} = \varrho \int\limits_{\Omega_e} h\,\Phi_i \Phi_j \,\mathrm{d}\,\Omega_e\,, \tag{5.42}$$

with Φ_i being a shape (interpolation) function, and Ω_e denoting a finite element domain. Since $\boldsymbol{m}^e_{ij} = \boldsymbol{m}^e_{ji}$, the element matrix $\boldsymbol{M}^e$ is symmetric, and, thus, the global matrix $\boldsymbol{M}$, assembled from the element matrices, is also symmetric.

The element matrix $\boldsymbol{K}^e$ consists of $N_e \times N_e$ component submatrices $\boldsymbol{K}^e_{ij}$ of the structure given by

$$\boldsymbol{k}^e_{ij} = \begin{bmatrix} k^{11}_{ij} & 0 & 0 & 0 \\ 0 & k^{22}_{ij} & 0 & 0 \\ 0 & 0 & k^{33}_{ij} & k^{34}_{ij} \\ 0 & 0 & k^{43}_{ij} & k^{44}_{ij} \end{bmatrix}. \tag{5.43}$$

The elements k^{11}_{ij} and k^{22}_{ij} of the matrix $\boldsymbol{k}^e_{ij}$ are calculated by integrating the evolution equations (5.34) and (5.35) for the ice concentration A and thickness h, and are

$$k^{11}_{ij} = \int\limits_{\Omega_e} R_A \Phi_i \Phi_j \,\mathrm{d}\,\Omega_e, \quad k^{22}_{ij} = \int\limits_{\Omega_e} R_h \Phi_i \Phi_j \,\mathrm{d}\,\Omega_e, \tag{5.44}$$

with the terms R_A and R_h defined by

$$R_A = \eta\,[1 - \alpha(A) H(-\eta)]\,, \quad R_h = \eta \alpha(A) H(-\eta). \tag{5.45}$$

The entries k^{33}_{ij}, k^{34}_{ij}, k^{43}_{ij} and k^{44}_{ij} in the matrix $\boldsymbol{k}^e_{ij}$ in (5.43) are obtained by integrating the two momentum equations (5.36). Each of these entries can be expressed as a sum of two terms, $k^{\beta\xi\sigma}_{ij}$ and $k^{\beta\xi c}_{ij}$, which are related to internal forces due to the ice stress gradients and the Coriolis acceleration, respectively; that is

$$k^{\beta\xi}_{ij} = k^{\beta\xi\sigma}_{ij} + k^{\beta\xi c}_{ij}, \quad \beta, \xi = 3, 4. \tag{5.46}$$

On account of the general form of the constitutive law (5.23) on p. 142, describing the stress in ice in terms of the three response functions ϕ_1, ϕ_2 and ϕ_3, the above components $k^{\beta\xi\sigma}_{ij}$ of the element matrix $\boldsymbol{k}^e_{ij}$ are given by

$$\begin{aligned} k^{\beta\xi\sigma}_{ij} = \int\limits_{\Omega_e} & A h f(A) H(-\eta)\, \frac{\partial \Phi_i}{\partial X_n} \frac{\partial \Phi_j}{\partial X_l} \times \\ & \times \left[\tfrac{1}{2}\phi_2\,(G_{rn} G_{sl} + \delta_{rs} G_{ml} G_{mn}) + \phi_3\, G_{rl} G_{sn}\right] \mathrm{d}\,\Omega_e\,, \end{aligned} \tag{5.47}$$

where δ_{rs} is the Kronecker symbol, and the indices are

$$l, m, n, r, s = 1, 2; \quad \beta = r + 2, \quad \xi = s + 2. \tag{5.48}$$

The terms $k_{ij}^{\beta\xi c}$, arising from the Coriolis forces, are, in turn, defined by

$$k_{ij}^{\beta\xi c} = -\epsilon_{rs3}\, \varrho f_c \int_{\Omega_e} h\, \Phi_i \Phi_j \,\mathrm{d}\, \Omega_e\,, \quad \beta, \xi = 3, 4, \tag{5.49}$$

with the index connections (5.48) for β and ξ. Recall that ϵ_{rs3} is the permutation symbol and f_c is the Coriolis coefficient defined by relation (5.16) on p. 139. Note that $k_{ij}^{\beta\xi\sigma} = k_{ji}^{\xi\beta\sigma}$ and $k_{ij}^{\beta\xi c} = -k_{ji}^{\xi\beta c}$. Thus, the element matrix $\boldsymbol{K}^e$, and hence the global matrix $\boldsymbol{K}$, is symmetric if the Coriolis forces are ignored; otherwise the symmetry of $\boldsymbol{K}$ is lost, which has some numerical consequences (more computer memory is needed and the time of computations is longer).

Finally, the right-hand side vector $\boldsymbol{f}$ in Eq. (5.37) is aggregated from element vectors $\boldsymbol{f}^e$:

$$\boldsymbol{f}^e = (\boldsymbol{f}_1^e,\ \ldots,\ \boldsymbol{f}_i^e,\ \ldots,\ \boldsymbol{f}_{N_e}^e)^T, \tag{5.50}$$

with $\boldsymbol{f}_i^e$ being a four-element vector containing forces acting at i-th node of a finite element; that is

$$\boldsymbol{f}_i^e = (0,\ 0,\ f_{i3},\ f_{i4})^T. \tag{5.51}$$

The nodal forces $f_{i\beta}$ ($\beta = 3, 4$) result from both the ice surface stresses generated by wind and water drag, $f_{i\beta}^w$, and the pressure term associated with the function ϕ_1 in the constitutive law (5.23), $f_{i\beta}^\sigma$. Hence,

$$f_{i\beta} = f_{i\beta}^w + f_{i\beta}^\sigma, \tag{5.52}$$

with

$$\begin{aligned} f_{i\beta}^w &= \int_{\Omega_e} A\, \Phi_i (\tau_{ar} + \tau_{wr})\,\mathrm{d}\, \Omega_e\,, \\ f_{i\beta}^\sigma &= -\int_{\Omega_e} \phi_1 A h f(A) H(-\eta)\, \frac{\partial \Phi_i}{\partial X_l}\, G_{rl}\,\mathrm{d}\, \Omega_e\,, \end{aligned} \tag{5.53}$$

where $\beta = r + 2$ and $r = 1, 2$, see the index connections (5.48).

The above relations for the coefficients of the matrices $\boldsymbol{M}$ and $\boldsymbol{K}$ and the forcing vector $\boldsymbol{f}$ are general in the sense that they hold for any set of the shape functions Φ_j. The latter set of functions depends on the way a given continuous domain is discretized, that is, on the geometry of finite elements, the number of nodal points within each element, and a particular form of shape functions adopted. In the model, the results of which are presented in the following Sect. 5.5, triangular elements with six nodal points (three vertices and three mid-side points) have been applied, with the shape functions Φ_j being bi–quadratic polynomials. All surface integrals have been calculated numerically by applying Gauss–Legendre quadrature formulae with seven sampling points within triangular elements (Zienkiewicz et al. 2005).

The system of first-order in time differential equations (5.37) has been integrated in the time domain by employing a single-step implicit algorithm. Within each time step, due to the non-linearity of equations and the dependence of the matrices $\boldsymbol{M}$ and $\boldsymbol{K}$ and the vector $\boldsymbol{f}$ on the solution vector $\boldsymbol{w}$, the system of equations has been solved iteratively by using a direct (Picard) iteration scheme to achieve convergence. The time stepping has been carried out by the weighted residual θ-method (Zienkiewicz et al. 2005). Application of this method reduces (5.37) to the solution of the following system of algebraic equations:

$$(\boldsymbol{M} + \Delta t\, \theta \boldsymbol{K})\, \boldsymbol{w}_{n+1} = [\boldsymbol{M} - \Delta t\, (1-\theta)\boldsymbol{K}]\, \boldsymbol{w}_n + \Delta t\, \bar{\boldsymbol{f}}, \tag{5.54}$$

which relates the solutions $\boldsymbol{w}_n$ and $\boldsymbol{w}_{n+1}$ at two instants t_n and t_{n+1}. In (5.54), $\Delta t = t_{n+1} - t_n$ is the time step length, θ is the weighting parameter, and $\bar{\boldsymbol{f}}$ is the time-averaged forcing vector. For a linear variation of $\boldsymbol{f}$ between t_n and t_{n+1}, the forcing vector $\bar{\boldsymbol{f}}$ is given by

$$\bar{\boldsymbol{f}} = (1-\theta)\boldsymbol{f}_n + \theta \boldsymbol{f}_{n+1}. \tag{5.55}$$

In the calculation, the value $\theta = 0.7$ has been adopted, though it has been found that using other values θ from the range $0.5 \le \theta \le 1$ does not significantly affect the numerical performance of the above time-integration scheme (Morland and Staroszczyk 1998).

5.5 Finite-Element Numerical Simulations

The accuracy and numerical stability of the finite-element algorithm presented in the preceding section has been tested by comparing its predictions with a class of exact analytical solutions to specific one-dimensional ice flow problems (Morland and Staroszczyk 1998). For these problems, involving the formation and subsequent motion through an ice pack of a moving interface separating regions of converging and diverging flows, analytic solutions have been constructed by applying an inverse method. It has turned out that a very good agreement between the discrete and analytic results has been achieved for three particular cases considered. These preliminary tests also enabled the analysis of the effect of the mesh resolution on the accuracy of finite-element results in the context of two-dimensional simulations described below.

The first two-dimensional ice pack flow problem solved numerically by the proposed finite-element model is depicted in Fig. 5.5a. In this problem, the ice pack initially occupies a rectangular domain $0 \le x_1 \le 25$ km, $0 \le x_2 \le 50$ km. Two adjacent sides of the rectangle, denoted by Γ_1 and Γ_2, are solid boundaries, and the other two, Γ_3 and Γ_4, are moving open water boundaries, the position of which must be tracked with the solution. Both converging and diverging zones arise during the ice

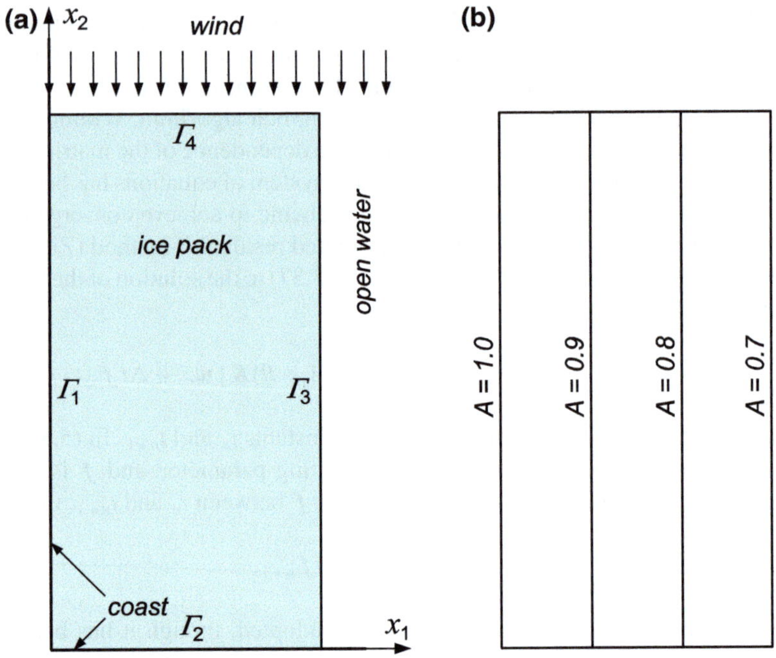

Fig. 5.5 **a** Rectangular ice pack with two solid (Γ_1, Γ_2) and two open water (Γ_3, Γ_4) boundaries, driven by a uniform wind stress applied in the negative direction of the x_2-axis. **b** Initial distribution of ice concentration A

pack flow, therefore this problem contains all the features which occur in realistic sea ice flows, and is an instructive test for the stability of a numerical model.

The ice pack is assumed to be at rest at time $t = 0$, and is driven by a steady wind stress of the constant magnitude τ_a acting along the negative x_2-axis direction. Initially, the ice has a uniform mean thickness h_0 and the ice area fraction A_0 is assumed to vary linearly in the x_1-axis direction, from $A_0 = 1$ at the coast Γ_1 to $A_0 = 0.7$ at the ice edge Γ_3 (see Fig. 5.5b). Accordingly, the initial conditions can be expressed by

$$t = 0: \quad v_1 = v_2 = 0, \quad h = h_0, \quad A = A_0 = 1 - 1.2 \times 10^{-5} x_1. \tag{5.56}$$

At the solid boundaries $\Gamma_v = \Gamma_1 \cup \Gamma_2$, either bonded (no-slip and zero normal velocity) or free-slip conditions are adopted. In the former case it is assumed that both velocity components are zero, while in the latter case the velocity component normal to the coast and the ice traction component tangential to the coast vanish. The open water boundaries $\Gamma_\sigma = \Gamma_3 \cup \Gamma_4$ are assumed to be stress-free. Hence, the no-slip boundary conditions are expressed by

$$(X_1, X_2) \in \Gamma_v: \quad \boldsymbol{v} = \boldsymbol{0}, \tag{5.57}$$

the free-slip conditions are defined by

$$(X_1, X_2) \in \Gamma_v: \quad \boldsymbol{v} \cdot \boldsymbol{n} = 0, \quad \boldsymbol{s} \cdot (\boldsymbol{\sigma n}) = 0, \tag{5.58}$$

and the stress-free conditions are given by

$$(X_1, X_2) \in \Gamma_\sigma: \quad \boldsymbol{n} \cdot (\boldsymbol{\sigma n}) = 0, \quad \boldsymbol{s} \cdot (\boldsymbol{\sigma n}) = 0. \tag{5.59}$$

In the above expressions $\boldsymbol{n}$ and $\boldsymbol{s}$ denote unit vectors normal and tangential to the respective boundary (which, in general, moves and deforms). Recall that these vectors are related to their counterparts $\bar{\boldsymbol{n}}$ and $\bar{\boldsymbol{s}}$ in the referential (initial) configuration by the formulae (5.29) on p. 145.

Below are shown the results of simulations carried out for a rectangular ice pack driven by the uniform wind stress of a magnitude $\tau_a = 0.05$ Pa acting over a period of three days. The calculations were performed for a linearly viscous flow law (4.39), with the ice viscosities $\mu = \zeta = 5 \times 10^8$ Pa s. The initial ice thickness was $h_0 = 2$ m, and the parameter $A_f = 0.5$ was adopted to define the critical ice concentration at which the ice ridging process starts, see the definition of the ridging function $\alpha(A)$ given by Eq. (5.6) on p. 137. The time integration of finite-element equations was carried out with the time step length $\Delta t = 360$ s.

The plots in Fig. 5.6 show the results of simulations obtained for the free-slip conditions (5.58) at the solid boundaries. The figure illustrates the evolution of the ice velocity field $\boldsymbol{v}$, the ice concentration A and the ice mean thickness h, and the respective plots correspond to the integration times of one, two and three days. It is seen that over the period of three days the ice pack undergoes significant deformation in the horizontal plane, with parts of the open water boundaries (Γ_3 and Γ_4) moving by a distance of about 20 km in the x_2-axis direction, and by about 8 km in the lateral x_1-direction. It can be observed in the velocity vector plots that the pack flow slows as the ice concentration gradually increases (Fig. 5.6b) and the pack becomes thicker (Fig. 5.6c) due to the ice ridging process. While the ice concentration A changes slowly throughout the whole pack as the ice field deforms, the ice thickness h increases considerably (by more than 50%) near the coastline $x_2 = 0$.

The corresponding results obtained for the case of the no-slip conditions at the solid boundaries Γ_1 and Γ_2 can be found in the paper by Morland and Staroszczyk (1998). In this, probably less realistic, situation than the above-presented free-slip case, the flow of ice is more affected by the presence of the rigid boundaries, and the predicted distributions of the ice concentration A and the ice thickness h are less uniform than those displayed in Fig. 5.6.

The problem depicted in Fig. 5.5 has been also solved by adopting the viscous-plastic rheological model for sea ice. The results of simulations, again for the free-slip conditions assumed at the solid boundaries Γ_1 and Γ_2, are presented in Fig. 5.7. The simulations have been conducted for the compressive strength parameter $P = 5 \times 10^3$ Pa, the rheological parameter $e = 2$ and the critical dilatation-rate $\Delta_c = 2 \times 10^{-9}$ s^{-1} (Schulkes et al. 1998).

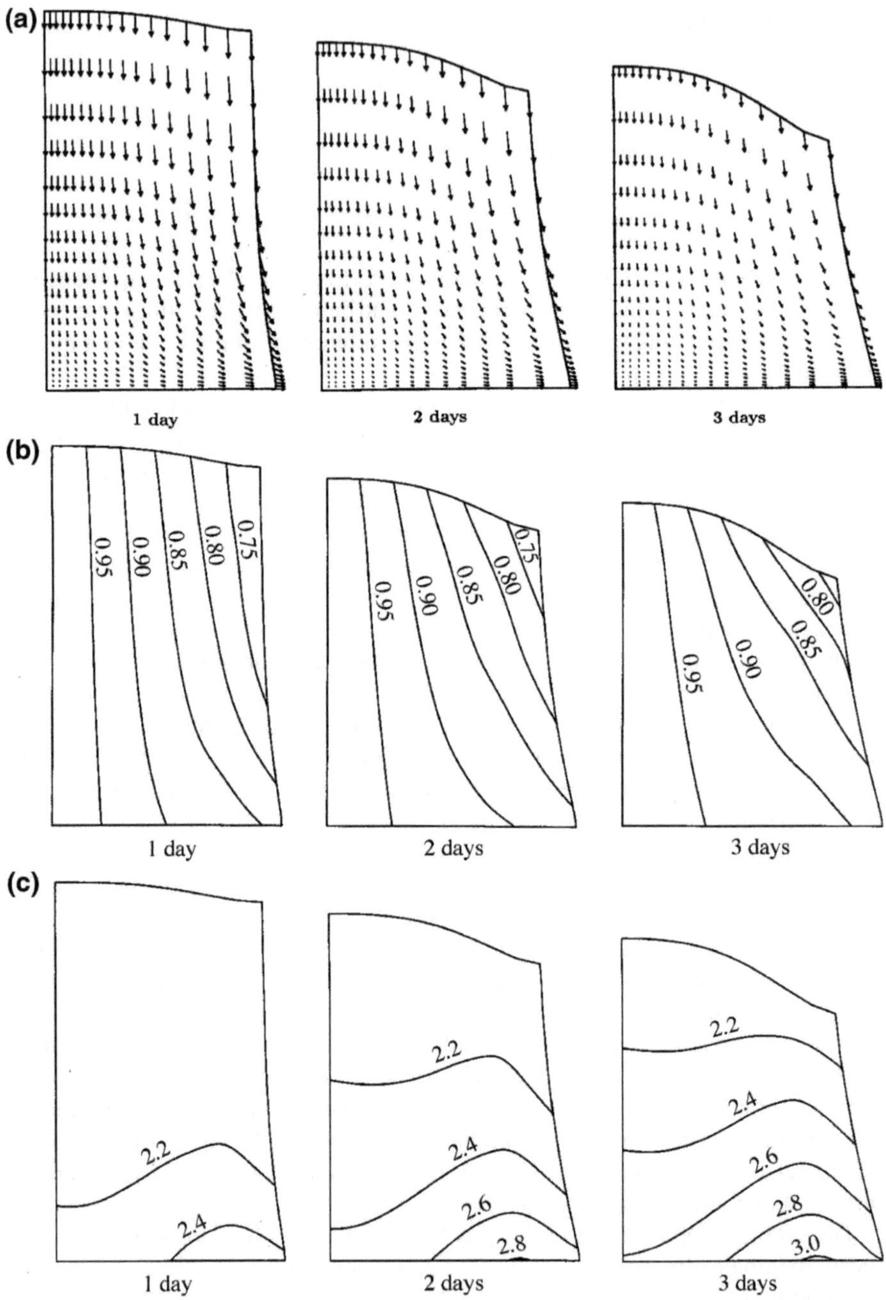

Fig. 5.6 Evolution of a rectangular ice pack driven by wind, for the viscous fluid rheology. Free-slip conditions are adopted at the solid boundaries Γ_1 and Γ_2. Plots **a** the velocity vector field $\boldsymbol{v}$ after one, two and three days; **b** the ice area fraction A after one, two and three days; and **c** the ice thickness h after one, two and three days. Reprinted from Morland and Staroszczyk (1998), Fig. 6, with permission of the Royal Society of London

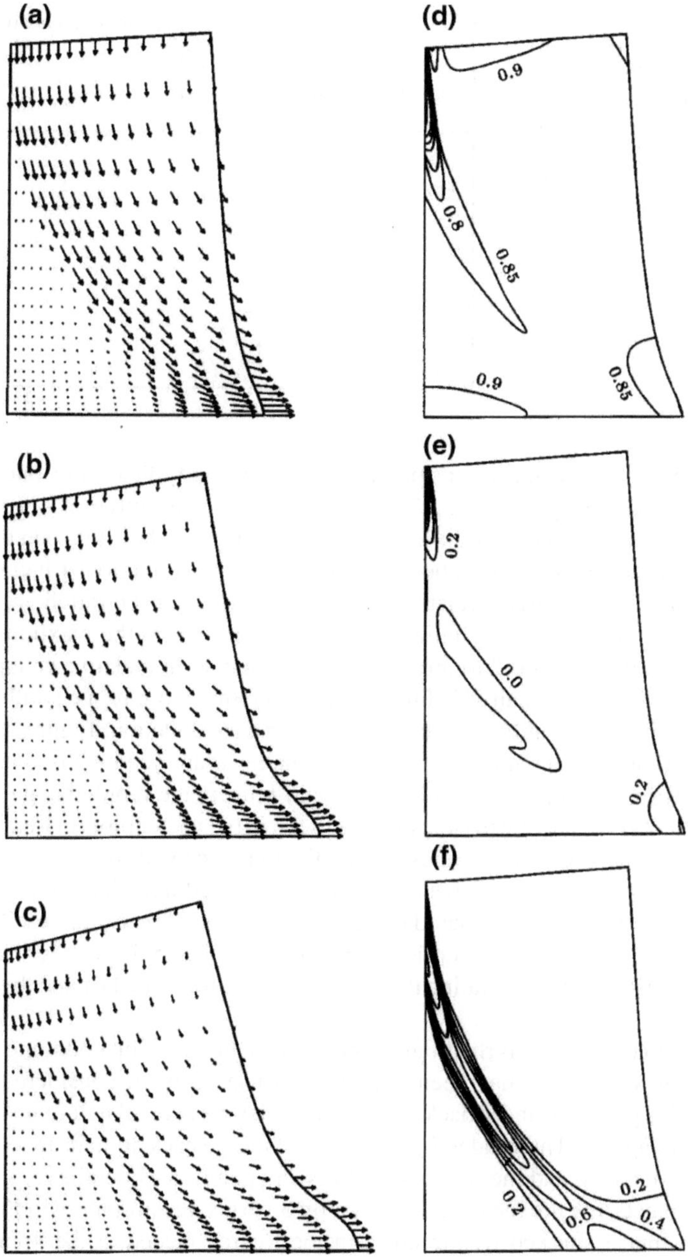

Fig. 5.7 Evolution of a rectangular ice pack driven by wind, for the viscous-plastic rheology. Free-slip conditions are adopted at the solid boundaries Γ_1 and Γ_2. Plots of the velocity vector field $\boldsymbol{v}$ **a** after one day, **b** after two days, **c** after three days. Contour plots of **d** the ice concentration A, **e** the velocity divergence η (in units 10^{-5} s^{-1}) and **f** the shear strain-rate invariant γ (in units 10^{-5} s^{-1}), after one day. Reprinted with permission from Schulkes et al. (1998), Fig. 3. Copyright 1998 by John Wiley and Sons

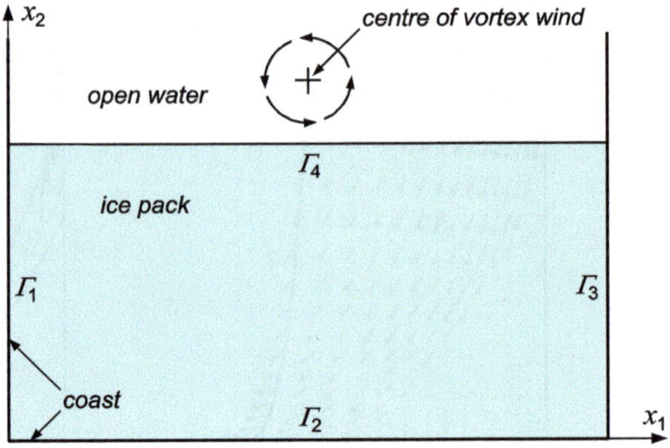

Fig. 5.8 Rectangular ice pack with three solid boundaries (Γ_1, Γ_2 and Γ_3) and one open water boundary (Γ_4), driven by a geostrophic vortex wind field, with the vortex centre, denoted by the cross, located at the open sea

Figure 5.7a–c show the vector plots of the ice velocity field after one, two and three days, respectively, and Fig. 5.7d–f display the ice concentration A, the velocity divergence η, and the shear strain-rate invariant γ after one day. It is seen that, as the pack is driven towards the solid boundary Γ_2, the ice area fraction A increases near this boundary, but the largest relative changes in the ice concentration are observed near the open water edge Γ_3, initially at $x_1 = 25$ km, where the maximum change equals about 30% (from 0.7 to around 0.9). In contrast, the relative changes in the ice thickness h are less than 1%. Further, it is seen that close to the solid boundary Γ_2 the pack is significantly stretched in the lateral x_1-direction. Figure 5.7f shows the concentration region of high shear strain-rate invariant γ, which is characteristic of materials in which plastic yield occurs, and reflects the rapid change in the ice velocities across the narrow plastic slip band seen in the vector plots in Fig. 5.7a–c. In regions where the stress magnitudes are smaller than the yield stress level defined by the parameter P, the ice cover behaves as a viscous medium which undergoes relatively small deformations compared to the plastic strains. More examples, illustrating the behaviour of the wind-driven ice pack shown in Fig. 5.5 for other rheological models (for instance, the cavitating fluid model) can be found in the paper by Schulkes et al. (1998).

The non-linearly viscous rheological model with the tear-drop bounding envelope, described in Sect. 5.2.4, has been applied to a test problem considered by Flato (1993). In this problem, an ice pack initially occupies a rectangular domain 250 km × 500 km, see Fig. 5.8. Three sides Γ_1, Γ_2 and Γ_3 of the ice pack domain are constrained by solid boundaries, and the fourth side Γ_4 is an open water boundary.

It is assumed that the ice pack is driven by a vortex geostrophic wind field, with the vortex centre, marked by the cross in the figure, located at the open sea at a distance of 50 km from the initial line of the ice edge Γ_4. The wind velocity field is defined by (Flato 1993)

$$\boldsymbol{u}_a(\boldsymbol{R}) = \min\left(\omega_a R,\ \frac{\Lambda}{R}\right) \boldsymbol{k} \times \frac{\boldsymbol{R}}{R}, \tag{5.60}$$

where $\boldsymbol{u}_a$ is the geostrophic wind velocity vector, $\boldsymbol{R}$ is the vector radius from the vortex centre and R is its magnitude, $\omega_a = 5 \times 10^{-4}\ \mathrm{s}^{-1}$ is an angular velocity, $\Lambda = 8 \times 10^5\ \mathrm{m}^2\ \mathrm{s}^{-1}$, and $\boldsymbol{k}$ denotes an upward vertical unit vector. With these parameters, the wind velocity u_a increases linearly form zero at the vortex centre to its maximum value 20 m s^{-1} at a distance of $R = 40$ km from the vortex centre, and then gradually decreases with increasing R, so that $u_a = 4$ m s^{-1} at $R = 200$ km, and $u_a = 2$ m s^{-1} at $R = 400$ km, etc. In order to calculate the ice surface and basal tractions, τ_a and τ_w respectively, caused by the wind and water current action, see Eq. (4.77) on p. 98, the wind, $\boldsymbol{u}_a$, and ice, $\boldsymbol{v}$, velocity vectors must be rotated on the horizontal plane by applying the relations

$$\bar{\boldsymbol{u}}_a = \boldsymbol{u}_a \cos\theta_a + \boldsymbol{k} \times \boldsymbol{u}_a \sin\theta_a\,, \tag{5.61}$$

$$\bar{\boldsymbol{v}} = \boldsymbol{v} \cos\theta_w + \boldsymbol{k} \times \boldsymbol{v} \sin\theta_w\,. \tag{5.62}$$

In the latter expressions, θ_a and θ_w are the wind and water turning angles, associated with the planetary boundary layer; both these angles are assumed to be equal to 25°.

The finite-element simulations were carried out for the initially uniform ice pack thickness and concentration, adopting $h_0 = 1$ m and $A_0 = 0.9$. The rheological properties of the ice were defined by the compressive strength parameter 2.75×10^4 Pa (Flato 1993), the critical dilation-rate parameter $\eta_c = 10^{-5}\ \mathrm{s}^{-1}$ (Morland and Staroszczyk 1998), and the critical ice concentration $A_f = 0.5$ that defines the onset of the ice ridging process. The finite-element mesh consisted of 400 triangular elements with six nodes each, so that the discrete system had 3444 degrees of freedom. The free-slip conditions (5.58) were adopted at the solid boundaries Γ_1, Γ_2 and Γ_3. The results of a five-day integration, performed with the time step length Δt varying from 36 to 900 s, are presented in Fig. 5.9. The plots in Fig. 5.9a–d illustrate the evolution of the finite-element grid in space, showing the changes in the deformation and position of material elements of the ice pack. These plots demonstrate well the capacity of the applied material coordinate approach of dealing with the problems in which an extensive evolution of free boundaries occurs. Figure 5.9e displays the distribution of the mean ice thickness h after five days, and shows that the ice thickness growth due to the ridging process is confined to relatively small regions near the coastlines, with the most significant increase, by about 80% of the initial value h_0, taking place near the boundary Γ_3. Finally, Fig. 5.9f illustrates the ice velocity field $\boldsymbol{v}$ after the five-day simulation.

5.6 Smoothed Particle Hydrodynamics Model

In Sect. 5.4, a discrete model based on the conventional finite-element model is developed to solve the sea-ice dynamics equations formulated in the material (Lagrangian) coordinates. The model requires the calculation of the material deformation and velocity gradients and their evolution throughout the whole time of simulation,

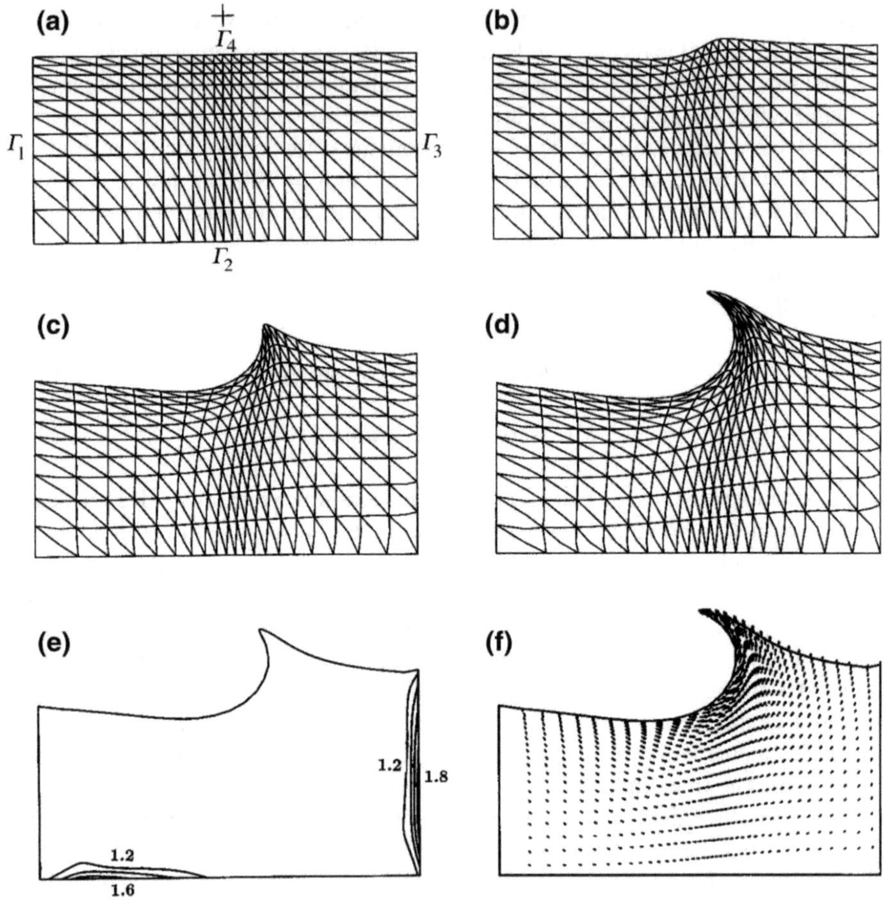

Fig. 5.9 Evolution of a rectangular ice pack driven by a geostrophic vortex wind, for the nonlinearly viscous fluid rheology with a bounding envelope. Free-slip conditions are adopted at the solid boundaries Γ_1, Γ_2 and Γ_3. **a** Finite-element mesh at the start of simulation, with the cross representing the wind vortex centre; **b**, **c** and **d** the mesh after one, three and five days, respectively; **e** the ice thickness distribution after five days; **f** the ice velocity vector field after five days. Reprinted from Morland and Staroszczyk (1998), Fig. 10, with permission of the Royal Society of London

beginning from the initial, undeformed configuration of an ice pack. In case of large ice displacements and deformations, the finite-element mesh can become highly distorted, as is seen in Fig. 5.9, and this, in turn, results in inaccurate evaluation of the deformation gradient components and the accumulation of numerical errors, eventually leading to the loss of stability of a numerical scheme. In order to effectively solve such numerically challenging problems, in which large deformations of a material occur, or fronts of material discontinuity develop and propagate, a family of so-called mesh-free discrete methods has been invented. In this section, a discrete model based on the approach known as the *smoothed particle hydrodynamics* (SPH) method is formulated. In recent years, this is one of the most extensively developed mesh-free methods.

The SPH method was invented, independently, by Lucy (1977) and Gingold and Monaghan (1977), but for nearly two decades its use was solely restricted to the field of astrophysics. Only in the mid 1990s, when the full potential of this method was recognized, the SPH method brought attention of the solid mechanics community, and ever since the interest in the method has been steadily growing. Consequently, it has found successful applications in many disciplines of physics, applied mechanics and engineering—multiple examples of its applications can be found in two review papers by Monaghan (2005, 2012). Compared to other fields of applied mechanics, the application of the SPH method to problems involving sea ice has been rare. The first applications of the SPH approach to the sea-ice modelling were due to Gutfraind and Savage (1997, 1998), who simulated the behaviour of ice in the open sea; that is, in the absence of solid boundaries (coasts), which significantly simplifies numerical calculations. Another example was the work by Shen et al. (2000), in which the SPH approach was employed to simulate the flow of river ice, with a focus on its jamming, in which case the movement of ice is severely restricted by solid boundaries (river banks) and is, essentially, nearly one-dimensional along a river. A typical sea-ice pack evolution problem, which is analysed here, can be considered as an in-between case, involving both extensive boundaries along which the ice interacts with open sea, and extensive boundaries along which the ice interacts with a coast.

The major idea of the SPH approach consists in representing a continuum by a collection of discrete material particles, each of which carries, in a fully Lagrangian sense, all information on local physical properties (such as mass, velocity, temperature, etc.) of the body under consideration. Since no pre-defined connections between discrete particles are required in the SPH approach, the method has a great flexibility in dealing with large deformations, material fragmentation, propagation of discontinuity surfaces, etc. In order to approximate field variables in terms of their values given at discrete particles, special interpolating functions, often referred to as *smoothing kernels*, are applied. Typically, a smoothing kernel has non-zero values only in a small domain, called the *kernel support* (usually the latter has a shape of a circle in two-dimensional problems). The characteristic size of the kernel support is defined by a kernel radius R. These basic SPH features and ideas are sketched in Fig. 5.10. Details on the SPH methodology can be found in the literature, for instance in the papers by Monaghan (1992, 2005, 2012), or in the books by Li and Liu (2004) and Violeau and Issa (2007).

The values of field variables at any point $\boldsymbol{x}$ of a continuum are calculated by summations of weighted contributions from all particles contained within the kernel support domain of radius R, with the weights determined by the values of the smoothing kernel function, W. Similarly, the spatial derivatives of field functions are evaluated by summation formulae involving spatial derivatives of the smoothing kernel functions (both the kernel functions and their derivatives are defined by analytical expressions). Accordingly, the value of a function f at position $\boldsymbol{x}_a$ is evaluated by means of a kernel function W centred at this particle, by applying the formula

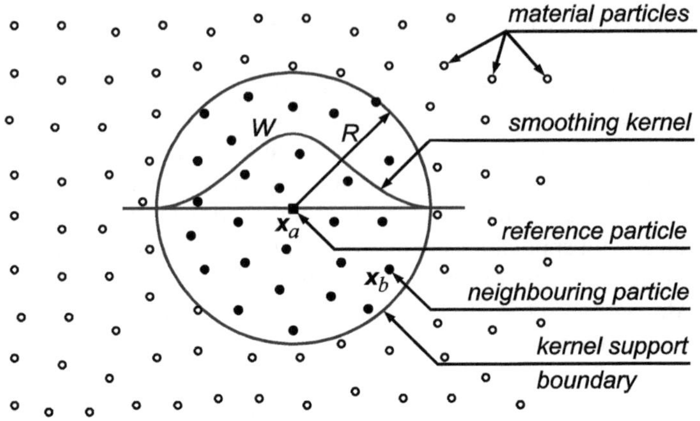

Fig. 5.10 The method of approximation of field variables in the SPH method. The approximation at a reference particle $\boldsymbol{x}_a$ (solid square) involves material particles $\boldsymbol{x}_b$ (solid circles) situated within a circular support domain of radius R centred at $\boldsymbol{x}_a$. Particles outside the support domain (empty circles) do not contribute to the approximation of field functions at the reference particle $\boldsymbol{x}_a$. W denotes a smoothing kernel function centred at $\boldsymbol{x}_a$.

$$f_a = f(\boldsymbol{x}_a) = \sum_{b=1}^{N} V_b\, f_b\, W(r_{ab})\,. \tag{5.63}$$

In Eq. (5.63), and in the remaining part of this section, the symbols a and b are used to denote discrete particle labels. $f_a = f(\boldsymbol{x}_a)$ is a discrete value of f at particle a, N denotes the number of discrete particles currently located within the kernel support domain of particle a, V_b is the volume of particle b, and $r_{ab} = |\boldsymbol{x}_{ab}| = |\boldsymbol{x}_a - \boldsymbol{x}_b|$ is the distance between particles a and b.

In order to express the differential equations describing the problem in discrete forms, SPH approximations of differential operators are required. Inspection of the mass conservation balance equations (5.1) and (5.2) on p. 136 and the momentum equation (5.14) on p. 139 shows that only the approximations of the divergence operators for vector and tensor fields are needed in the analysis. These approximations are adopted in the forms recommended by Monaghan (1992) and Gray et al. (2001):

$$(\text{div } \boldsymbol{f})_a = -\frac{1}{\varrho_a} \sum_{b=1}^{N} m_b\, \boldsymbol{f}_{ab} \cdot \nabla_a W_{ab} \tag{5.64}$$

and

$$(\text{div } \boldsymbol{A})_a = \varrho_a \sum_{b=1}^{N} m_b \left(\frac{\boldsymbol{A}_a}{\varrho_a^2} + \frac{\boldsymbol{A}_b}{\varrho_b^2} \right) \cdot \nabla_a W_{ab}\,. \tag{5.65}$$

In these expressions, $\boldsymbol{f}$ and $\boldsymbol{A}$ denote, respectively, a vector and a two-dimensional tensor fields, ϱ_a is the density of particle a, m_b is the mass of particle b, and

$f_{ab} = f_a - f_b$. $\nabla_a W_{ab}$ denotes the gradient of the kernel function W centred at particle a and calculated at particle b. This gradient is defined by

$$\nabla_a W_{ab} = \frac{\boldsymbol{x}_{ab}}{r_{ab}} \frac{\partial W(r_{ab})}{\partial r_{ab}} . \tag{5.66}$$

Application of the divergence operator approximations (5.64) and (5.65) to differential equations (5.1), (5.2) and (5.14) yields

$$\frac{\mathrm{d}\, A_a}{\mathrm{d}\, t} = -A_a \eta_a \left[1 - \alpha(A_a) H(-\eta_a)\right] , \tag{5.67}$$

$$\frac{\mathrm{d}\, h_a}{\mathrm{d}\, t} = -h_a \eta_a \alpha(A_a) H(-\eta_a), \tag{5.68}$$

$$\frac{\mathrm{d}\, \boldsymbol{v}_a}{\mathrm{d}\, t} = \frac{\varrho_a}{\varrho} A_a\, f(A_a) \sum_{b=1}^{N} m_b \left(\frac{\boldsymbol{\sigma}_a}{\varrho_a^2} + \frac{\boldsymbol{\sigma}_b}{\varrho_b^2}\right) \cdot \nabla_a W_{ab} + \frac{A_a}{\varrho h_a} (\boldsymbol{\tau}_a + \boldsymbol{\tau}_w)_a . \tag{5.69}$$

The above equations must be supplemented by constitutive relations for the stress $\boldsymbol{\sigma}$; for the illustrations presented in the next section, the viscous-plastic flow law defined by Eq. (4.49) on p. 81 has been used. In the momentum equation (5.69), the Coriolis acceleration term has been ignored, as the Coriolis effect can be neglected on the moderate length scales of hundred kilometres considered here. In view of (5.64), the ice dilatation-rate $\eta = \operatorname{div} \boldsymbol{v}$, when approximated at particle a, is given by

$$\eta_a = (\operatorname{div} \boldsymbol{v})_a = -\frac{1}{\varrho_a} \sum_{b=1}^{N} m_b\, \boldsymbol{v}_{ab} \cdot \nabla_a W_{ab} , \tag{5.70}$$

where $\boldsymbol{v}_{ab} = \boldsymbol{v}_a - \boldsymbol{v}_b$. In addition to relations (5.67)–(5.69), in order to track the motion of ice on the sea surface, a trajectory equation has also to be solved for each discrete particle:

$$\frac{\mathrm{d}\, \boldsymbol{x}_a}{\mathrm{d}\, t} = \boldsymbol{v}_a . \tag{5.71}$$

It should be noted at this point that there is an important qualitative difference between the ice intrinsic density ϱ, which is constant in time, and the densities ϱ_a and ϱ_b of discrete material particles, which change in time. The latter densities connect discrete particle masses m_a (which remain constant when there are no thermodynamic processes involved, or vary in time otherwise) with discrete particle volumes V_a (which change in time as the ice pack deforms) through the equation

$$m_a = \varrho_a V_a . \tag{5.72}$$

The particle density ϱ_a, in turn, is related to the local ice concentration A_a and the local ice thickness h_a by the formula

$$\varrho_a = \varrho A_a h_a . \tag{5.73}$$

Thus, the particle density ϱ_a expresses the mass of ice per unit surface of the sea, so that it has the meaning of a partial density, and the physical unit $\mathrm{kg\,m^{-2}}$.

The system of equations (5.67)–(5.69) and (5.71) set for each particle a of the discrete representation of a sea-ice pack is equivalent to six scalar relations for six unknown field functions: ice concentration A, ice thickness h, two components of the ice velocity vector $\boldsymbol{v}$, and two components of the position vector $\boldsymbol{x}$. This system of six equations was integrated in the time domain by applying a predictor-corrector method (Staroszczyk 2011), with a Courant-Friedrichs-Levy (CFL) condition used for controlling the stability of computations. The smoothing kernel function W was adopted in the form of a quintic spline function proposed by Morris (1996). In order to increase the accuracy of the approximation of field variables at discrete particles near the ice pack boundaries, standard kernels and their gradients were modified by following a method proposed by Belytschko et al. (1998) (this technique is known in the literature as the corrected smoothed particle hydrodynamics, C-SPH). One of the issues which has not been yet resolved satisfactorily in the SPH is the implementation of boundary conditions at solid boundaries (here given by Eqs. (5.57) and (5.58) on p. 150). From among a number of techniques that are known in the literature, a method proposed by Cummins and Rudman (1999) was applied, in which so-called ghost (virtual) particles are generated outside solid boundaries to mirror physical properties of corresponding particles in the ice flow domain.

5.7 SPH Numerical Simulations

The SPH model presented in the previous section was applied to simulate the dynamic behaviour of a sea-ice pack driven by a vortex wind field in the geometric configuration illustrated in Fig. 5.8 in Sect. 5.5. A very similar flow configuration was adopted by Li et al. (2014) for ice pack evolution simulations by applying another particle method, namely the discrete-element method (DEM).

In the SPH simulations, the ice response was described by the viscous-plastic flow law (4.49) on p. 81, assuming zero tensile strength of ice ($P_2 = 0$). More results, obtained for a different ice rheology and other flow configurations, can be found in the paper by Staroszczyk (2017). The initial ice flow conditions at time $t = 0$ were prescribed by

$$A(\boldsymbol{x}, t_0) = A_0(\boldsymbol{x}), \quad h(\boldsymbol{x}, t_0) = h_0(\boldsymbol{x}), \quad \boldsymbol{v}(\boldsymbol{x}, t_0) = \boldsymbol{0}. \tag{5.74}$$

The boundary conditions at the ice pack–open sea interfaces were assumed to be stress-free, and the conditions at the ice–solid boundaries were taken as either bonded (zero tangential and normal velocities), see (5.57), or slip-free (zero tangential traction and zero normal velocity), see (5.58).

As in Sect. 5.5, the numerical computations were performed for the initially 500 km × 250 km rectangular pack, with the wind vortex centre located 50 km off the initial open sea–ice pack edge Γ_4 (see Fig. 5.8). The results presented below have been obtained for the maximum vortex wind velocity $u_a = 10$ m s^{-1}. The initial ice thickness was $h_0 = 1$ m and the initial ice concentration was $A_0 = 0.7$, with the critical concentration level $A_f = 0.5$. The viscous-plastic rheological model parameters were: $P_1 = 5$ kPa, $e = 2$ and $\Delta_c = 2 \times 10^{-9}$ s^{-1}. The computational grid in the initial configuration consisted of $80 \times 40 = 3200$ discrete particles uniformly distributed along the directions of both coordinate axes, with the initial inter-particle spacings 6.25 km. Free-slip conditions were adopted along the three coastlines Γ_1, Γ_2 and Γ_3 for the illustrations presented below.

The results of simulations have shown that the SPH model easily treats large ice pack deformations and large displacements of the ice pack–open sea boundary. This is illustrated by the plots in Fig. 5.11, showing the evolution of the ice pack domain and displaying discrete particle distributions at the start of the flow and after five, ten and fifteen days of the ice flow. The colours of the particles have no physical meaning

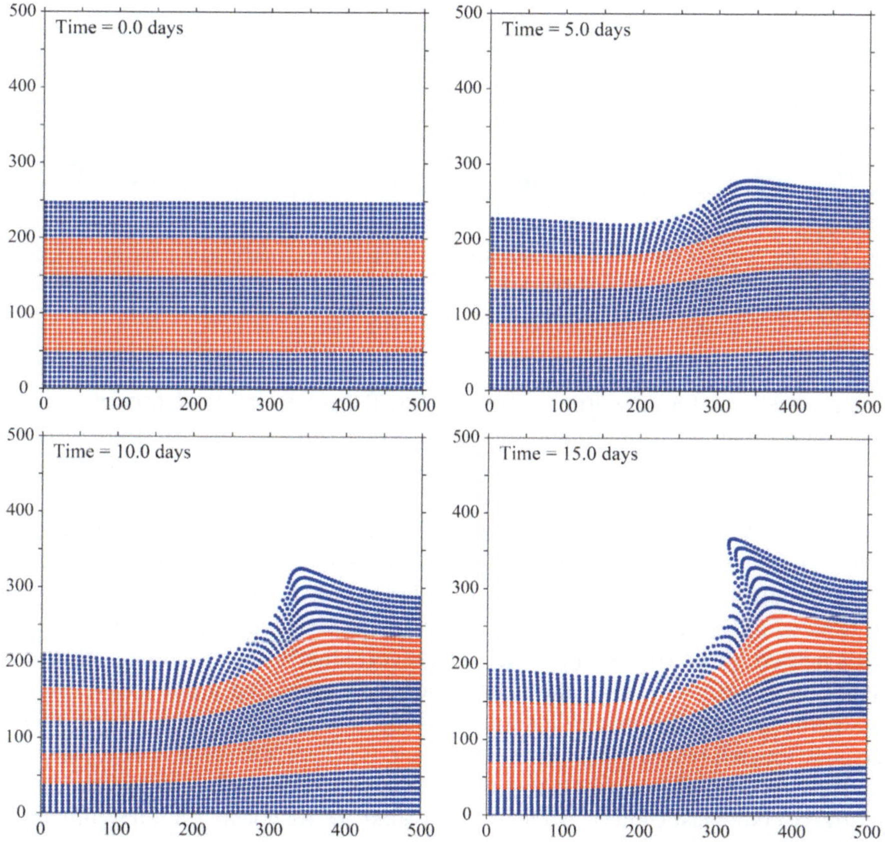

Fig. 5.11 Evolution of the initially rectangular ice pack in the flow configuration shown in Fig. 5.8, with free-slip conditions along coastlines. Discrete particle distributions at the start of flow and after 5, 10 and 15 days

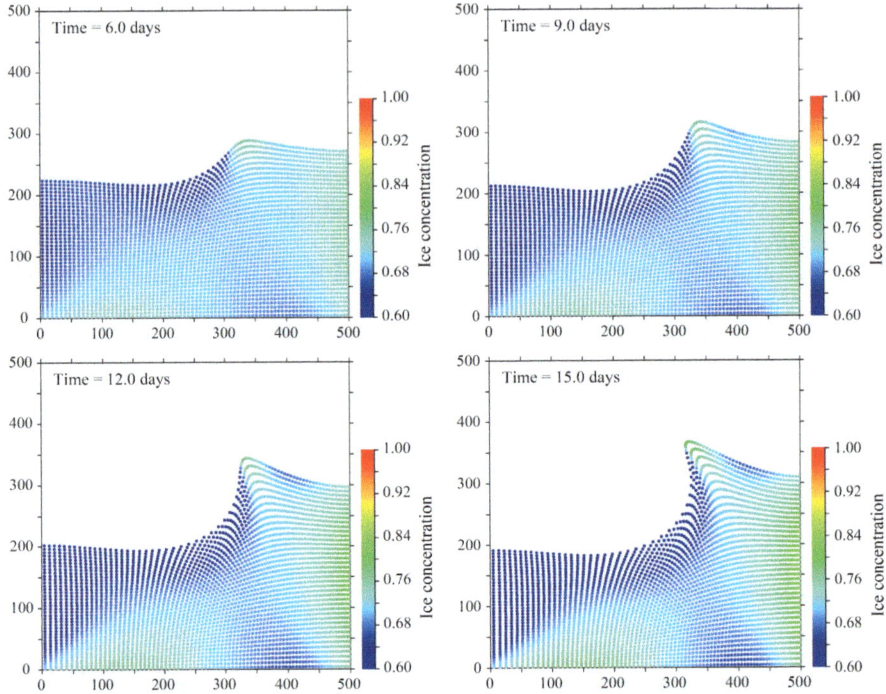

Fig. 5.12 Evolution of the initially rectangular ice pack in the flow configuration shown in Fig. 5.8, with free-slip conditions along coastlines. Ice concentration distributions after 6, 9, 12 and 15 days ($A_0 = 0.7$)

and are used only for the purpose of illustrating the pattern of the deformation field. One can see that the displacements of the material particles of ice, after fifteen days, exceed 120 km (the 'tongue' of the ice pack), with smaller displacements of around 60 km occurring along the solid boundaries Γ_1 and Γ_3.

The following Fig. 5.12 illustrates the evolution of the ice concentration field $A(x, y, t)$, displaying the ice area fractions after six, nine, twelve and fifteen days of the simulations. It can be noted that the largest increase in the ice area fraction A, from its initial value $A_0 = 0.7$, occurs along the coastline Γ_3, towards which the ice is pushed by the vortex wind field. A relatively smaller increase in A is observed along the coastline Γ_2, with the maximum values occurring around $x \approx 150$ km. A significant increase in the ice concentration also takes place at the tip of the ice pack tongue, which may appear surprising; however, the same feature was also predicted by the DEM calculations (Li et al. 2014) in which sea ice was modelled as an elastic-viscous-plastic material. On the other hand, the ice concentration significantly decreases near the coastline Γ_1 and the ice–open sea boundary Γ_1 (except for the ice tongue), where the ice pack flow is divergent. In contrast to the ice area fraction field, changes in the ice thickness $h(x, y, t)$ are essentially confined to the vicinity

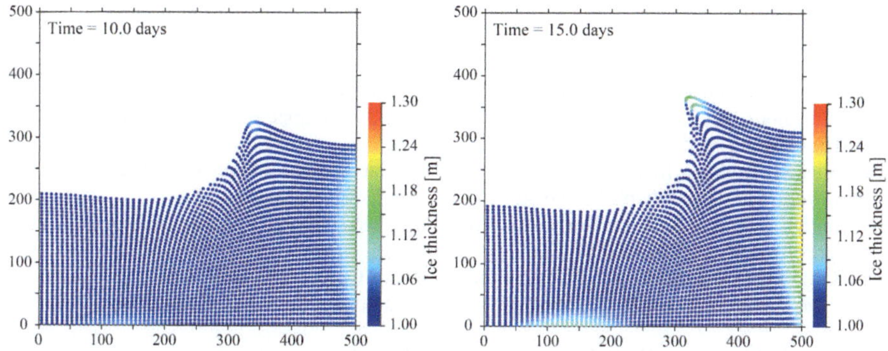

Fig. 5.13 Evolution of the initially rectangular ice pack in the flow configuration shown in Fig. 5.8, with free-slip conditions along coastlines. Ice thickness distributions after 10 and 15 days ($h_0 =$ 1.0 m)

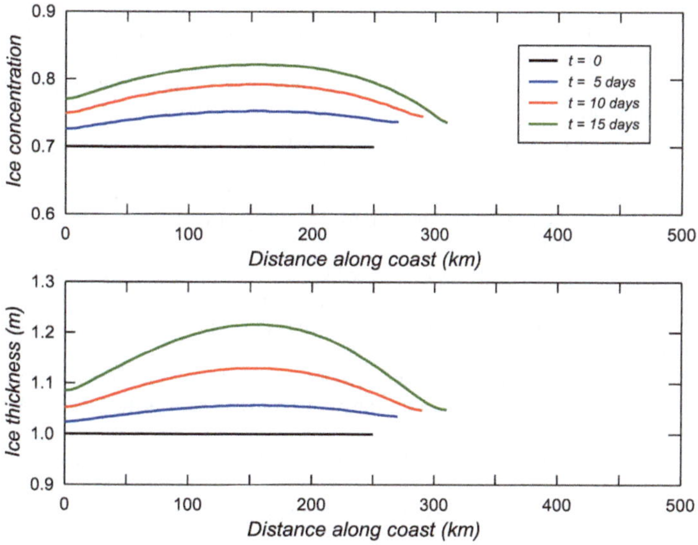

Fig. 5.14 Evolution of the profiles of ice concentration and ice thickness along the coastline at $x = 500$ km in the ice pack flow problem defined in Fig. 5.8. The same labelling applies to both plots

of the solid boundary Γ_3, as can be seen in Fig. 5.13, displaying the ice thickness distributions after ten and fifteen days of the pack flow. Corresponding to the results shown in Figs. 5.12 and 5.13 are the plots in Fig. 5.14, illustrating changes in time of the ice concentration and ice thickness along the coastline Γ_3 at $x = 500$ km (that is, along the y-axis direction). The right-hand ends of the curves show the changing y-position of the ice pack–open sea boundary along the coastline Γ_3.

References

Belytschko T, Krongauz Y, Dolbow J, Gerlach C (1998) On the completeness of meshfree particle methods. Int J Numer Method Eng 43(5):785–819. https://doi.org/10.1002/(SICI)1097-0207(19981115)43:5

Cummins SJ, Rudman M (1999) An SPH projection method. J Comput Phys 152(2):584–607. https://doi.org/10.1006/jcph.1999.6246

Feltham DL (2008) Sea ice rheology. Annu Rev Fluid Mech 40:91–112. https://doi.org/10.1146/annurev.fluid.40.111406.102151

Flato GM (1993) A particle-in-cell sea-ice model. Atmos Ocean 31(3):339–358

Flato GM, Hibler WD (1992) Modeling pack ice as a cavitating fluid. J Phys Oceanogr 22(6):626–651

Gingold RA, Monaghan JJ (1977) Smoothed particle hydrodynamics: theory and application to non-spherical stars. Mon Not R Astron Soc 181:375–389

Gray JMNT (1999) Loss of hyperbolicity and ill-posedness of the viscous-plastic sea ice rheology in uniaxial divergent flow. J Phys Oceanogr 29:2920–2929

Gray JMNT, Killworth PD (1995) Stability of the viscous-plastic sea ice rheology. J Phys Oceanogr 25:971–978

Gray JMNT, Morland LW (1994) A two-dimensional model for the dynamics of sea ice. Phil Trans R Soc Lond A 347(1682):219–290. https://doi.org/10.1098/rsta.1994.0045

Gray JP, Monaghan JJ, Swift RP (2001) SPH elastic dynamics. Comput Methods Appl Mech Eng 190(49–50):6641–6662. https://doi.org/10.1016/S0045-7825(01)00254-7

Guba O, Lorenz J, Sulsky D (2013) On well-posedness of the viscous-plastic sea ice model. J Phys Oceanogr 43:2185–2199. https://doi.org/10.1175/JPO-D-13-014.1

Gutfraind R, Savage SB (1997) Marginal ice zone rheology: comparison of results from continuum-plastic models and discrete-particle simulations. J Geophys Res 102(C6):12647–12661. https://doi.org/10.1029/97JC00124

Gutfraind R, Savage SB (1998) Flow of fractured ice through wedge-shaped channels: smoothed particle hydrodynamics and discrete-element simulations. Mech Mater 29(1):1–17

Hibler WD (1977) A viscous sea ice law as a stochastic average of plasticity. J Geophys Res 82(27):3932–3938

Hibler WD (1979) A dynamic thermodynamic sea ice model. J Phys Oceanogr 9(4):815–846

Hibler WD, Ip CF (1995) The effect of sea ice rheology on Arctic buoy drift. ASME AMD 207:255–263

Hunke EC, Dukowicz JK (1997) An elastic-viscous-plastic model for sea ice dynamics. J Phys Oceanogr 27(9):1849–1867

Ip CF, Hibler WD, Flato GM (1991) On the effect of rheology on seasonal sea-ice simulations. Ann Glaciol 15:17–25

Kara AB, Wallcraft AJ, Metzger EJ, Hurlburt HE, Fairall CW (2007) Wind stress drag coefficient over the global ocean. J Clim 20(23):5856–5864. https://doi.org/10.1175/2007JCLI1825.1

Leppäranta M, Hibler WD (1985) The role of plastic interaction in marginal ice zone dynamics. J Geophys Res 90(C6):11 899–11 909. https://doi.org/10.1029/JC090iC06p11899

Li B, Li H, Liu Y, Wang A, Ji S (2014) A modified discrete element model for sea ice dynamics. Acta Oceanol Sin 33(1):56–63. https://doi.org/10.1007/s13131-014-0428-3

Li S, Liu WK (2004) Meshfree particle methods. Springer, Berlin

Lu P, Li Z, Cheng B, Leppäranta M (2011) A parameterization of the ice-ocean drag coefficient. J Geophys Res 116(C07):C07019. https://doi.org/10.1029/2010JC006878

Lucy LB (1977) A numerical approach to the testing of the fission hypothesis. Astron J 82(12):1013–1024

Monaghan JJ (1992) Smoothed particle hydrodynamics. Annu Rev Astron Astrophys 30:543–574. https://doi.org/10.1146/annurev.aa.30.090192.002551

Monaghan JJ (2005) Smoothed particle hydrodynamics. Rep Prog Phys 68(8):1703–1759. https://doi.org/10.1088/0034-4885/68/8/R01

Monaghan JJ (2012) Smoothed particle hydrodynamics and its diverse applications. Annu Rev Fluid Mech 44:323–346. https://doi.org/10.1146/annurev-fluid-120710-101220
Morland LW, Staroszczyk R (1998) A material coordinate treatment of the sea-ice dynamics equations. Proc R Soc Lond A 454(1979):2819–2857. https://doi.org/10.1098/rspa.1998.0283
Morris JP (1996) Analysis of smoothed particle hydrodynamics with applications. PhD thesis, Monash University, Melbourne, Australia
Overland JE, Pease CH (1988) Modeling ice dynamics of coastal seas. J Geophys Res 93(C12):15619–15637. https://doi.org/10.1029/JC093iC12p15619
Parkinson CL, Washington WM (1979) A large-scale numerical model of sea-ice. J Geophys Res 84(C1):311–337. https://doi.org/10.1029/JC084iC01p00311
Pritchard RS (1975) An elastic-plastic constitutive law for sea ice. J Appl Mech Ser E 42(2):379–384
Sanderson TJO (1988) Ice mechanics. Risks to offshore structures, Graham and Trotman, London
Schreyer HL, Sulsky DL, Munday LB, Coon M, Kwok R (2006) Elastic-decohesive constitutive model for sea ice. J Geophys Res 111(C11):C11S26 https://doi.org/10.1029/2005JC003334
Schulkes RMSM, Morland LW, Staroszczyk R (1998) A finite-element treatment of sea ice dynamics for different ice rheologies. Int J Numer Anal Methods Geomech 22(3):153–174
Shen HT, Su J, Liu L (2000) SPH simulation of river ice dynamics. J Comput Phys 165(2):752–770 https://doi.org/10.1006/jcph.2000.6639
Staroszczyk R (2011) Simulation of solitary wave mechanics by a corrected smoothed particle hydrodynamics method. Arch Hydro-Eng Environ Mech 58(1–4):23–45. https://doi.org/10.2478/v10203-011-0002-9
Staroszczyk R (2017) SPH modelling of sea-ice pack dynamics. Arch Hydro-Eng Environ Mech 64(2):115–137. https://doi.org/10.1515/heem-2017-0008
Sulsky D, Peterson K (2011) Toward a new elastic-decohesive model of Arctic sea ice. Phys D 240(20):1674–1683. https://doi.org/10.1016/j.physd.2011.07.005
Sulsky D, Schreyer H, Peterson K, Kwok R, Coon M (2007) Using the material-point method to model sea ice dynamics. J Geophys Res 112(C2):C02S90. https://doi.org/10.1029/2005JC003329
Tsamados M, Feltham DL, Wilchinsky AV (2013) Impact of a new anisotropic rheology on simulations of Arctic sea ice. J Geophys Res 118:91–107. https://doi.org/10.1029/2012JC007990
Violeau D, Issa R (2007) Numerical modelling of complex turbulent free-surface flows with the SPH method: an overview. Int J Numer Methods Fluids 53(2):277–304. https://doi.org/10.1002/fld.1292
Wilchinsky AV, Feltham DL (2004) A continuum anisotropic model of sea-ice dynamics. Proc R Soc Lond A 460(2047):2105–2140. https://doi.org/10.1098/rspa.2004.1282
Wilchinsky AV, Feltham DL (2011) Modeling Coulombic failure of sea ice with leads. J Geophys Res 116(C08040). https://doi.org/10.1029/2011JC007071
Zienkiewicz OC, Taylor RL, Zhu JZ (2005) The finite element method: its basis and fundamentals, 6th edn. Elsevier Butterworth-Heinemann, Amsterdam

Chapter 6
Micro-mechanical Models for Polar Ice

In the problems discussed in Chaps. 4 and 5 devoted to the mechanics of sea ice, the macroscopic anisotropy of ice plays a limited role, and its only manifestation of some practical importance is the depth variation of mechanical properties of land-fast ice that was formed under still water conditions, as illustrated in Fig. 2.1 on p. 10. For this reason, the anisotropy of ice is usually ignored in sea ice applications, maybe except for some specific small-scale cases in which the elastic, not creep or brittle, response of floating ice is important.

A completely different situation takes place when the mechanical behaviour of grounded polar ice is considered. In general, such ice is macroscopically strongly anisotropic, and only the upper, relatively thin (on the typical depth scales of large polar ice caps) layers of ice can be treated as an isotropic medium. Moreover, on the long geophysical time scales characteristic of polar ice, the macroscopic anisotropy evolves during the passage of ice through the depth of an ice sheet, as described in Sect. 2.2. This evolution of the macroscopic properties of ice and its microstructure (ice fabric) is due to the response of the polycrystalline material to changing stress and strains configurations by the mechanism of *induced anisotropy*. All these macroscopic changes have its source in the processes occurring at the level of a single crystal of ice. Thus, to understand and properly describe the behaviour of ice on the macroscopic level, one must first understand and describe the behaviour of a single crystal of ice embedded in a polycrystalline aggregate. Only then, on the basis of microscopic constitutive laws for a single crystal, can we attempt, by applying homogenization methods, to derive macroscopic constitutive equations relating macroscopic stresses to macroscopic strains and strain-rates in ice. The formulation of such micro- and macroscopic laws for polar ice is the subject of this chapter.

As already discussed in Chap. 3, the main mode of deformation of polycrystalline ice on long time scales is its creep. Therefore, in order to formulate constitutive laws describing the creep flow of polar ice, only creep properties of a single crystal need to be considered, and hence the elastic and viscoelastic microscopic properties can be ignored in the analysis as being unimportant. The creep of a single crystal of ice is the irreversible deformation which results from the movements of dislocations

R. Staroszczyk, *Ice Mechanics for Geophysical and Civil Engineering Applications*, GeoPlanet: Earth and Planetary Sciences, https://doi.org/10.1007/978-3-030-03038-4_6

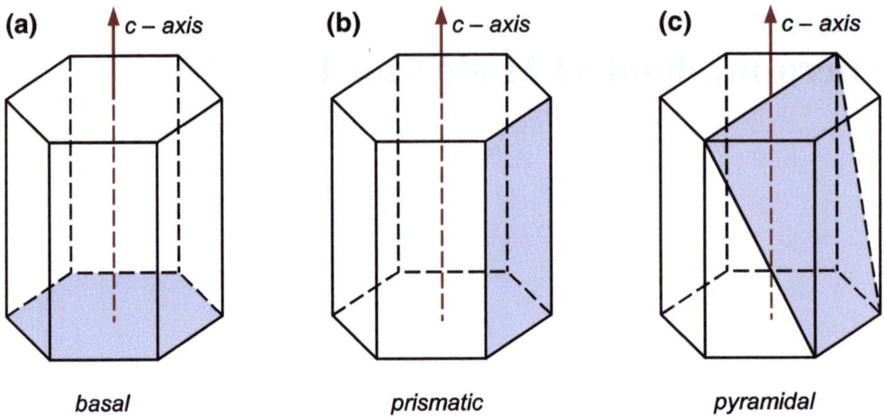

Fig. 6.1 Three crystallographic slip planes in ice Ih crystals

(*dislocation glides*) within the crystal (refer to Sect. 3.2). These dislocation glides occur along a set of characteristic crystallographic planes, forming a *slip system*. Basically, in the hexagonal ice Ih crystals (see Fig. 3.3 on p. 24) there are three preferential slip planes, illustrated in Fig. 6.1.

As follows from the diagram in Fig. 3.6 on p. 29, the creep deformation by basal slip requires, at a given strain-rate, stress levels which are by up to two orders of magnitude smaller than those needed for creep in other, non-basal, slip planes, such as the prismatic or pyramidal planes. Therefore, one can assume that nearly all the creep deformation in an ice crystal takes place by slip on its basal planes, called for this reason *easy glide planes*. The presence of the easy glide planes in individual crystals has the consequences for the deformation mechanisms occurring in a polycrystal. This is illustrated, in an idealized manner, in Fig. 6.2, showing the behaviour of an aggregate composed of a set of crystallites (represented by thin sheets, with their normal vectors being the crystal c-axes), and sliding relative to each other along the crystal basal planes. It is seen that when the polycrystal is subjected to tensile stresses acting along a certain direction, then the individual crystal c-axes rotate away from this direction. And vice versa, when compressive stresses act on a polycrystal along some direction, then the crystal c-axes rotate towards this direction. This 'deck of cards' analogue is certainly a simplification of the real behaviour of a polycrystalline ice aggregate, but illustrates well the major features of the deformation due to the mechanism of the *crystal lattice rotation*, also known as the mechanism of *rotation recrystallization*. This mechanism is responsible for the development of the oriented microstructure of polycrystalline polar ice, or the *ice fabric*, observed in ice samples extracted from bore cores drilled in Antarctica and Greenland (Alley 1992; Gow et al. 1997; Thorsteinsson et al. 1997; Gow and Meese 2007; Durand et al. 2009; Faria et al. 2014).

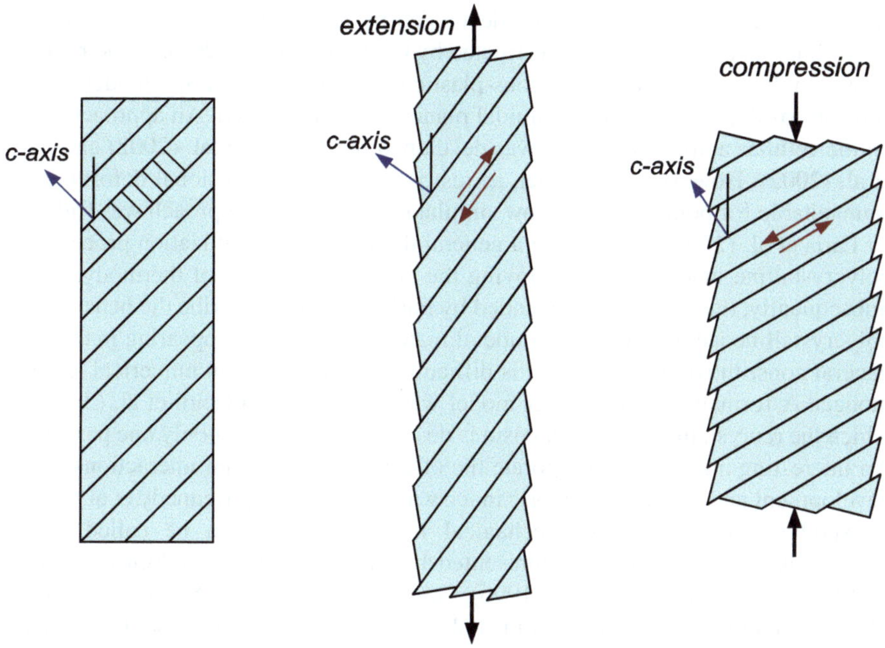

Fig. 6.2 Idealization of the crystal lattice rotation mechanism

In what follows in this chapter, a number of micro-mechanical models describing the behaviour of polycrystalline ice are formulated. These models have been developed with the aim to capture the main micro-mechanisms observed in natural ice which have effect on the formation and evolution of the anisotropic properties of the material. First of all, the mechanism of the crystal lattice rotation is incorporated, by which the crystal *c*-axes gradually rotate towards the principal axes of compression and away from the principal axes of tension, giving rise to strong fabrics found in deep regions of large polar ice sheets. The other important micro-process that should be accounted for in the models is the mechanism of dynamic recrystallization (see Sect. 2.2). By this mechanism, in turn, due to high shear stresses and high temperatures in near-bottom regions of polar ice sheets, the strong fabrics that have earlier developed in ice are destroyed in an abrupt (on geophysical time scale) manner.

The first micro-mechanical models for polycrystalline ice were formulated in the mid-1990s. Azuma (1994, 1995) proposed a model, in which an individual crystal is assumed to deform only by basal slip, and the microscopic stress acting on each crystal is related to the bulk macroscopic stress in a way determined on the basis of experimental results. Van der Veen and Whillans (1994) adopted a similar approach, by supposing that the only active slip system during the creep deformation of a grain is that associated with basal gliding, but, following Lliboutry (1993), they made an assumption of a uniform stress in a polycrystal, requiring that the stress in each grain is equal to the macroscopic stress applied to the polycrystal. A more general model, based on the theory developed by Hutchinson (1976) and extended by Molinari et al. (1987), was constructed by Castelnau et al. (1996). In their model, known in the literature as the viscous-plastic self-consistent (VPSC) model, crystal slips on basal, prismatic and pyramidal planes are incorporated. An approach based on the cellular automata method was developed by Ktitarev et al. (2002) and Faria et al. (2002). However, this model, as restricted to one-dimensional deformations, is unsuitable for polar ice sheet flow simulations. A distinct approach was followed by Faria et al. (2003) to construct a general theory of recrystallization processes in polycrystalline materials by employing the general principles of thermodynamics. Subsequently, this theory was extended by Faria (2006) to describe the behaviour of polycrystalline ice. Due to a multitude of material parameters appearing in this very general constitutive description, it is difficult to implement it in numerical ice flow models. A formally much simpler model was proposed by Placidi et al. (2010), in which the recrystallization mechanism is described by means of only one parameter. An interesting attempt to incorporate the effects of inter-crystal interactions on the development and evolution of fabric in ice was undertaken by Kennedy et al. (2013).

Another group of micro-mechanical models, which can be called *micro-macroscopic* formulations, is represented by the papers by Svendsen and Hutter (1996), Meyssonnier and Philip (1996), Gödert and Hutter (1998), Gagliardini and Meyssonnier (1999) and Gagliardini et al. (2001). All these models are based on the assumption that each material point of a polycrystalline aggregate contains crystals of all possible orientations, and the distribution of these orientations is defined by a continuous function, the evolution of which describes the changes in the anisotropic properties of the medium. Various aspects of the modelling of polycrystalline ice mechanics are discussed in a review paper by Placidi et al. (2006).

The models presented below belong to the class of so-called *multi-grain* (or *discrete-grain*) models (Staroszczyk 2001, 2002, 2004, 2009, 2011). Such models are based on an approach that seems to be physically reasonably well motivated, in which a material point of the polycrystalline material is represented by a finite number of discrete grains. Each individual grain in the aggregate is treated as a transversely isotropic and linearly viscous body, and its behaviour is followed separately from other grains. Non-linearity of the material behaviour can be accounted for by relating the ice viscosity to a function of strain-rate or stress invariants, in a manner typical of theoretical glaciology. The macroscopic response of the whole polycrystal is derived from the responses of all constituent crystals by applying one of homogenization methods; here two methods, known as the uniform-stress and uniform-strain approaches, are employed. The parameters of the presented micro-mechanical models have been determined by correlations with available experimental data for the observed limit behaviour of natural ice at large strains. The predictions of the models illustrate the evolution of anisotropic fabrics in ice and the variation of macroscopic viscosities of polycrystalline ice in simple flow simulations.

An alternative method to that described in this chapter, in which the constitutive laws for polar ice are constructed by applying a fundamentally different, namely phenomenological approach, is presented in Chap. 7.

6.1 Single Crystal Kinematics

Due to the transverse isotropy of the elementary hexagonal crystal of ice, with the axis of rotational symmetry coinciding with the crystal c-axis, the changing position of the crystal in space can be uniquely defined by the orientation of a unit vector aligned along the c-axis. Hence, we introduce a unit vector $\boldsymbol{c}$ along the crystal c-axis, and two angles: the zenith angle θ ($0 \leq \theta \leq \pi/2$) and the longitude φ ($0 \leq \varphi \leq 2\pi$), which define the orientation of the crystal in a fixed rectangular Cartesian reference frame Ox_i ($i = 1, 2, 3$), see Fig. 6.3. Since in some instances it is more convenient to describe the microscopic behaviour and material properties of a single crystal in a reference frame associated with that crystal, rather than in the global co-ordinates Ox_i, we also adopt a local rectangular frame Ox_i^c ($i = 1, 2, 3$), moving together with the crystal. The axes of the local frame are chosen in such a way that x_3^c coincides with the direction of the c-axis (the vector $\boldsymbol{c}$), x_1^c lies in the plane $Ox_3x_3^c$, and x_2^c has the direction that preserves the right-handedness of the local coordinate system. All tensor quantities, the components of which are expressed in the moving local frame will be indicated by the superscript ‘c’, and those expressed in the fixed global frame will be left without any suffix.

The transformation of components of non-scalar quantities from the local to the global reference frame is described by means of the *rotation tensor* $\boldsymbol{R}$, the components of which are defined by

$$\boldsymbol{R} = \begin{pmatrix} \cos\theta\cos\varphi & -\sin\varphi & \sin\theta\cos\varphi \\ \cos\theta\sin\varphi & \cos\varphi & \sin\theta\sin\varphi \\ -\sin\theta & 0 & \cos\theta \end{pmatrix}. \tag{6.1}$$

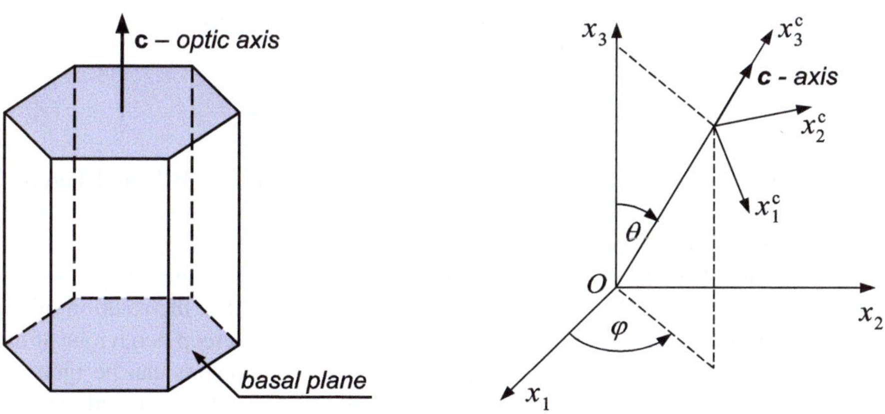

Fig. 6.3 Transversely isotropic hexagonal ice crystal and local and global coordinate frames, with the longitude (azimuth) angle φ and the zenith angle θ defining the changing crystal c-axis orientation in space

Since the local frame Ox_i^c moves relative to the fixed global frame Ox_i, the above (orthogonal) rotation matrix is time-dependent; that is, $\boldsymbol{R} = \boldsymbol{R}(t)$. Hence, the position vectors in both coordinate systems, $\boldsymbol{x}$ and $\boldsymbol{x}^c$, are related by

$$\boldsymbol{x} = \boldsymbol{R}(t)\,\boldsymbol{x}^c, \quad \boldsymbol{x}^c = \boldsymbol{R}^T(t)\,\boldsymbol{x}, \tag{6.2}$$

where $\boldsymbol{R}^T$ is the transpose of $\boldsymbol{R}$. Further, in view of Eq. (6.2)$_1$, the velocity fields observed in the global and local frames, $\boldsymbol{v}$ and $\boldsymbol{v}^c$ respectively, transform by

$$\boldsymbol{v} = \dot{\boldsymbol{x}} = \dot{\boldsymbol{R}}\boldsymbol{x}^c + \boldsymbol{R}\boldsymbol{v}^c, \tag{6.3}$$

where the superposed dots denote time derivatives. Differentiation of the velocities $\boldsymbol{v}$ and $\boldsymbol{v}^c$ with respect to the spatial coordinates yields the velocity gradients $\boldsymbol{L}$ and $\boldsymbol{L}^c$ measured in both frames, with components defined by

$$L_{ij} = \frac{\partial v_i}{\partial x_j}, \quad L_{ij}^c = \frac{\partial v_i^c}{\partial x_j^c} \quad (i, j = 1, 2, 3). \tag{6.4}$$

On account of Eq. (6.3), the velocity gradients $\boldsymbol{L}$ and $\boldsymbol{L}^c$ are connected by the relation

$$\boldsymbol{L} = \dot{\boldsymbol{R}}\boldsymbol{R}^T + \boldsymbol{R}\boldsymbol{L}^c\boldsymbol{R}^T. \tag{6.5}$$

Decomposition of the tensor $\boldsymbol{L}$ into its symmetric and anti-symmetric parts defines the strain-rate (or stretching) tensor $\boldsymbol{D}$ and the spin (or the rotation-rate) tensor $\boldsymbol{W}$:

$$\boldsymbol{D} = \tfrac{1}{2}\left(\boldsymbol{L} + \boldsymbol{L}^T\right), \quad \boldsymbol{W} = \tfrac{1}{2}\left(\boldsymbol{L} - \boldsymbol{L}^T\right). \tag{6.6}$$

These definitions, together with relation (6.5), furnish the transformation rules for the strain-rate and spin tensors expressed by

$$\boldsymbol{D} = \boldsymbol{R}\boldsymbol{D}^c\boldsymbol{R}^T, \quad \boldsymbol{W} = \dot{\boldsymbol{R}}\boldsymbol{R}^T + \boldsymbol{R}\boldsymbol{W}^c\boldsymbol{R}^T. \tag{6.7}$$

Since, by definition, the spin tensors are skew-symmetric ($\boldsymbol{W} = -\boldsymbol{W}^T$) and therefore have only three non-trivial components (Chadwick 1999), the tensor quation (6.7)$_2$ is equivalent to three scalar equations which express three non-zero components of $\boldsymbol{W}$ in terms of three non-zero components of $\boldsymbol{W}^c$: W_{12}^c, W_{13}^c and W_{23}^c. The first component, W_{12}^c, is irrelevant to this analysis, since it describes the rotation of the crystal about its axis of symmetry (which does not affect the creep behaviour of the crystal); hence, the respective equation for the $(\cdot)_{12}$ components can be ignored. The remaining two spin tensor components, W_{13}^c and W_{23}^c, can be expressed in terms of the strain-rate tensor components by assuming that the grain basal planes remain parallel to each other during the creep deformation of the crystal. This provides two kinematic relations (Meyssonnier and Philip 1996):

$$W_{13}^c = D_{13}^c, \quad W_{23}^c = D_{23}^c. \tag{6.8}$$

Accordingly, in view of (6.8), the tensor expression $(6.7)_2$ yields the following two evolution equations for the angles θ and φ:

$$\dot{\theta} = -\, D^c_{13} + W_{13}\cos\varphi + W_{23}\sin\varphi, \tag{6.9}$$

$$\dot{\varphi}\sin\theta = -\, D^c_{23} - W_{12}\sin\theta - (W_{13}\sin\varphi - W_{23}\cos\varphi)\cos\theta. \tag{6.10}$$

The above two kinematic equations describe uniquely the rotation of the crystal c-axis, as long as the microscopic strain-rates D^c_{13} and D^c_{23}, expressed in the lattice frame Ox^c_i, and the microscopic spins W_{12}, W_{13} and W_{23}, given in the fixed frame Ox_i, are known. The microscopic strain-rates are related to stresses by constitutive laws, while the microscopic spins are determined by the macroscopic deformations of a whole polycrystal. The formulation of the microscopic constitutive equations is the subject of the next section.

6.2 Constitutive Laws for an Ice Crystal

The results of experiments carried out by Rigsby (1958) and Jones (1982) show that the creep behaviour of polycrystalline ice can be regarded as independent of confining pressure. Therefore, one can assume that the creep of ice is entirely determined by the deviatoric stress (which is a common approximation of the creep behaviour in general), and treat the material as an incompressible medium, with the mean pressure not prescribed by a constitutive law, but determined by the momentum equation and boundary conditions in a particular problem considered. The same assumptions as for the polycrystalline ice aggregate are applied to the viscous behaviour of a single crystal of ice. Kamb (1961) proved that the creep response of an individual hexagonal crystal is entirely independent of the glide direction on the crystal basal plane in the case of its linear behaviour, and is weakly dependent on the glide direction in the case of non-linear behaviour. Thus, irrespective whether it is linear or non-linear range of the creep response of ice to stress, the single crystal of ice is universally treated as a material exhibiting transverse isotropy, with the crystal c-axis being the axis of rotational symmetry, and the crystal basal plane being the plane of isotropy.

Accordingly, the constitutive law for a crystal involves the deviatoric Cauchy stress, $\boldsymbol{S}$, and the strain-rate, $\boldsymbol{D}$. In addition, to account for the anisotropy of the material, a so-called *structure tensor*, denoted by $\boldsymbol{M}$, is used to account for the transverse symmetry of the body. The latter tensor is defined by means of the crystal c-axis unit vector $\boldsymbol{c}$ as follows:

$$\boldsymbol{M} = \boldsymbol{c}\otimes\boldsymbol{c}, \quad M_{ij} = c_i c_j \quad (i, j = 1, 2, 3), \tag{6.11}$$

where the symbol $\otimes$ denotes an outer product. The structure tensor has the properties $\operatorname{tr}\boldsymbol{M} = 1$ and $\boldsymbol{M}^2 = \boldsymbol{M}$. The deviatoric stress $\boldsymbol{S}$ is given in terms of the Cauchy stress $\boldsymbol{\sigma}$ and the hydrostatic pressure p by

$$\boldsymbol{S} = \boldsymbol{\sigma} - \tfrac{1}{3}\text{tr}\,\boldsymbol{\sigma}\boldsymbol{I} = \boldsymbol{\sigma} + p\,\boldsymbol{I}, \quad p = -\tfrac{1}{3}\text{tr}\,\boldsymbol{\sigma}. \tag{6.12}$$

The constitutive laws for the anisotropic ice are derived within the framework of the general theory of *frame-indifferent* (*objective*) constitutive equations (Chadwick 1999; Liu 2002; Truesdell and Noll 2004). In accordance with this theory, admissible constitutive relations must satisfy a set of conditions which ensure that the observed behaviour of a material is the same for any pair of (moving) observers (for details see Appendix B). By applying this theory, an irreducible (canonical) form of a non-linear constitutive equation can be derived for any type of material symmetry, such as isotropy, transverse isotropy, orthotropy, etc. (Boehler 1987). The irreducible form of a frame-indifferent constitutive law for a transversely isotropic medium, which relates two symmetric second-order tensors, $\boldsymbol{A}_1$ and $\boldsymbol{A}_2$ say, and which is linear in these tensors, is expressed by

$$\boldsymbol{A}_1 = \alpha_1\boldsymbol{I} + \alpha_2\boldsymbol{M} + \alpha_3\boldsymbol{A}_2 + \alpha_4\,(\boldsymbol{M}\boldsymbol{A}_2 + \boldsymbol{A}_2\boldsymbol{M})\,. \tag{6.13}$$

The above form is obtained from a general non-linear Eq. (B.7) on p. 325. In (6.13), α_1 and α_2 are functions of the invariants $\text{tr}\,\boldsymbol{A}_2$ and $\text{tr}\,(\boldsymbol{M}\boldsymbol{A}_2)$, and α_3 and α_4 are constants. The invariant $\text{tr}\,(\boldsymbol{M}\boldsymbol{A}_2) = \boldsymbol{c}^T\boldsymbol{A}_2\,\boldsymbol{c}$ describes the component of the tensor $\boldsymbol{A}_2$ in the privileged material direction represented by the vector $\boldsymbol{c}$, parallel to the crystal c-axis. In view of Eqs. (B.8), the coefficients α_k $(k = 1, \ldots, 4)$ are given by

$$\begin{aligned} \alpha_1 &= a_1 + a_2\text{tr}\,\boldsymbol{A}_2 + a_3\text{tr}\,(\boldsymbol{M}\boldsymbol{A}_2), \\ \alpha_2 &= a_4 + a_5\text{tr}\,\boldsymbol{A}_2 + a_6\text{tr}\,(\boldsymbol{M}\boldsymbol{A}_2), \\ \alpha_3 &= a_7, \quad \alpha_4 = a_8. \end{aligned} \tag{6.14}$$

Thus, there are eight constants $(a_1, \ldots, a_8)$ in the most general form of a linear law. Since our aim is to construct a constitutive law for the creep behaviour of ice, one of the above two tensors, $\boldsymbol{A}_1$ or $\boldsymbol{A}_2$, will be the deviatoric stress tensor $\boldsymbol{S}$, and the other will be the strain-rate tensor $\boldsymbol{D}$, with a specific choice depending on whether the stress will be expressed in terms of the strain-rate, or vice versa.

The number of non-vanishing material constants a_i appearing in (6.13) and (6.14) can be reduced from eight to three by following the procedure described in detail by Staroszczyk (2004). Here we only note that two constants, namely a_1 and a_4, become zero if one assumes that the material is stress-free if it does not deform, that is, $\boldsymbol{S} = \boldsymbol{O}$ when $\boldsymbol{D} = \boldsymbol{O}$, where $\boldsymbol{O}$ is the zero tensor. Further, by taking advantage of the fact that both tensors $\boldsymbol{S}$ and $\boldsymbol{D}$ are traceless ($\text{tr}\,\boldsymbol{S} = 0$ by definition and $\text{tr}\,\boldsymbol{D} = 0$ by ice incompressibility), we notice that the two terms involving $\text{tr}\,\boldsymbol{A}_2$ disappear in (6.14). Hence, the constants a_2 and a_5 have no effect on the creep response of the material and therefore can be neglected. Thus,

$$\alpha_1 = a_3\text{tr}\,(\boldsymbol{M}\boldsymbol{A}_2), \quad \alpha_2 = a_6\text{tr}\,(\boldsymbol{M}\boldsymbol{A}_2), \tag{6.15}$$

and the law (6.13) reduces to

$$\boldsymbol{A}_1 = a_3 \text{tr}\,(\boldsymbol{M}\boldsymbol{A}_2)\boldsymbol{I} + a_6 \text{tr}\,(\boldsymbol{M}\boldsymbol{A}_2)\boldsymbol{M} + a_7 \boldsymbol{A}_2 + a_8\,(\boldsymbol{M}\boldsymbol{A}_2 + \boldsymbol{A}_2\boldsymbol{M})\,. \tag{6.16}$$

By equating now the deviatoric parts of both sides of (6.16) one can eliminate the first term on the RHS of (6.16) (since $\boldsymbol{I}$ is a spherical tensor), which implies that the constant a_3 does not enter the constitutive equation. Accordingly, the flow law for a transversely isotropic and incompressible ice crystal takes the form

$$\boldsymbol{A}_1 = a_6 \text{tr}\,(\boldsymbol{M}\boldsymbol{A}_2)\left(\boldsymbol{M} - \tfrac{1}{3}\,\boldsymbol{I}\right) + a_7 \boldsymbol{A}_2 + a_8\left[\boldsymbol{A}_2\boldsymbol{M} + \boldsymbol{M}\boldsymbol{A}_2 - \tfrac{2}{3}\text{tr}\,(\boldsymbol{M}\boldsymbol{A}_2)\,\boldsymbol{I}\right]. \tag{6.17}$$

which involves only three material constant: a_6, a_7 and a_8. These three constants, defining the creep behaviour of the material, should be (ideally) determined from the results of simple laboratory tests.

6.2.1 Strain-Rate – Stress Formulation

The frame-indifferent microscopic constitutive law relating strain-rate to deviatoric stress is obtained by taking $\boldsymbol{A}_1 = \boldsymbol{D}$ and $\boldsymbol{A}_2 = \boldsymbol{S}$ in Eq. (6.17), which then becomes:

$$\boldsymbol{D} = a_6 \text{tr}\,(\boldsymbol{M}\boldsymbol{S})\left(\boldsymbol{M} - \tfrac{1}{3}\,\boldsymbol{I}\right) + a_7 \boldsymbol{S} + a_8\left[\boldsymbol{S}\boldsymbol{M} + \boldsymbol{M}\boldsymbol{S} - \tfrac{2}{3}\text{tr}\,(\boldsymbol{M}\boldsymbol{S})\,\boldsymbol{I}\right]. \tag{6.18}$$

The above law describes the crystal behaviour in the global reference frame Ox_i ($i = 1, 2, 3$). In order to physically interpret and then determine in experiments the three material constants a_6, a_7 and a_8, it is convenient to adopt this frame to coincide with the local frame Ox_i^c associated with the single crystal lattice. In this case, $\boldsymbol{D} = \boldsymbol{D}^c$ and $\boldsymbol{S} = \boldsymbol{S}^c$, and the tensors and their combinations entering Eq. (6.18) are defined by

$$\boldsymbol{M} = \begin{pmatrix} 0 & 0 & 0 \\ 0 & 0 & 0 \\ 0 & 0 & 1 \end{pmatrix}, \quad \boldsymbol{M}\boldsymbol{S}^c = \begin{pmatrix} 0 & 0 & 0 \\ 0 & 0 & 0 \\ S_{31}^c & S_{32}^c & S_{33}^c \end{pmatrix}, \quad \text{tr}\,(\boldsymbol{M}\boldsymbol{S}^c) = S_{33}^c\,, \tag{6.19}$$

$$\boldsymbol{S}^c\boldsymbol{M} + \boldsymbol{M}\boldsymbol{S}^c - \tfrac{2}{3}\text{tr}\,(\boldsymbol{M}\boldsymbol{S}^c)\,\boldsymbol{I} = \begin{pmatrix} -\frac{2}{3}S_{33}^c & 0 & S_{13}^c \\ 0 & -\frac{2}{3}S_{33}^c & S_{23}^c \\ S_{31}^c & S_{32}^c & \frac{4}{3}S_{33}^c \end{pmatrix}. \tag{6.20}$$

With expressions (6.19) and (6.20), the law (6.18) yields the relations which connect the microscopic strain-rate components D_{ij}^c to the corresponding microscopic deviatoric stress components S_{ij}^c by

$$D_{12}^c = a_7 S_{12}^c, \quad D_{13}^c = (a_7 + a_8)\,S_{13}^c, \quad D_{33}^c = \tfrac{1}{3}\,(2a_6 + 3a_7 + 4a_8)\,S_{33}^c\,. \tag{6.21}$$

Now let us introduce the following three *fluidities* (reciprocal viscosities) η_{ij}: η_{13} for shearing in the plane parallel to the crystal c-axis (basal shearing), η_{12} for shearing in the plane normal to the c-axis (prismatic shearing), and η_{33} for unconfined axial compression along the c-axis. These fluidities are defined in terms of the strain-rates and deviatoric stresses expressed in the local coordinate system Ox_i^c by

$$\eta_{ij} = \frac{2D_{ij}^c}{S_{ij}^c}. \tag{6.22}$$

The factor 2 is used in the above definition to ensure conformity with the standard form of the viscous flow law for isotropic fluids $\boldsymbol{S} = 2\mu_0 \boldsymbol{D}$ (then $S_{ij} = 2D_{ij}/\eta_0$ due to $\mu_0 = 1/\eta_0$), with μ_0 and η_0 denoting, respectively, the isotropic fluid viscosity and fluidity. From relations (6.21) and (6.22) it follows that

$$a_7 = \tfrac{1}{2}\,\eta_{12}, \quad a_7 + a_8 = \tfrac{1}{2}\,\eta_{13}, \quad 2a_6 + 3a_7 + 4a_8 = \tfrac{1}{2}\,\eta_{33}. \tag{6.23}$$

The solution of the above three equations determines the material constants in terms of the fluidities as follows:

$$a_6 = \left(\tfrac{3}{4}\,\alpha + \tfrac{1}{4}\,\beta - 1\right)\eta\,, \quad a_7 = \tfrac{1}{2}\,\beta\eta\,, \quad a_8 = \tfrac{1}{2}\,(1-\beta)\,\eta\,, \tag{6.24}$$

where $\eta = \eta_{13}$ is the largest fluidity amongst η_{12}, η_{13} and η_{33}. α and β are two dimensionless rheological parameters, defining the axial and prismatic shear fluidities in terms of the basal shear fluidity $\eta = \eta_{13}$:

$$\alpha = \frac{\eta_{33}}{\eta_{13}}, \quad \beta = \frac{\eta_{12}}{\eta_{13}}, \tag{6.25}$$

with the inequalities $0 \leq \alpha \leq 1$ and $0 \leq \beta \leq 1$. Substitution of relations (6.24) in (6.18) leads to the constitutive equation

$$\begin{aligned} \boldsymbol{D} = \tfrac{1}{2}\eta \Big\{ &\tfrac{1}{2}\,(3\alpha + \beta - 4)\,\mathrm{tr}\,(\boldsymbol{MS})\left(\boldsymbol{M} - \tfrac{1}{3}\,\boldsymbol{I}\right) + \beta\,\boldsymbol{S} + \\ &+ (1-\beta)\left[\boldsymbol{SM} + \boldsymbol{MS} - \tfrac{2}{3}\mathrm{tr}\,(\boldsymbol{MS})\,\boldsymbol{I}\,\right]\Big\} \end{aligned} \tag{6.26}$$

for the viscous deformation of the transversely isotropic crystal of ice. In particular, the case $\alpha = \beta = 1$ describes an isotropic crystal, and the case $\alpha = \beta = 0$ corresponds to the situation in which the crystal deforms only by basal slip. It can be easily verified that for the isotropic grain the law (6.26) reduces to the standard equation $\boldsymbol{D} = \eta_0 \boldsymbol{S}/2$, with $\eta = \eta_0$, or, equivalently, to the relation $\boldsymbol{S} = 2\mu_0 \boldsymbol{D}$. In the Voigt notation, in which tensor components are expressed as elements of vectors, the flow law (6.26) can be expressed in the crystal coordinate system in an alternative form as

$$\begin{pmatrix} D^c_{11} \\ D^c_{22} \\ D^c_{33} \\ D^c_{12} \\ D^c_{13} \\ D^c_{23} \end{pmatrix} = \frac{\eta}{2} \begin{pmatrix} \frac{1}{2}(\alpha+\beta) & \frac{1}{2}(\alpha-\beta) & & & & \\ \frac{1}{2}(\alpha-\beta) & \frac{1}{2}(\alpha+\beta) & & (0) & & \\ & & \alpha & & & \\ & & & \beta & & \\ & (0) & & & 1 & \\ & & & & & 1 \end{pmatrix} \begin{pmatrix} S^c_{11} \\ S^c_{22} \\ S^c_{33} \\ S^c_{12} \\ S^c_{13} \\ S^c_{23} \end{pmatrix}, \tag{6.27}$$

with the symmetric fluidity matrix depending on the three rheological parameters: the basal shear fluidity η and the two dimensionless parameters α and β describing the degree of anisotropy of the crystal. An alternative fluidity matrix was derived by Gagliardini and Meyssonnier (1999); however, they considered a simpler case in which $\alpha = \beta$ (that is, the fluidities for prismatic shear and axial compression are equal), which results in a diagonal form of the matrix in (6.27).

6.2.2 *Stress – Strain-Rate Formulation*

In the particular case of the coordinate frame associated with the crystal, the constitutive law in the form expressing deviatoric stress in terms of strain-rate can be easily derived by inversion of (6.27). However, such an inversion of the general form (6.26) is not so straightforward. Therefore, to derive a counterpart of the viscous flow (6.26) in which stresses are now given in terms of strain-rates, we resort to the same method as that employed to obtain (6.26). Hence, we start again from the generic equation (6.17), and assume that now $\boldsymbol{A}_1 = \boldsymbol{S}$ and $\boldsymbol{A}_2 = \boldsymbol{D}$. Accordingly, the stress—strain-rate form of the viscous flow law is expressed by

$$\boldsymbol{S} = a_6 \mathrm{tr}\,(\boldsymbol{M}\boldsymbol{D})\left(\boldsymbol{M} - \tfrac{1}{3}\boldsymbol{I}\right) + a_7 \boldsymbol{D} + a_8\left[\boldsymbol{M}\boldsymbol{D} + \boldsymbol{D}\boldsymbol{M} - \tfrac{2}{3}\mathrm{tr}\,(\boldsymbol{M}\boldsymbol{D})\,\boldsymbol{I}\right], \tag{6.28}$$

where the material constants a_6, a_7 and a_8 have now different meanings than in the strain-rate – stress equation given by (6.18). In order to relate these constant to measurable quantities, we introduce three viscosities: μ_{13}, μ_{12} and μ_{33}, which are, respectively, the shear viscosities for basal and prismatic slips, and the axial viscosity for compression along the c-axis. These viscosities are the reciprocals of the fluidities defined by (6.22), thus

$$\mu_{ij} = \eta_{ij}^{-1}. \tag{6.29}$$

Furthermore, we also introduce two dimensionless rheological parameters A and B, defining the degree of anisotropy of the single crystal, defined by

$$A = \frac{\mu_{33}}{\mu_{13}}, \quad B = \frac{\mu_{12}}{\mu_{13}}, \tag{6.30}$$

which, in view of (6.29) and (6.25), are related to the previously used parameters α and β by

$$A = \alpha^{-1}, \quad B = \beta^{-1}. \tag{6.31}$$

By following the same argument as before for the strain-rate – stress formulation of the constitutive equation, the three constants a_6, a_7 and a_8 can be related to the three viscosities μ_{12}, μ_{13} and μ_{33}, or, equivalently, to the two dimensionless parameters A and B and the basal shear viscosity $\mu = \mu_{13}$ (the smallest viscosity amongst μ_{12}, μ_{13} and μ_{33}) (Staroszczyk 2002). After inserting the resulting expressions for a_6, a_7 and a_8 into (6.28), one obtains the following frame-indifferent constitutive law describing the viscous creep behaviour of an incompressible transversely isotropic crystal of ice:

$$\begin{aligned} \boldsymbol{S} = 2\mu \Big\{ & \tfrac{1}{2}\,(3A + B - 4)\,\mathrm{tr}\,(\boldsymbol{M}\boldsymbol{D})\big(\boldsymbol{M} - \tfrac{1}{3}\,\boldsymbol{I}\big) + B\boldsymbol{D} + \\ & + (1 - B)\big[\boldsymbol{M}\boldsymbol{D} + \boldsymbol{D}\boldsymbol{M} - \tfrac{2}{3}\mathrm{tr}\,(\boldsymbol{M}\boldsymbol{D})\,\boldsymbol{I}\,\big]\Big\}, \end{aligned} \tag{6.32}$$

where $A \geq 1$ and $B \geq 1$. The case $A = B = 1$ corresponds to an isotropic crystal, whereas the limit case $A \to \infty$ and $B \to \infty$ describes a crystal which deforms only by basal glide. In the local coordinate system associated with the crystal lattice, the law (6.32) can be expressed as

$$\begin{pmatrix} S^c_{11} \\ S^c_{22} \\ S^c_{33} \\ S^c_{12} \\ S^c_{13} \\ S^c_{23} \end{pmatrix} = 2\mu \begin{pmatrix} \tfrac{1}{2}(A+B) & \tfrac{1}{2}(A-B) & & & & \\ \tfrac{1}{2}(A-B) & \tfrac{1}{2}(A+B) & & & (0) & \\ & & A & & & \\ & & & B & & \\ & (0) & & & 1 & \\ & & & & & 1 \end{pmatrix} \begin{pmatrix} D^c_{11} \\ D^c_{22} \\ D^c_{33} \\ D^c_{12} \\ D^c_{13} \\ D^c_{23} \end{pmatrix}. \tag{6.33}$$

It can be easily shown, recalling the identities (6.31) and $\mu = 1/\eta$, that the viscosity matrix in Eq. (6.33) is indeed an inverse of the fluidity matrix in Eq. (6.27).

6.3 Macroscopic Behaviour of an Ice Polycrystal

The microscopic constitutive laws formulated in Sect. 6.2 describe the viscous creep of a single crystal of ice. In order to derive macroscopic constitutive equations for a polycrystalline aggregate, composed, in general, of crystals of all possible orientations in space, it is necessary to apply one of homogenization methods. These methods, based on various simplifications adopted in them, enable the estimation of the macroscopic properties of a polycrystal.

A number of approaches, which all originate from the field of metallurgy, have been developed to connect the properties of single crystals to those of polycrystals.

The historically first such an approach, known as the *uniform stress method*, also referred to as the Sachs-Reuss (SR) method, is based on the assumption that all crystals in an aggregate are subject to the same stresses. This approximation was first introduced to ice mechanics by Lliboutry (1993) who argued that the differences in micro-stresses between adjacent ice grains are negligibly small due to the process of continuous migration of grain boundaries. The other approach, commonly applied in metallurgy and structural geology, is known as the *uniform strain method*, or the Taylor-Voigt (TV) method, and implies that uniform macroscopic deformation applied to an aggregate induces the same uniform deformations in all constituent grains. Both homogenization methods significantly simplify the real intrinsic behaviour of the material, since they do not account for the local interactions occurring between individual crystals. Furthermore, in the uniform stress approach, the local (that is, across the grain boundaries) compatibility conditions are, generally, not satisfied, whereas in the uniform strain approximation the local equilibrium conditions are, usually, violated. The question as to which of the two methods is more suitable for polycrystalline ice remains open, since, as yet, there is no obvious indication as to why one of the approaches should be better than the other. It appears that the uniform stress method is more common in theoretical glaciology, though the results of numerical comparisons made by Meyssonnier and Philip (1999) (by using a self-consistent model similar to that of Castelnau et al. (1996), in which no assumptions on local stresses and strains are made) do not clearly favour either of the models. Nevertheless, both Sachs-Reuss and Taylor-Voigt models are useful for practice, since these two extreme approximations of the stress and the strain homogeneity in a polycrystalline aggregate give, respectively, a lower and an upper bound for the macroscopic stress at a given strain-rate (Bishop and Hill 1951; Hill 1952; Arminjon 1991); hence, they provide bounds on the macroscopic viscosity of polar ice. Thus, it is of interest to develop the models for polycrystalline ice which make use of both limit approximations. Apart from the Sachs-Reuss, Taylor-Voigt and self-consistent model (Molinari et al. 1987), several other, formally more complex approaches, are known in materials science, including, for instance, the Hashin-Shtrikman (1963) and Mori-Tanaka (1973) methods, which enable more accurate estimations of the real behaviour of polycrystalline bodies. Such models, though, have not been used yet in theoretical glaciology. At this stage it seems that a good practical solution in geophysical applications is to take an arithmetic mean of the lower and upper bounds predicted by the Sachs-Reuss and Tailor-Voigt methods, as giving a sufficiently accurate approximation to the macroscopic ice viscosity.

Once the microscopic properties of all individual grains in a polycrystal have been established by applying microscopic constitutive equations, kinematic constraints and simplifications imposed by a homogenization technique employed, the macroscopic response of the polycrystalline aggregate can be determined. In the multi-grain formulation presented here, a simple averaging method is adopted, in which the components of any tensor quantity describing the whole polycrystal are defined as weighted arithmetic averages of the components of the corresponding microscopic tensor entities. In what follows, all macroscopic quantities will be indicated by a superposed bar. Accordingly, the components of a macroscopic tensor $\bar{\boldsymbol{T}}$ are evaluated in terms of the components of microscopic tensors $\boldsymbol{T}^{(k)}$ $(k = 1, 2, \ldots, N_c)$ associated with constituent crystals by the formula

$$\bar{T}_{ij} = \frac{1}{V_0} \sum_{k=1}^{N_c} V_k \, T_{ij}^{(k)}, \quad V_0 = \sum_{k=1}^{N_c} V_k \,. \tag{6.34}$$

In (6.34), V_k denotes the volume of a k-th discrete grain, V_0 is the total volume of all grains in the aggregate, and N_c is the number of crystals at a given material point and at a given time instant. If only the mechanism of rotation recrystallization is considered, then the number of crystals can be treated as constant. However, when the mechanism of the dynamic (migration) recrystallization is active, then N_c varies in time due to the process of nucleation of new crystals and the disappearance of old ones. Moreover, the volumes V_k of the crystals which undergo dynamic recrystallization change either. To obtain statistically satisfactory results of the above averaging procedure, the number of discrete grains used in the calculations should exceed a minimum value of 230, found by Elvin (1996) on the basis of numerical simulations carried out for polycrystalline S2 ice. In all simulations, the results of which are presented in the further part of this chapter, at least 800 discrete grains have been used.

The majority of the micro-mechanical models that have been developed to describe induced anisotropy of ice are based on the Sachs-Reuss approximation of the stress homogeneity within a polycrystalline aggregate (Lile 1978; Lliboutry 1993; Van der Veen and Whillans 1994; Gödert and Hutter 1998; Gagliardini and Meyssonnier 1999; Staroszczyk 2001), implying that each crystal, irrespective of its orientation, bears the same microscopic stress, equal to the overall macroscopic stress applied to the aggregate. Formally, this is expressed by

$$\boldsymbol{S} = \bar{\boldsymbol{S}}. \tag{6.35}$$

The tensor equation (6.35) is equivalent to six relations between the microscopic and macroscopic deviatoric stresses. The six microscopic stress components determine, through the constitutive law (6.26), six microscopic strain-rate components. However, these six strain-rates D_{ij} are insufficient to trace the evolution of the oriented structure of the material, described by (6.9) and (6.10), since microscopic spins W_{12}, W_{13} and W_{23} are needed as well. Therefore, three additional micro-macroscopic relations are required, and those, following Gödert and Hutter (1998) and Gagliardini and Meyssonnier (1999), are chosen to be

$$\boldsymbol{W} = \bar{\boldsymbol{W}}, \tag{6.36}$$

connecting the microscopic and macroscopic spins. Hence, the uniform stress model considered here is, in fact, based on the combination of six stress and three kinematic conditions.

In other branches of materials science, especially in metallurgy, more common than the uniform stress models are the methods based on the Taylor-Voigt approximation of the strain uniformity in the polycrystalline aggregate. An example is a family of theories referred to as the Taylor-Bishop-Hill (TBH) models (Bishop and Hill 1951; Wenk et al. 1989; Zhang and Jenkins 1993). In terms of the velocity gradients, the Taylor-Voigt condition takes the form:

$$\boldsymbol{L} = \bar{\boldsymbol{L}}, \tag{6.37}$$

which is equivalent to

$$\boldsymbol{D} = \bar{\boldsymbol{D}} \quad \text{and} \quad \boldsymbol{W} = \bar{\boldsymbol{W}}. \tag{6.38}$$

Thus, the uniform strain model is based on nine kinematic relations, connecting the single crystal and polycrystalline aggregate deformation variables.

6.3.1 Uniform Stress Model

The stress uniformity condition (6.35) implies that, in general, the local strain-rates $\boldsymbol{D}$ are different in each crystal. These strain-rates, expressed in the global coordinates Ox_i, are prescribed by the constitutive equation (6.26) in which the microscopic stress $\boldsymbol{S}$ is replaced by the macroscopic stress $\bar{\boldsymbol{S}}$. In order to follow the evolution of the crystal c-axes orientations in space, described by Eqs. (6.9) and (6.10), one needs the strain-rate components D^c_{13} and D^c_{23} given in the local frame Ox^c_i. The latter components can be derived from the transformation relation $(6.7)_1$ to yield

$$\begin{aligned} D^c_{13} &= \frac{\eta}{4}\sin 2\theta\left[\bar{S}_{11}(1+\cos^2\varphi) + \bar{S}_{22}(1+\sin^2\varphi) + \bar{S}_{12}\sin 2\varphi\right] + \\ &\quad + \frac{\eta}{2}\cos 2\theta\,(\bar{S}_{13}\cos\varphi + \bar{S}_{23}\sin\varphi), \end{aligned} \tag{6.39}$$

$$\begin{aligned} D^c_{23} &= \frac{\eta}{4}\sin\theta\left[(\bar{S}_{22} - \bar{S}_{11})\sin 2\varphi + 2\bar{S}_{12}\cos 2\varphi\right] + \\ &\quad - \frac{\eta}{2}\cos\theta\,(\bar{S}_{13}\sin\varphi - \bar{S}_{23}\cos\varphi). \end{aligned} \tag{6.40}$$

The insertion of the above two relations in Eqs. (6.9) and (6.10), in which the microscopic spin tensor components W_{ij} are replaced by the corresponding macroscopic quantities $\bar{W}_{ij}$, see the condition (6.36), allows us to follow the evolution of the angles θ and φ defining the orientation of the c-axis of an individual crystal. By doing so for all the crystals in the aggregate one can trace the evolution of the ice fabric (the oriented microstructure of ice). By applying then the averaging formula (6.34), it is possible to evaluate the macroscopic properties of ice associated with its current fabric.

Limit Viscosities

The macroscopic property which is crucial in the context of polar ice sheet flow simulations, is the viscosity of ice. Therefore, it is of practical importance to determine the dependence of the macroscopic viscosities of anisotropic ice on its microstructure (fabric), starting from an initially isotropic fabric typical of ice found near the free surface of an ice sheet, to strongly anisotropic fabrics characteristic of the bottom layers of polar glaciers.

First, determine the macroscopic viscosity of the isotropic polycrystalline ice. Clearly, this must be a function of the microscopic rheological parameters defining the viscous properties of a single crystal, and the distribution of the crystal orientations in the fabric. In the case of macroscopic isotropy of the material, it can be assumed that the c-axes of all the crystals in the aggregate are randomly distributed in space. Suppose that the number of constituent grains is sufficiently large, and that each crystal (with its orientation defined by the angles θ and φ) has the same size and, thus, 'occupies' the same area on a unit hemisphere, equal to $N_c = 2\pi/\sin\theta\, d\theta\, d\varphi$. Then, at the limit $N_c \to \infty$, the summation in Eq. (6.34) can be replaced by surface integration, which, when applied to the strain-rate components, transforms the averaging formula $(6.34)_1$ into:

$$\bar{D}_{ij} = \frac{1}{2\pi} \int_0^{2\pi} \int_0^{\pi/2} D_{ij}(\theta, \varphi)\, \sin\theta\, \mathrm{d}\theta\, \mathrm{d}\varphi. \tag{6.41}$$

To derive a relation between the macroscopic and microscopic fluidities (reciprocal viscosities) of the isotropic ice, consider a simple stress configuration, namely that of pure shear, in which the only non-vanishing deviatoric stress components are, say, $\bar{S}_{13} = \bar{S}_{31}$. For these two components, the flow law (6.26), with $S_{13} = \bar{S}_{13}$, yields the microscopic strain-rate D_{13}

$$\begin{aligned} D_{13} = \frac{\eta}{2} \bar{S}_{13} \big[&(3\alpha + \beta - 4) \sin^2\theta \cos^2\theta \cos^2\varphi + \\ &+ \beta + (1 - \beta)(\sin^2\theta \cos^2\varphi + \cos^2\theta) \big]. \end{aligned} \tag{6.42}$$

After substituting the above equation into (6.41) and performing the prescribed integration, one finds the macroscopic strain-rate $\bar{D}_{13}$ to be given by

$$\bar{D}_{13} = \frac{\eta}{10} \bar{S}_{13} \, (\alpha + 2\beta + 2). \tag{6.43}$$

By the analogy with the viscous flow law $\bar{\boldsymbol{S}} = 2\mu_0 \bar{\boldsymbol{D}}$, the latter expression determines the macroscopic viscosity of the isotropic ice, μ_0, as

$$\mu_0 = \mu \, \frac{5}{\alpha + 2\beta + 2}. \tag{6.44}$$

Recall that $\mu = \mu_{13}$ is the microscopic shear viscosity for the crystal basal slip, and the dimensionless rheological parameters α and β describe the strength of crystal anisotropy. For the isotropic crystals ($\alpha = \beta = 1$), relation (6.44) gives $\mu_0 = \mu$, implying that the microscopic and macroscopic viscous properties of the material are the same.

As the polycrystalline ice deforms under uniaxial compression or simple shear, the individual crystals gradually rotate in such a way that, eventually, all the crystal c-axes become aligned in parallel, forming a very strong fabric. Such highly anisotropic fabrics develop at very large axial or shear strains, and are typical of near-bottom regions of polar ice sheets (Gow et al. 1997; Thorsteinsson et al. 1997; Gow and Meese 2007; Faria et al. 2014). In a limit situation, when all the crystal c-axes are aligned in one direction, the macroscopic properties of the polycrystal are essentially those of a single monocrystal, with the rheological properties described by the constitutive Eqs. (6.26) or (6.27). Hence, the macroscopic viscosities of the polycrystal are the same as the microscopic viscosities of a single crystal. Therefore, with the macroscopic viscosity of the isotropic polycrystalline ice given by (6.44), the ratio of the limit macroscopic viscosity for shearing on the plane parallel to the direction of all crystal c-axes is expressed by the relation

$$\frac{\mu_{13}}{\mu_0} = \frac{\alpha + 2\beta + 2}{5} . \tag{6.45}$$

Similarly, the limit macroscopic viscosities for shear in the plane normal to the crystal c-axes (prismatic shearing), for uniaxial compression along the c-axes, and for uniaxial compression in the direction normal to the c-axes, are given, respectively, by

$$\frac{\mu_{12}}{\mu_0} = \frac{\alpha + 2\beta + 2}{5\beta} , \quad \frac{\mu_{33}}{\mu_0} = \frac{\alpha + 2\beta + 2}{5\alpha} , \quad \frac{\mu_{11}}{\mu_0} = \frac{4(\alpha + 2\beta + 2)}{5(\alpha + 3\beta)} . \tag{6.46}$$

The limit viscosity ratios μ_{13}/μ_0 and μ_{33}/μ_0, defined by the above expressions (6.45) and $(6.46)_2$, are usually measured in simple shear and uniaxial compression tests carried out on samples of polycrystalline ice. It follows from relation (6.45) that in the case of the most anisotropic single crystals, defined by the material parameters $\alpha = \beta = 0$, when the only mode of the crystal deformation is the basal glide, the limit viscosity ratio $\mu_{13}/\mu_0 = 0.4$ holds. On the other hand, equation $(6.46)_2$ gives for $\alpha = \beta = 0$ an unbounded value of the limit viscosity ratio μ_{33}/μ_0. This means that when all the crystal c-axes are aligned in one direction, then the polycrystal cannot deform in this direction by axial compression. Since for physical reasons this is unrealistic, a non-zero value of the rheological parameter α must be adopted in the uniform stress constitutive model.

The reciprocals of the limit viscosity ratios given by Eqs. (6.45) and $(6.46)_2$, μ_0/μ_{13} and μ_0/μ_{33}, are commonly described in glaciology as the *enhancement factors* for shear and compression, E_s and E_a, respectively (Budd and Jacka 1989). By correlating the viscosity relations (6.45) and $(6.46)_2$ with empirical data, one can attempt to determine the single crystal rheological parameters in micro-mechanical constitutive models. Budd and Jacka (1989) measured the values of the enhancement factors $E_a = 3$ and $E_s = 8$ for ice at temperatures near the melting point, meaning that both axial and shear viscosities of strongly anisotropic ice are considerably smaller than those for isotropic ice at the start of deformation. However, the data for ice at temperatures below −20 °C (Pimienta et al. 1987; Thorsteinsson et al. 1997) indicate that the axial viscosity increases with the development of the anisotropic fabric in ice; that is, $E_a < 1$. Hence, two different types of polar ice can be distinguished, referred to as *warm ice* and *cold ice*. There is no universal agreement between researchers as to the specific values of the enhancement factors for both types of polar ice, and different values are reported on the basis of experimental data (Pimienta et al. 1987; Ma et al. 2010; Treverrow et al. 2012), or estimated on the basis of numerical simulations (Mangeney et al. 1996; Thorsteinsson et al. 1999). Certainly, the differences among the estimates for E_a and E_s must be attributed to the fact that different stress, strain and temperature regimes were explored. Generally, the values of the shear enhancement factor E_s can range from about 4 to about 10, for both warm and cold ice. As regards the enhancement factor for compression, the values of E_a varying from 2.8 to even 6.3 are reported (Treverrow et al. 2012) for warm ice, and E_a ranging from about 1/10 to about 1/3 for cold ice appear in the literature. Throughout this book, the following pairs of the compression and shear enhancement factors will be used in numerical simulations: $E_a = 3$ and $E_s = 8$ for warm ice, and $E_a = 1/3$ and $E_s = 5$ for cold ice.

Simple Flow Simulations

The proposed micro-mechanical model has been used to simulate the viscous creep behaviour of polycrystalline ice in simple flow configurations, with the purpose to examine, and illustrate, the mechanism of the development of anisotropic fabrics with increasing macroscopic deformation of ice, and to investigate the associated process of the evolution of the macroscopic viscosities of ice with the increasing degree of its anisotropy.

Two configurations which were considered in the simulations included uniaxial compression and simple shear flow regimes (see Fig. 6.4). The first regime is typical of the central, near-divide, regions of polar ice caps, whereas the other flow regime is dominant in near-bottom layers of large polar ice sheets, especially in regions which are far from ice divides (refer to Fig. 2.3 on p. 15).

Let adopt the material rectangular coordinates OX_i $(i = 1, 2, 3)$ and the spatial coordinates Ox_i $(i = 1, 2, 3)$, with the same origin O and the directions of the coordinate axes. Then, the deformation field describing unconfined uniaxial compression along the X_3-axis is defined by

$$x_1 = \lambda_1 X_1, \quad x_2 = \lambda_2 X_2, \quad x_3 = \lambda_3 X_3, \quad \lambda_1 = \lambda_2 = \lambda_3^{-1/2}, \tag{6.47}$$

where λ_i $(i = 1, 2, 3)$ are the principal stretches along the X_i axes, all equal to unity at the start of flow at time $t = 0$ from an initially isotropic state, and $\lambda_3 < 1$ in compression and $\lambda_3 > 1$ in extension at $t > 0$. The last relation in (6.47) results from the ice incompressibility condition $\lambda_1\lambda_2\lambda_3 = 1$. The velocity field, obtained by time-differentiation of relations (6.47), is described by

$$v_1 = -\tfrac{1}{2}x_1\dot{\lambda}_3/\lambda_3, \quad v_2 = -\tfrac{1}{2}x_2\dot{\lambda}_3/\lambda_3, \quad v_3 = x_3\dot{\lambda}_3/\lambda_3. \tag{6.48}$$

The macroscopic deviatoric stress and velocity gradient tensors have diagonal forms given by

$$\bar{\boldsymbol{S}} = \begin{pmatrix} -\tfrac{1}{2}\bar{S}_{33} & 0 & 0 \\ 0 & -\tfrac{1}{2}\bar{S}_{33} & 0 \\ 0 & 0 & \bar{S}_{33} \end{pmatrix}, \quad \bar{\boldsymbol{L}} = \begin{pmatrix} -\tfrac{1}{2}\dot{\lambda}_3/\lambda_3 & 0 & 0 \\ 0 & -\tfrac{1}{2}\dot{\lambda}_3/\lambda_3 & 0 \\ 0 & 0 & \dot{\lambda}_3/\lambda_3 \end{pmatrix}, \tag{6.49}$$

where $\bar{S}_{33} = \tfrac{2}{3}\bar{\sigma}_{33}$, with $\bar{\sigma}_{33}$ being the macroscopic axial stress (negative in compression) applied to ice. The diagonal form of $\bar{\boldsymbol{L}}$ implies that $\bar{\boldsymbol{D}} = \bar{\boldsymbol{L}}$ and the spin is zero, $\bar{\boldsymbol{W}} = \boldsymbol{O}$. The axial deformation along the X_3-axis is given by the stretch λ_3, but, in subsequent plots, a more common measure, expressed by the axial strain ε_{33}, will be used

$$\varepsilon_{33} = \lambda_3 - 1, \tag{6.50}$$

with negative values for compression, and positive for tension.

In the simple shear configuration, see Fig. 6.4b, it is assumed that the deformation takes place in the plane OX_1X_3 and the viscous flow starts from an initially isotropic state. The deformation field is then described by

$$x_1 = X_1 + \kappa X_3, \quad x_2 = X_2, \quad x_3 = X_3, \tag{6.51}$$

where κ is a shear strain increasing from zero. The associated velocity field is given by

$$v_1 = \dot{\kappa}x_3, \quad v_2 = v_3 = 0, \tag{6.52}$$

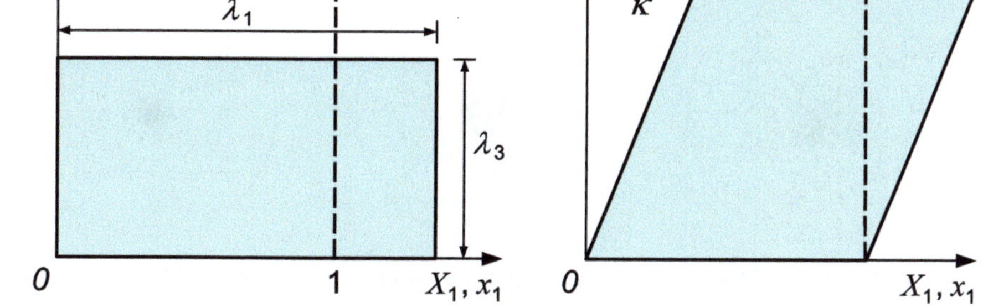

Fig. 6.4 Simple flow configurations: **a** uniaxial compression along the X_3-axis and **b** simple shear in the plane OX_1X_3

giving the macroscopic velocity gradient, strain-rate, and spin tensors

$$\bar{\boldsymbol{L}} = \begin{pmatrix} 0 & 0 & \dot{\kappa} \\ 0 & 0 & 0 \\ 0 & 0 & 0 \end{pmatrix}, \quad \bar{\boldsymbol{D}} = \begin{pmatrix} 0 & 0 & \frac{1}{2}\dot{\kappa} \\ 0 & 0 & 0 \\ \frac{1}{2}\dot{\kappa} & 0 & 0 \end{pmatrix}, \quad \bar{\boldsymbol{W}} = \begin{pmatrix} 0 & 0 & \frac{1}{2}\dot{\kappa} \\ 0 & 0 & 0 \\ -\frac{1}{2}\dot{\kappa} & 0 & 0 \end{pmatrix}. \tag{6.53}$$

The numerical simulations were carried by using $N_c = 800$ discrete grains in the uniform stress model. The initial crystal c-axes orientations were distributed at random. The calculations were performed for the isotropic ice viscosity $\mu_0 =$ 10 MPa$\cdot$a (where the unit 'a' denotes year), and the axial deviatoric stress $|\bar{S}_{33}| =$ 0.1 MPa, kept constant during the flow. The latter two values can be regarded as typical magnitudes of both quantities in large polar ice sheets (Staroszczyk and Morland 2000).

Figures 6.5 and 6.6 illustrate the development of anisotropic ice fabrics with increasing axial deformation ε_{33} under compression and tension. The fabrics are presented by means of equal-area *Schmid diagrams*, in which the dots show the positions of crystal c-axes on the unit hemisphere projected on the plane Ox_1x_2 (normal to the direction of axial loading). The plots illustrate the predictions of the uniform stress (Sachs-Reuss) model, in which the microscopic rheological properties describing the

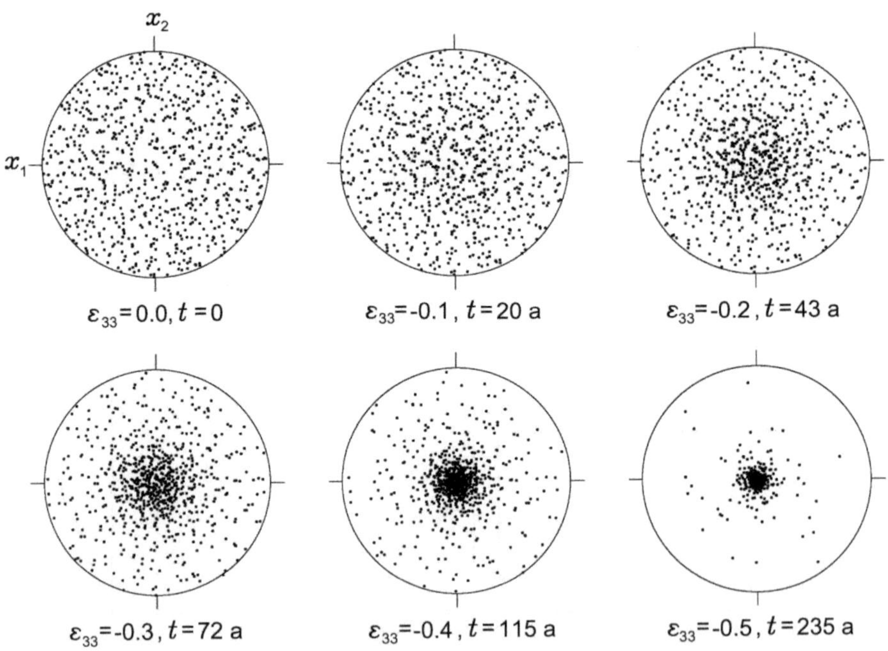

Fig. 6.5 Evolution of anisotropic ice fabric in uniaxial compression along the x_3-axis as a function of the macroscopic axial strain ε_{33} and time t (given in years), for the crystal rheological parameters $\alpha = \beta = 0.1$ in the uniform stress (Sachs-Reuss) model

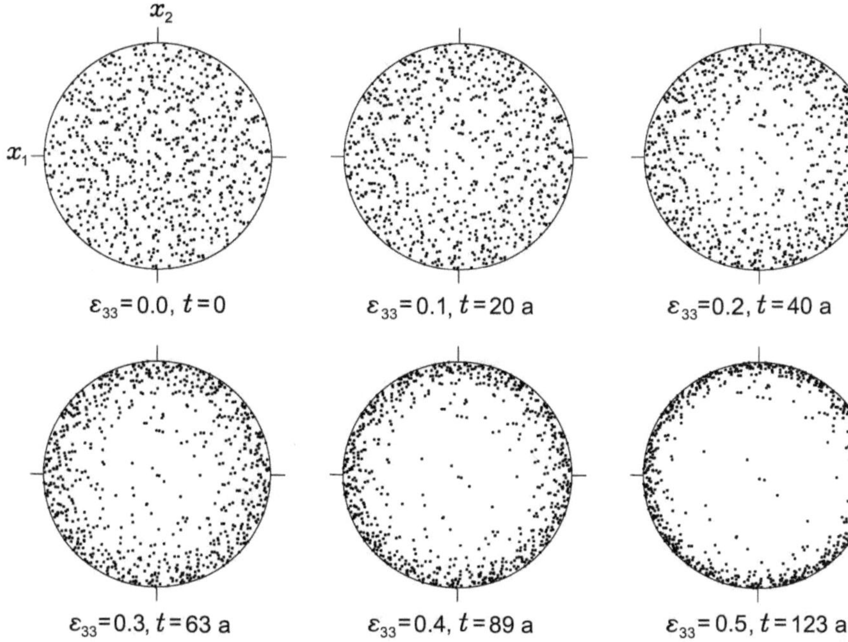

Fig. 6.6 Evolution of anisotropic ice fabric in uniaxial tension along the x_3-axis as a function of the macroscopic axial strain ε_{33} and time t (given in years) for the crystal rheological parameters $\alpha = \beta = 0.1$ in the uniform stress (Sachs-Reuss) model

anisotropy of a single crystal are defined by the parameters $\alpha = \beta = 0.1$. Figure 6.5 demonstrates the mechanism of crystal lattice rotation under compression. The diagrams show how the crystals gradually align in the direction of the principal axis of compression x_3, which at large strains gives rise to a single maximum fabric, with nearly all the crystal c-axes clustered around the x_3-axis. The plots in Fig. 6.6, in turn, illustrate the behaviour of polycrystalline ice under uniaxial tension. In this stress configuration, the progressive movement of the crystal c-axes away from the principal axis of tension x_3 can be observed, so that at large axial strains ε_{33} all the crystals tend to align near the plane Ox_1x_2, normal to the direction of tension. Such behaviour of ice, predicted by the model and shown in the presented plots, is in good qualitative agreement with the behaviour of polar ice observed in the field (Budd and Jacka 1989; Alley 1992; Thorsteinsson et al. 1997).

The results displayed in Figs. 6.7 and 6.8 illustrate the variation of macroscopic axial viscosities with increasing strains; that is, with increasing anisotropy of the polycrystalline ice aggregate. The axial viscosity μ_{33} is defined in terms of the macroscopic stress $\bar{S}_{33}$ (kept constant) and the current macroscopic strain-rate $\bar{D}_{33}$ by the relation $\mu_{33} = \bar{S}_{33}/(2\bar{D}_{33})$, and is normalized by the isotropic ice viscosity μ_0; hence, the ratios μ_{33}/μ_0 are plotted in the figures. It is clearly seen that the viscous behaviour of ice, both in compression and tension, is very sensitive to the values of

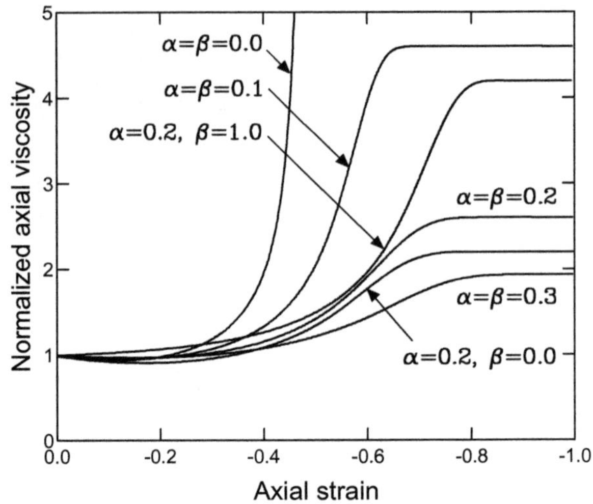

Fig. 6.7 Variation of the normalized macroscopic axial viscosity μ_{33}/μ_0 with the strain ε_{33} in uniaxial compression, for different values of the rheological parameters α and β in the uniform stress (Sachs-Reuss) model

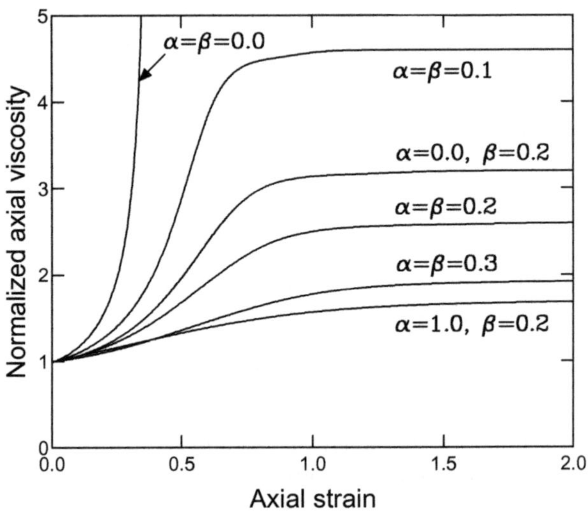

Fig. 6.8 Variation of the normalized macroscopic axial viscosity μ_{33}/μ_0 with the strain ε_{33} in uniaxial tension, for different values of the rheological parameters α and β in the uniform stress (Sachs-Reuss) model

the microscopic rheological parameters α and β. As already mentioned, the limit axial viscosities attain infinite values in the case of crystals deforming only by basal glide ($\alpha = \beta = 0$). It can also be observed that, in general, the limit axial viscosities are different for the ice subjected to compression and tension, with the exception of cases in which $\alpha = \beta$ (this property follows from the second and third relations (6.46)).

A characteristic feature which can be seen in Fig. 6.7 is that polycrystalline ice softens during the first phase of its compression, with a slight decrease in the normalized viscosity μ_{33}/μ_0 from a unit value for isotropic ice at the start of deformation, to the values which are equal to about 0.95 for the combinations of $\alpha \le 0.2$ and $\beta \le 0.2$. This type of behaviour was observed in experiments (Mellor and Cole 1982; Budd and Jacka 1989), and can be explained by the fact that the majority of crystals in an initially isotropic polycrystal have their c-axes oriented at the zenith angles θ greater than 45°. Thus, as ice undergoes compression, then immediately after the start of deformation the c-axes of many crystals pass through the 'soft orientation zone'; that is, the surface of the cone defined by the zenith angle $\theta = 45°$. Since at the latter angle the resolved shear stress acting on the crystal basal plane has the largest magnitude, and, in turn, the gliding on the basal plane is the easiest way for a crystal to deform, the overall result is that the macroscopic viscous resistance of the polycrystal to stress decreases in the first stage of flow. Such a softening mechanism does not occur under tension, since then the majority of crystals, during their rotation away from the principal direction of tensile stress, do not pass through the cone $\theta = 45°$. Hence, the monotonic increase of the macroscopic viscosities with increasing strain seen in Fig. 6.8.

More results, and their more detailed discussion, can be found in the paper by Staroszczyk (2001).

6.3.2 *Uniform Strain Model*

In the uniform strain model, based on the approximation (6.37), the local deviatoric stress $\boldsymbol{S}$ is, generally, different in each crystal in the aggregate. This stress is defined by the constitutive law (6.32), with the microscopic strain-rate $\boldsymbol{D}$ replaced by the macroscopic strain-rate $\bar{\boldsymbol{D}}$ due to the condition $(6.38)_1$. The components of the macroscopic stress $\bar{\boldsymbol{S}}$ are then evaluated by applying the averaging formula (6.34), which results in the relation between the macroscopic strain-rate and macroscopic stress. In order to follow the evolution of the crystal c-axes orientations, described by Eqs. (6.9) and (6.10) on p. 173, the strain-rate components expressed in the global coordinates Ox_i are transformed to those in the local lattice frame Ox_i^c by using $(6.7)_1$ on p. 172. This yields:

$$
\begin{aligned}
D_{13}^c &= \tfrac{1}{2}\sin 2\theta\left[\bar{D}_{11}(1+\cos^2\varphi) + \bar{D}_{22}(1+\sin^2\varphi) + \bar{D}_{12}\sin 2\varphi\right] + \\
&\quad + \cos 2\theta\,(\bar{D}_{13}\cos\varphi + \bar{D}_{23}\sin\varphi), \qquad (6.54)\\
D_{23}^c &= \tfrac{1}{2}\sin\theta\left[(\bar{D}_{22} - \bar{D}_{11})\sin 2\varphi + 2\bar{D}_{12}\cos 2\varphi\right] + \\
&\quad - \cos\theta\,(\bar{D}_{13}\sin\varphi - \bar{D}_{23}\cos\varphi). \qquad (6.55)
\end{aligned}
$$

The above relations, together with the local spin tensor components W_{ij} replaced by $\bar{W}_{ij}$, after the substitution in the kinematic Eqs. (6.9) and (6.10), determine the current orientations of crystals in the global coordinate frame. Hence, Eqs. (6.54) and (6.55) describe the evolution of the ice fabric in the uniform strain model.

Limit Viscosities

To determine the macroscopic viscosity of isotropic ice in terms of the microscopic rheological parameters, a method (Staroszczyk 2002, 2004) which is very similar to that applied earlier in Sect. 6.3.1 is followed. Hence, a simple shear flow configuration (see Fig. 6.4b) is adopted, in which, for prescribed macroscopic strain-rates, the microscopic deviatoric stresses for a single crystal are calculated from the constitutive law (6.32) on p. 178. These stresses are then used to derive the macroscopic stresses by applying the averaging formula (6.41) (with the strain-rates replaced by the stresses) for a random fabric. As a result, the following expression for the macroscopic isotropic ice viscosity μ_0 is obtained:

$$\mu_0 = \frac{\mu}{5}\,(A + 2B + 2). \tag{6.56}$$

The assumption of the isotropy of the single crystal, expressed by $A = B = 1$, leads, obviously, to $\mu_0 = \mu$. On the other hand, when the crystals are supposed to deforms only by basal glide, which is described by $A \to \infty$ and $B \to \infty$, then (6.56) yields an unbounded isotropic viscosity of the polycrystal. Recall that the uniform stress model predicts for this case a bounded value of the macroscopic viscosity, $\mu_0 = 2.5\,\mu$, see Eq. (6.44) on p. 182.

In the limit situation of all crystal c-axes aligned in parallel, forming thus a single maximum fabric, the constitutive relation (6.32), combined with the definition (6.56), yields the limit macroscopic shear viscosity μ_{13} defined by

$$\frac{\mu_{13}}{\mu_0} = \frac{5}{A + 2B + 2}. \tag{6.57}$$

Further, the limit macroscopic viscosities for shear in the plane normal to the crystal c-axes, for uniaxial compression along the c-axes, and for uniaxial compression in the direction perpendicular to the c-axes, are given by

$$\frac{\mu_{12}}{\mu_0} = \frac{5B}{A + 2B + 2}, \qquad \frac{\mu_{33}}{\mu_0} = \frac{5A}{A + 2B + 2}, \qquad \frac{\mu_{11}}{\mu_0} = \frac{5(A + 3B)}{4(A + 2B + 2)}. \tag{6.58}$$

The reciprocal viscosity ratios (6.57) and $(6.58)_2$ are equal to the enhancement factors for shear and compression, E_s and E_a, respectively. This enables the correlation of the microscopic rheological constants A and B with the observed macroscopic behaviour of polycrystalline ice at indefinitely large deformations. Accordingly, we have

$$A = \frac{E_s}{E_a}, \qquad B = \frac{5}{2}\,E_s - \frac{1}{2}\frac{E_s}{E_a} - 1. \tag{6.59}$$

For the specific values of the enhancement factors corresponding to cold and warm ice adopted in this book for numerical simulations, expressions (6.59) give

$$\text{cold ice} \quad (E_a = \tfrac{1}{3},\ E_s = 5): \quad A = 15, \quad B = 4, \tag{6.60}$$

$$\text{warm ice} \quad (E_a = 3,\ E_s = 8): \quad A = \tfrac{8}{3}, \quad B = \tfrac{53}{3}. \tag{6.61}$$

It should be pointed out that the above values of the enhancement factors cannot be attained in the uniform stress model considered earlier, for neither cold nor warm ice. In this respect, the uniform strain model proves to be more flexible, since it enables the correlation with the observed behaviour of polycrystalline ice, whereas the uniform stress model does not.

Finally, let us determine the relationship between the isotropic ice viscosities predicted by the uniform strain and the uniform stress models for the same properties of constituent crystals in the aggregate. Denoting by μ_0^{SR} and μ_0^{TV} the isotropic ice viscosities given by the Sachs-Reuss and Taylor-Voigt approximations, respectively, it can be found from definitions (6.44) and (6.56) that

$$\frac{\mu_0^{TV}}{\mu_0^{SR}} = \frac{1}{25}(A + 2B + 2)(\alpha + 2\beta + 2), \tag{6.62}$$

with the connections $A = \alpha^{-1}$ and $B = \beta^{-1}$ introduced by (6.31). For the isotropic constituent crystals, when $A = B = \alpha = \beta = 1$, the μ_0^{TV}/μ_0^{SR} ratio is, obviously, unity. With increasing anisotropy of the crystals in the aggregate, the viscosity ratio (6.62) also increases, indicating a growing difference between the upper and lower bounds on the polycrystalline ice viscosity predicted by the two approximations. For the particular values of the constants A and B for cold and warm ice, given by (6.60) and (6.61) respectively, relation (6.62) yields:

$$A = 15, \quad B = 4: \quad \mu_0^{TV}/\mu_0^{SR} = 2.5667, \tag{6.63}$$

$$A = \tfrac{8}{3}, \quad B = \tfrac{53}{3}: \quad \mu_0^{TV}/\mu_0^{SR} = 3.8906. \tag{6.64}$$

Simple Flow Simulations

The uniform strain model has been applied to simulate the viscous response of polycrystalline ice in the same uniaxial and simple shear configurations (see Fig. 6.4 on p. 185) as those considered in Sect. 6.3.1 for the uniform stress model calculations. The discrete-grain model consisted again of $N_c = 800$ crystals with equal volumes. The isotropic ice viscosity $\mu_0 = 10$ MPa$\cdot$ a was used in the simulations, and the microscopic rheological parameters $A = 15$ and $B = 4$ were chosen as those describing the cold ice characterized by the viscosity enhancement factors $E_a = 1/3$ and $E_s = 5$, see relations (6.59).

The simulations for the simple shear flow regime, described by Eqs. (6.51)–(6.53), were conducted under the assumption that ice is sheared at a constant strain-rate $\bar{D}_{13} = 5 \times 10^{-3}$ a^{-1}. The process of formation and subsequent evolution of ice fabric during simple shear is illustrated in Fig. 6.9. The distributions of the crystal c-axes are plotted on the plane Ox_1x_2, parallel to the glide plane, and normal to the shear plane Ox_1x_3 in which the deformation takes place. It can be observed that in this flow

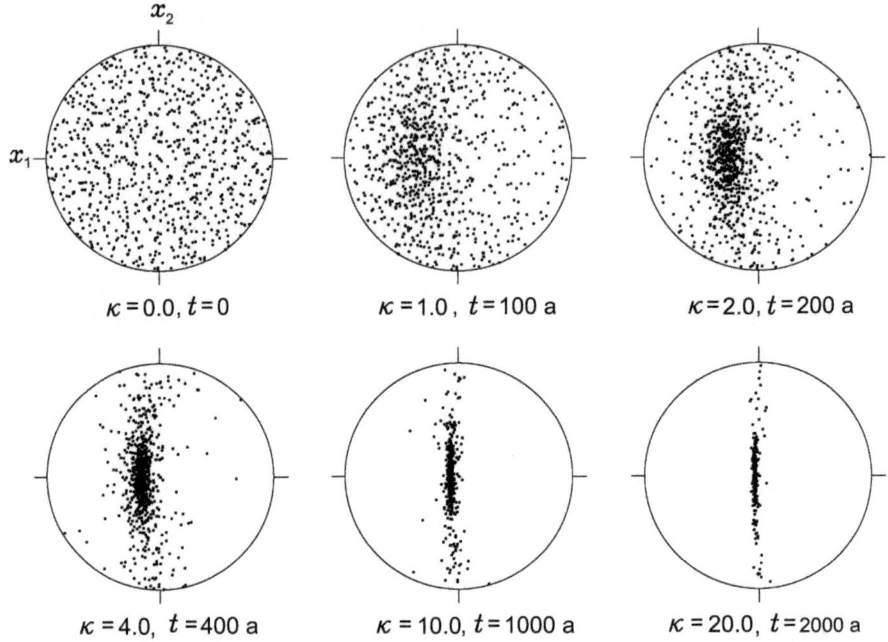

Fig. 6.9 Evolution of anisotropic ice fabric in simple shear in the Ox_1x_3 plane as a function of the macroscopic shear strain κ and time t (in years), for the crystal rheological parameters $A = 15$ and $B = 4$ (cold ice) in the uniform strain (Taylor-Voigt) model. Reprinted from Staroszczyk (2002), Fig. 5. Copyright 2002 by the Institute of Fundamental Technological Research of the Polish Academy of Sciences

configuration the majority of crystals first rotate towards the plane Ox_2x_3, and only at large shear strains κ do their c-axes start to cluster around the axis x_3 (which becomes the principal axis of compression for $\kappa \to \infty$). Eventually, a single maximum fabric forms which is very similar to that developing under uniaxial compression (see Fig. 6.5 on p. 186). The characteristic features of the fabrics illustrated in Figs. 6.5 and 6.9 are consistent with the results obtained by Morland and Staroszczyk (2009), who deduced the general properties of the mechanism of fabric evolution in polycrystalline bodies without assuming any particular constitutive relation describing the material.

The quantitative features of the process of anisotropic fabric evolution are presented in Fig. 6.10, showing the variation of the macroscopic axial and shear viscosities with the increasing shear deformation κ. The viscosities are plotted in the dimensionless forms μ_{ij}/μ_0, and the initial isotropic state corresponds to $\kappa = 0$. It can be observed that, in the case of the cold ice represented in the figure, the limit values (at shear strain $\kappa \to \infty$) of all the axial viscosities are larger than those for isotropic ice, whereas all the limit shear viscosities are smaller than those for the isotropic ice. It can be also seen that the shear viscosity μ_{13} initially increases with κ at relatively small shear deformations (with a maximum value occurring at $\kappa \sim 2$), which indicates the initial hardening of the material in this flow regime, before the

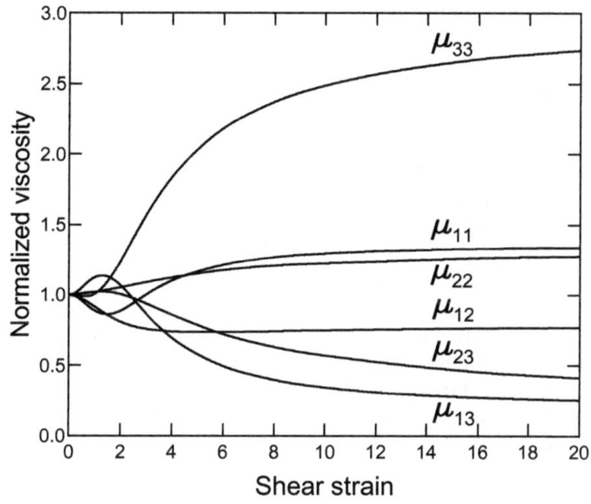

Fig. 6.10 Variation of the normalized axial and shear viscosities μ_{ij}/μ_0 with the strain κ in simple shear for the crystal rheological parameters $A = 15$ and $B = 4$ (cold ice) in the uniform strain (Taylor-Voigt) model. Reprinted from Staroszczyk (2002), Fig. 7. Copyright 2002 by the Institute of Fundamental Technological Research of the Polish Academy of Sciences

phase of progressive softening of the ice fabric at larger shear strains. Since the fabric that forms at the limit $\kappa \to \infty$ coincides with that created at large deformations during uniaxial compression, the full anisotropy gradually transforms into transverse isotropy, with $\mu_{13} \to \mu_{23}$ and $\mu_{11} \to \mu_{22}$. Moreover, the viscosities plotted in Fig. 6.10 approach, as $\kappa \to \infty$, the limit values defined by the enhancement factors $E_a = 1/3$ and $E_s = 5$ (that is, $\mu_{33}/\mu_0 \to 1/E_a = 3$ and $\mu_{13}/\mu_0 \to 1/E_s = 1/5$).

The uniform strain model has been also used to simulate the creep behaviour of polycrystalline ice in the uniaxial flow configuration. The ice fabrics predicted by the model are qualitatively very similar to those presented in Figs. 6.5 and 6.6 and obtained by applying the uniform stress model. For this reason, these fabrics are not shown here; relevant illustrations can be found in the paper by Staroszczyk (2002). The only difference between the predictions of the two alternative models described in this section is that the uniform strain model predicts 'stiffer' behaviour of polycrystalline ice than its uniform stress counterpart: the same degree of crystal c-axes concentration around the principal axis of compression requires a larger compressive strain along that axis.

It is of interest to compare the predictions of the two proposed models regarding the evolution of the strength of anisotropy of a polycrystalline aggregate with increasing macroscopic deformation. Such comparisons are presented in Fig. 6.11, showing the variation of the axial and shear viscosities with the axial strain ε_{33} under compression. Again, the viscosities μ_{ij} are normalized by the isotropic ice viscosity μ_0. We note that (1) the isotropic ice viscosities (occurring at $\varepsilon_{33} = 0$) predicted by the uniform strain (Taylor-Voigt) and uniform stress (Sachs-Reuss) models are, for the chosen single crystal parameters $A = 1/\alpha$ and $B = 1/\beta$, in the ratio $\mu_0^{TV}/\mu_0^{SR} = 2.5667$ defined by (6.63), and (2) the same limit viscosities μ_{12}, μ_{13} and μ_{33}, when $\varepsilon_{33} \to -1$, for the Sachs-Reuss model given by (6.45) and (6.46), and for the Taylor-Voigt model given by (6.57) and (6.58), are predicted by the two approximations. However, it is seen

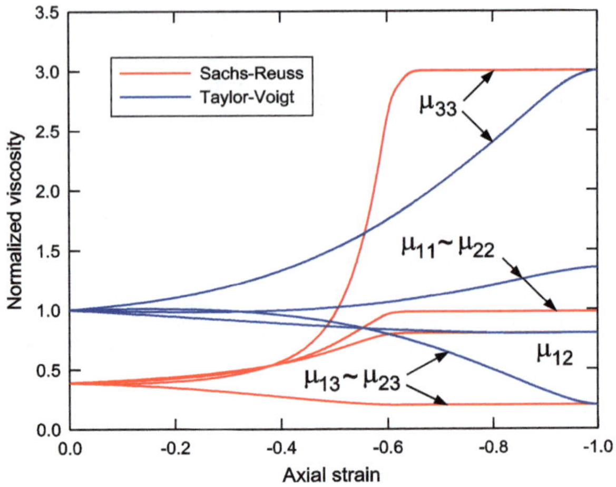

Fig. 6.11 Variation of the normalized axial and shear viscosities with the strain ε_{33} in uniaxial compression for the crystal rheological parameters $A = 15$ and $B = 4$ (cold ice). Predictions of the uniform stress (Sachs-Reuss) and the uniform strain (Taylor-Voigt) models. Adapted from Staroszczyk (2002), Fig. 4. Copyright 2002 by the Institute of Fundamental Technological Research of the Polish Academy of Sciences

that the limit viscosities given by the two models are approached in different ways. For the uniform stress model, the most significant increase in the fabric strength occurs for the axial strains ε_{33} changing from -0.4 to -0.6, with the limit values reached just after the latter value is exceeded, while for the uniform strain model, which predicts 'stiffer' behaviour of ice, the limit alignment of individual grains, and hence the limit viscosities, are attained much later, at the strains ε_{33} close to -1.

6.4 Dynamic Recrystallization Models

In the previous sections of this chapter two micro-mechanical models describing the mechanism of crystal lattice rotation are presented. The latter mechanism is active throughout the whole path of the ice descent from the free surface to depth in polar ice sheets, and plays a major role in the development of the macroscopic anisotropy of the polycrystalline material. This feature is demonstrated by the results presented in Sect. 6.3, showing progressive reorientation of crystal *c*-axes and the formation of strong single-maximum fabrics under sustained compression and shear. Such characteristic fabrics are found in ice cores retrieved from large depths of polar ice sheets. However, as ice particles during their downward motion enter near-bottom regions of polar glaciers, the microstructure of ice often changes dramatically, and the strong fabrics developed earlier undergo an abrupt destruction. Typically, these near-base fabrics have crystal *c*-axes broadly scattered in irregular patterns around the vertical, or multi-maxima fabrics with very coarse and interlocking grains are formed (Duval 1981; Duval and Castelnau 1995; De La Chapelle et al. 1998; Duval et al. 2000; Faria et al. 2014). The micro-mechanism which is responsible for such behaviour of ice is known as the dynamic (also termed migration, discontinuous or annealing) recrystallization. This process usually takes place within the deepest 100 m of an ice sheet, at ice near-melting temperatures (that is, above $-10\,°C$). Not all the factors which initiate and control the dynamic recrystallization mechanism have been identified yet, but it seems that the most important are high deviatoric stress, strain, and strain-rate magnitudes, with some role also played by the local bedrock topography (De La Chapelle et al. 1998; Duval et al. 2000).

The recrystallization process is caused by rapid migration od grain boundaries between dislocation-free crystals and those with high density of dislocations, and leads to the nucleation of new grains at the expense of old ones (which eventually disappear). Thus, on the microscopic scale, the process is controlled by the amount of energy stored at crystal dislocations piling up on crystal boundaries, and hence is driven by the differences in concentrations of crystal defects in neighbouring grains. Spatial orientations of newly formed grains are such that the bulk deformation of a polycrystal is enhanced. Therefore, the macroscopic result of the process is significant weakening of the strength of anisotropy of the medium compared to non-recrystallizing ice. For this reason, the migration recrystallization significantly modifies the macroscopic response of ice to stress, and therefore has a pronounced effect on the overall behaviour of polar ice masses, since the latter deform mainly by shearing in near-base regions; that is, in the regions in which the recrystallization occurs. Hence, this mechanism is important and has to be accounted for in numerical simulations of ice flows if realistic results are to be obtained. Some examples illustrating the role of the dynamic recrystallization in the large-scale flows of polar ice sheets will be given in Chap. 8.

First attempts to describe theoretically the mechanism of dynamic recrystallization in ice took place in the mid-1990s, and a number of models have been formulated since then. Van der Veen and Whillans (1994) developed a micro-mechanical model in which the onset of the mechanism is defined in terms of macroscopic strains in the polycrystalline aggregate. Staroszczyk and Morland (2001) proposed a phenomenological model in which, ignoring most of the micro-processes underlying the ice fabric evolution, the onset of the migration recrystallization is described by means of a critical macroscopic strain-rate invariant. Subsequently, Morland (2002) formulated another phenomenological model in which the onset of recrystallization is related to a critical crystal lattice distortion parameter. An unconventional approach based on the cellular automata method was developed by Ktitarev et al. (2002) and Faria et al. (2002); their model, however, suffers from its restriction to one-dimensional deformations. Another micro-mechanical model was proposed by Thorsteinsson (2002), who considered the effects of crystal interactions on the migration recrystallization process. A more advanced model based on similar ideas was developed more recently by Kennedy et al. (2013). A distinct approach was followed by Faria et al. (2003) to construct a general theory of recrystallization processes in polycrystalline materials by employing general principles of thermodynamics. This theory, applied to polycrystalline ice, was subsequently extended by Faria (2006). Placidi et al. (2010) constructed a similar model, in which the recrystallization mechanism is described in terms of only one parameter. An alternative micro-mechanical approach was pursued by Staroszczyk (2009, 2011) to develop two models based on the Taylor-Voigt and Sachs-Reuss approximations, in which the recrystallization process is controlled, respectively, by microscopic stresses or microscopic strain-rates. A more exhaustive

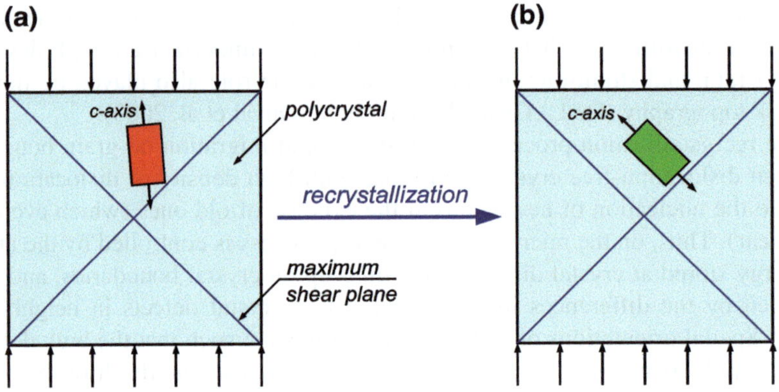

Fig. 6.12 Idealization of the dynamic recrystallization mechanism: hard crystals (**a**) are replaced by soft ones (**b**). Diagonal lines indicate the planes of maximum macroscopic shear stresses in a polycrystalline aggregate under uniaxial compression

discussion of various aspects of the modelling of recrystallization phenomena in polar ice can be found in a review paper by Placidi et al. (2006).

Two essential components which a recrystallization model must include are (1) the criterion for the onset of recrystallization of an existing grain, and (2) the determination of the initial spatial orientation of a newly nucleated grain. Following the observations of De La Chapelle et al. (1998), it is assumed in the models presented here that the mechanism of dynamic recrystallization affects the so-called *hard grains* in a polycrystalline aggregate; that is, the grains which are the most stressed in the present configuration and which can deform with difficulty (hence, are the least strained). Further, as a results of recrystallization, new *soft grains* are formed, which can deform easily in the present configuration and are least stressed. These ideas are schematically illustrated in Fig. 6.12.

It is commonly accepted (Budd and Jacka 1989; Alley 1992; Van der Veen and Whillans 1994; De La Chapelle et al. 1998) that new crystals of ice are oriented in a way that is most favourable for their further creep deformation (which is easiest by the slip on the crystal basal planes). That is, a newly grown grain is least stressed in a current stress configuration—this follows from the fact that the most stressed grains, due to their orientation in an ice polycrystal, are those which deform at the smallest rates, and vice versa, the least stressed grains are those deforming at largest rates. In the earlier discrete-grain models (Van der Veen and Whillans 1994; Ktitarev et al. 2002) it was assumed that the preferred direction of the *c*-axis of a newly nucleated crystal is such that its crystal basal planes are parallel to the plane of the maximum macroscopic shear stress in a polycrystal. In general, when all three principal macroscopic stresses are different from one another, the maximum shear planes are inclined at the angles of 45° to the principal axes of the maximum and minimum compressive stresses in the aggregate. Hence, the optimal orientations of the crystal *c*-axes of newly formed soft grains are at angles of 45° to the principal axes of maximum compression. It turns out, however, that the above is true only in the case of crystals deforming only by the basal glide. When other modes of crystal creep deformations are also possible, then the optimal orientations of newly formed crystals differ slightly (by several degrees) from the above direction of 45° (Staroszczyk 2009, 2011). This property will be demonstrated below.

In what follows, the discrete-grain models presented in Sects. 6.2 and 6.3, originally developed for describing the mechanism of crystal lattice rotation, are now extended to incorporate the mechanism of dynamic recrystallization as well. Three distinct models are formulated, in which the recrystallization process is driven, in turn, by microscopic deviatoric stresses, microscopic strain-rates and microscopic strains. Hence, three different criteria will be adopted to define the onset of the process of dynamic recrystallization for given macroscopic stress and deformation fields. For simplicity, the effect of temperature is neglected, by assuming that the process occurs at constant temperature. The three models described below will be used to simulate the evolution of the microstructure (fabric) of polycrystalline ice undergoing migration recrystallization, and to investigate the variation of macroscopic viscosities with increasing deformations in simple flow configurations. The results of these simulations are presented in the next Sect. 6.5.

6.4.1 Stress – controlled Recrystallization

The first variant of the dynamic recrystallization model is based on the assumption that the process is driven solely by the local stresses acting on individual crystals. Hence, it is assumed that recrystallize only those crystals in an aggregate which are most stressed. Accordingly, a microscopic stress magnitude measure, expressed by the second principal invariant of the deviatoric stress tensor $\boldsymbol{S}$, is adopted

$$S_{eq}^2 = J_2 = \tfrac{1}{2} S_{ij} S_{ij} \quad (i, j = 1, 2, 3)\,, \tag{6.65}$$

where S_{eq} is termed the equivalent micro-stress, and the summation convention for repeated indices applies. Further, a critical magnitude of the equivalent micro-stress, S_{eq}^{cr}, is introduced, and it is supposed that a given crystal k $(k = 1, 2, \ldots, N_c)$ undergoes recrystallization as long as the equivalent micro-stress $S_{eq}^{(k)}$ in that crystal exceeds, or equals, the critical equivalent stress magnitude. That is, the recrystallization criterion is expressed in the form:

$$S_{eq}^{(k)} \geq S_{eq}^{cr}. \tag{6.66}$$

In order to account in the model for the effect of temperature on the recrystallization process, one can follow an approach similar to that applied by Staroszczyk and Morland (2001) in their phenomenological formulation. In that approach, the critical stress measure S_{eq}^{cr} is assumed to be a function of the absolute temperature T in such a way that S_{eq}^{cr} decreases as ice temperature increases above some critical level; below the critical temperature level no recrystallization occurs.

The dynamic recrystallization model based on the condition (6.66) is constructed by extending the multi-grain Taylor-Voigt (uniform strain) model described in Sect. 6.3.2; by construction, the Sachs-Reuss model cannot be used for this purpose as it postulates the stress homogeneity in a polycrystalline aggregate. Recall that the Taylor-Voigt theory uses the constitutive law (6.32) on p. 178, involving two dimensionless rheological parameters defining the strength of anisotropy of a single crystal.

As noted earlier in this section, a crucial ingredient of a recrystallization model is the method of determination of optimal orientations of crystal c-axes of new grains which are formed as a result of the recrystallization process. It is a difficult task to accurately determine such a preferred crystal orientation (that is, the angles φ and θ defining a crystal c-axis orientation in space) for a general macroscopic viscous flow field. However, it is possible to achieve for simpler configurations: uniaxial compression and simple shear flows, illustrated in Fig. 6.4 on p. 185 and defined by Eqs. (6.47)–(6.53). The description of how to derive analytical formulae used for the determination of the favourable c-axes orientations of soft crystals can be found in the paper by Staroszczyk (2009). Here only the final results are presented.

First consider an unconfined uniaxial flow, assumed to occur under compressive stresses applied along the x_3-axis. In such a configuration, due to the rotational symmetry of the flow field, all the stresses acting on an individual crystal are functions of only the zenith angle θ and are independent of the crystal azimuth angle φ (refer to Fig. 6.3 on p. 171 for the definitions of the angles φ and θ). Thus, the equivalent stress S_{eq} at a given crystal, given by (6.65), is also a function of only the angle θ (and, of course, of the material parameters μ_0, A and B entering the constitutive equation). The value of the angle θ_{min} at which the function $S_{eq}(\theta)$ attains its minimum value defines the optimal orientation of a new crystal which is nucleated during the dynamic recrystallization process. For convenience, the microscopic equivalent stress S_{eq} is normalized by the magnitude of the macroscopic equivalent stress $\bar{S}_{eq}$ in a polycrystalline aggregate. Hence, a dimensionless function $\zeta_{eq} = S_{eq}/\bar{S}_{eq}$ is introduced to describe the microscopic stress state at a given crystal in terms of the c-axis orientation angle θ and the two dimensionless microscopic rheological parameters A and B. This function is expressed by

$$\begin{aligned} \zeta_{eq} = \frac{S_{eq}}{\bar{S}_{eq}} = \frac{5}{2(A+2B+2)} &\times \\ &\times \left[3(A^2+B^2)\sin^4\theta + 6A^2\cos^4\theta + 3\sin^2 2\theta - 2A^2\right]^{1/2}. \end{aligned} \tag{6.67}$$

It can be easily seen that for an isotropic crystal, described by $A = B = 1$, relation (6.67) gives $\zeta_{eq} = 1$ irrespective of θ, which means that all crystals in the aggregate are equally stressed. In the other limit case, of crystals deforming only by basal glide and described by $A \to \infty$ and $B \to \infty$, relation (6.67) gives

$$\zeta_{eq} = \frac{S_{eq}}{\bar{S}_{eq}} \to \frac{5}{6}\left[4 - 3\sin^2(2\theta)\right]^{1/2}. \tag{6.68}$$

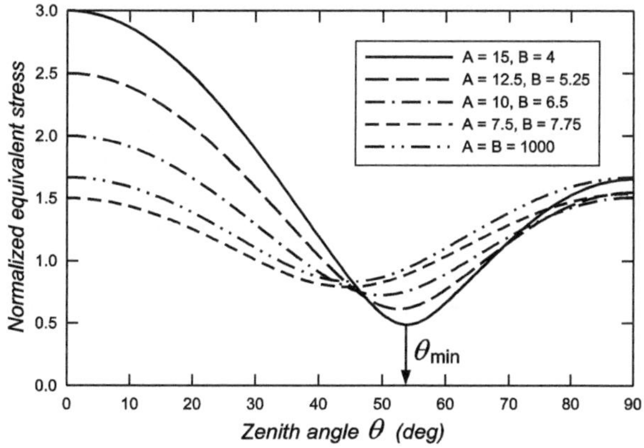

Fig. 6.13 Normalized equivalent stress ζ_{eq} in uniaxial compression as a function of the crystal c-axis zenith angle θ and the microscopic rheological parameters A and B. Reprinted from Staroszczyk (2009), Fig. 2. Copyright 2009 by the Institute of Fundamental Technological Research of the Polish Academy of Sciences

The latter expression gives the maximum normalized equivalent stress equal to $5/3$ for the crystal c-axis oriented at either $\theta = 0$ or $\theta = 90°$ (hard crystals), and the minimum value of this stress equal to $5/6$ for $\theta = 45°$ (soft crystals). These predictions are in good quantitative agreement with the results presented by De La Chapelle et al. (1998), obtained by applying a viscous-plastic self-consistent model by Castelnau et al. (1996), in which the resistance of non-basal slip systems was assumed to be 70 times larger than that of basal slip. Also, it follows from (6.67) that for $\theta = 0$, for any combinations of the rheological parameters A and B, the equivalent stress $S_{eq}/\bar{S}_{eq}$ is equal to $1/E_a$, the reciprocal of the axial enhancement factor.

The variation of the equivalent stress $\zeta_{eq} = S_{eq}/\bar{S}_{eq}$, given by (6.67), with the crystal c-axis zenith angle θ, for various combinations of the microscopic rheological parameters A and B, is illustrated in Fig. 6.13. The plots show that, depending on the strength of anisotropy of the crystal (defined by the parameters A and B), the most favourable orientation for the crystal c-axis in a polycrystalline aggregate under uniaxial compression is that defined by the zenith angles θ by a few degrees larger than 45°. For the particular values of the rheological parameters $A = 15$ and $B = 4$, best correlating with the observed limit macroscopic properties of cold polar ice, the optimal inclination of the crystal is at the angle of $\theta_{min} = 53.86°$ to the axis of the macroscopic compressive stress.

A more complex case is that of the simple shear flow regime, in which the microscopic stresses at a given grain depend on both crystal orientation angles, φ and θ. Assuming that the deformation takes place in the Ox_1x_3 plane, the dimensionless function ζ_{eq} is given by the expression (Staroszczyk 2009):

$$\zeta_{eq} = \frac{S_{eq}}{\bar{S}_{eq}} = \frac{5}{2(A+2B+2)} \left\{(3A^2 + B^2)\sin^2 2\theta \cos^2\varphi + \right. \\ \left. + 4\left[(B^2\sin^2\theta + \cos^2\theta)\sin^2\varphi + \cos^2 2\theta \cos^2\varphi\right]\right\}^{1/2}, \tag{6.69}$$

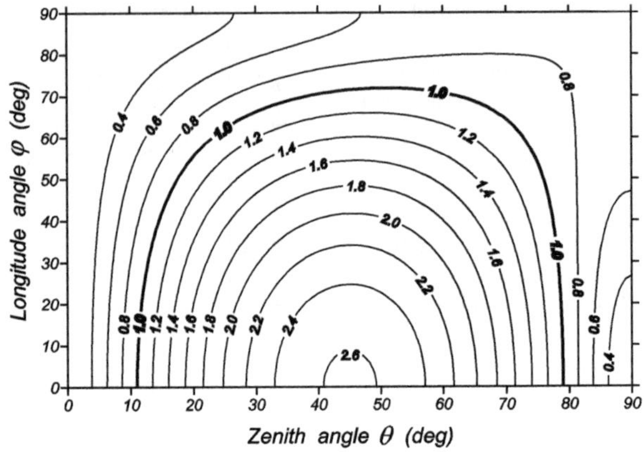

Fig. 6.14 Normalized equivalent stress ζ_{eq} in simple shear as a function of the crystal c-axis orientation angles θ and φ, for the microscopic rheological parameters $A = 15$ and $B = 4$ (cold ice). Reprinted from Staroszczyk (2009), Fig. 3. Copyright 2009 by the Institute of Fundamental Technological Research of the Polish Academy of Sciences

which describes the dependence of the microscopic equivalent stress in simple shear on the material parameters A and B and the crystal c-axis orientation angles θ and φ.

It follows from (6.69) that the stress ratio $S_{eq}/\bar{S}_{eq}$ reaches its minimum value, equal to $1/E_s$ (the reciprocal of the shear enhancement factor) for the crystal c-axis aligned either along the x_3-axis ($\theta = 0$) or along the x_1-axis ($\theta = 90°$ and either $\varphi = 0$ or $\varphi = 180°$). This means that in the simple shear configuration, irrespective of the values of the rheological constants A and B, the most favourably oriented crystals are those with their c-axes lying in the shear plane Ox_1x_3 and inclined at the angle 45° to the principal axis of compression. The dependence of the normalized equivalent stress ζ_{eq} on the c-axis orientation angles θ and φ, for the rheological parameters $A = 15$ and $B = 4$, is illustrated in Fig. 6.14. Only the range $0 \leq \varphi \leq \pi/2$ is considered, due to the symmetry property $\zeta_{eq}(\theta, \varphi) = \zeta_{eq}(\theta, \pi - \varphi)$ (the contour plot is symmetric about the co-ordinate line $\varphi = \pi/2$). It is seen in the figure that the most poorly oriented grains (that is, the most stressed) are those with the c-axis directions defined by $\theta = 45°$ and either $\varphi = 0$ or $\varphi = 180°$. The first of the latter two directions is normal, and the other is parallel, to the direction of the principal axis of compression in the assumed simple shear configuration. For the adopted values of A and B, the normalized equivalent stress has then the magnitude of about 2.63.

The plots in Figs. 6.13 and 6.14 show how the microscopic equivalent deviatoric stress depends on the crystal orientation in uniaxial and simple shear configurations. Certainly, in case of a general flow field, the dependence S_{eq} on φ and θ is more complex. Therefore, it would be very difficult (and computationally costly) to attempt to find an optimal orientation of the crystal c-axis for any particular (and changing in time) deformation configuration. Hence, for practical reasons, the c-axis orientation of a newly formed crystal is determined in the model in such a way that it lies on a conical surface at the angle θ_{min} to the current principal axis of the maximum compression, and the angle φ is such that the orientation direction of the new grain is closest to the orientation direction of the old grain, from which the new one is created. Some arbitrariness of this assumption is realized, but it is chosen in the proposed dynamic recrystallization model for the sake of simplicity of numerical calculations. Two alternative options for the determination of spatial orientations of newly formed grains were explored by Staroszczyk (2009). However, the numerical illustrations presented further in Sect. 6.5 show the results of the model based solely on the assumption described above.

6.4.2 Strain-Rate – Controlled Recrystallization

The second variant of the dynamic recrystallization model stems from the assumption that recrystallize those crystals in a polycrystalline aggregate which deform at smallest strain-rates, which is a characteristic feature of hard grains. Therefore, an invariant of the microscopic strain-rate tensor $\boldsymbol{D}$, defined by

$$D_{eq}^2 = I_2 = \tfrac{1}{2}\mathrm{tr}\,\boldsymbol{D}^2 = \tfrac{1}{2}D_{ij}D_{ij} \quad (i, j = 1, 2, 3)\,, \tag{6.70}$$

is introduced as a deformation measure for identifying hard crystals in the aggregate. D_{eq} is termed the equivalent strain-rate, and the summation convention for repeated indices applies in the definition (6.70). To establish a recrystallization criterion, a critical magnitude of the above invariant, D_{eq}^{cr}, is introduced, and it is assumed that a given grain k ($k = 1, 2, \ldots, N_c$) is subject to recrystallization when the equivalent microscopic strain-rate in that crystal is less, or equal to, the critical equivalent strain-rate level. Thus, a grain k recrystallizes if

$$D_{eq}^{(k)} \leq D_{eq}^{cr}. \tag{6.71}$$

As before, the temperature-dependence of the recrystallization process is not considered, but it can be easily incorporated in the model by assuming that D_{eq}^{cr} is a monotonically increasing function of the absolute temperature of ice.

The recrystallization model based on the criterion (6.71) is constructed as an extension of the Sachs-Reuss (uniform stress) approach presented in Sect. 6.3.1. In the latter approach, the creep behaviour of the material is described by the microscopic constitutive relation (6.26) on p. 176, defining the anisotropic rheological properties of a single crystal in terms of the dimensionless parameters α and β.

In order to determine the optimal orientations of new grains developing in the aggregate due to the migration recrystallization mechanism, a method analogous to that outlined in Sect. 6.4.1 is applied again. Hence, uniaxial and simple shear flow configurations are considered, for which the local strain-rate components D_{ij} are determined from the flow law (6.26) as functions of the macroscopic stress, the crystal orientation angles φ and θ, and the microscopic rheological parameters α and β. This yields the equivalent strain-rate D_{eq}^2 defined by (6.70). The latter is then normalized by means of the corresponding macroscopic equivalent strain-rate $\bar{D}_{eq}$ for a polycrystal under the same macroscopic stress. As a result, dimensionless relations similar to those given by (6.67) and (6.69) can be derived as functions of the crystal c-axis orientation angles. By finding then the values of φ and θ for which the latter functions attain maximum values in a given macroscopic stress field, one can determine the spatial orientations that are most favourable for newly nucleated grains.

The case of the uniaxial compression flow is again a simpler flow configuration than that of simple shear, because the resulting expression for the equivalent strain-rate involves only the zenith angle θ (Staroszczyk 2011):

$$\xi_{eq} = \frac{D_{eq}}{\bar{D}_{eq}} = \frac{5}{2(\alpha + 2\beta + 2)} \times \left[3\,(3\alpha^2 + \beta^2)\sin^4\theta - 12\alpha^2 \sin^2\theta + 3\sin^2 2\theta + 4\alpha^2\right]^{1/2} . \tag{6.72}$$

It follows from Eq. (6.72) that for an isotropic crystal, defined by $\alpha = \beta = 1$, we have $\xi_{eq} = 1$, which means that in this case all crystals in the aggregate deform at the same rates, irrespective of their spatial orientations. In the other limit case of crystals creeping only by basal glide, defined by $\alpha = \beta = 0$, relation (6.72) predicts $\xi_{eq} = \frac{5}{4}\sqrt{3}\,\sin 2\theta$. The latter expression is zero for $\theta = 0$ and $\theta = 90°$ (the crystals do not deform at all when their c-axes are parallel or normal to the principal axis of compression), and reaches its maximum value of ~ 2.16 for $\theta_{max} = 45°$. This means that when the crystal deforms only by basal slip, then its favourable orientation is such that the basal plane is exactly parallel to the plane of the macroscopic shear stress. When, however, other modes of the crystal creep deformation are also possible ($\alpha > 0$ and/or $\beta > 0$), then the optimal c-axis orientation slightly differs from $\theta = 45°$, but not by much. This can be seen in Fig. 6.15, illustrating the variation of the normalized invariant ξ_{eq} with the angle θ for several combinations of the rheological parameters α and β. For all the cases depicted in the figure, the azimuth angles at which the function ξ_{eq} reaches maximum values are within the range $45° \lesssim \theta_{max} \lesssim 47°$. Hence,

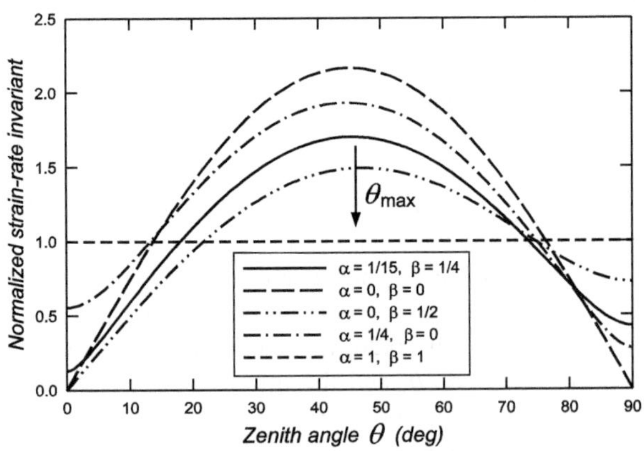

Fig. 6.15 Normalized strain-rate invariant $\xi_{eq} = D_{eq}/\bar{D}_{eq}$ in uniaxial compression as a function of the c-axis zenith angle θ and the crystal rheological parameters α and β. Hard (slowly-deforming) crystals are those with orientations close to either $\theta = 0$ or $\theta = 90°$, and soft (fast-deforming) crystals are those with orientations around $\theta = 45°$. Reprinted with permission from Staroszczyk (2011), Fig. 2. Copyright 2011 by Springer Nature

the effect of the crystal anisotropy on the value of the optimal grain orientation angle predicted by the Sachs-Reuss approximation can be regarded as small. Recall that in the case of the Taylor-Voigt model considered earlier in this section (see Fig. 6.13) the effect of the crystal anisotropy is more pronounced, with the predicted optimal values of the zenith angle θ ranging from about 45° to about 54°.

The case of the simple shear flow configuration is very difficult for analytical treatment, since then, to enforce the deformation field defined by the kinematic relations (6.51) on p. 185, not only the shear stress components $S_{13} = S_{31}$ (for the flow occurring in the Ox_1x_3 plane) need to be applied, but also the axial stress components S_{11} and S_{22} (with the inequality $S_{11} \neq S_{22}$ in general) have to be considered in the analysis (this property is known as the *Poynting effect* (Chadwick 1999)). The two axial stresses S_{11} and S_{22} depend on the current shear strain κ, which makes the analytical determination of the function $\xi_{eq}(\varphi, \theta)$ for a single crystal even more complicated. For these reason, the full analysis of the simple shear case is omitted here.

In the proposed uniform stress migration recrystallization model, the newly formed crystals, irrespective of the current flow configuration, are assumed to lie on a conical surface inclined at the angle θ_{max} to the current principal axis of compression, with θ_{max} determined from Eq. (6.72), derived for the uniaxial flow, as the angle for which the invariant function ξ_{eq} reaches its maximum. Further, it is assumed that the new crystals are distributed at random on this conical surface; that is, no value of the c-axis azimuth angle φ is favoured. The predictions of this model are presented in Sect. 6.5.

6.4.3 Strain – Controlled Recrystallization

The third version of the dynamic recrystallization model is based on the assumption that the process is induced by strain incompatibilities between individual crystals and a polycrystalline matrix in which they are imbedded. Hence, it is postulated that a given grain starts to recrystallize once a difference between a microscopic strain of the crystal and the macroscopic strain of the polycrystal, expressed in an invariant form, reaches a certain threshold value. The threshold strain parameter that defines the onset of recrystallization is adopted by correlating the model predictions with observed polar ice fabrics. The grains which undergo recrystallization are replaced by new grains, the spatial orientations of which are established from the condition that the initial strain-rates of the newly formed grains are maximized in the current macroscopic stress configuration (that is, the new crystals are aligned along easy-glide directions).

In order to describe the local deformations of individual crystals and the macroscopic deformation of the polycrystal, the previously introduced deformation gradient tensor $\boldsymbol{F}(\boldsymbol{X}, t)$ (5.24) and the velocity gradient $\boldsymbol{L}(\boldsymbol{x}, t)$ (6.4) are employed, with their components given by

$$F_{ij} = \frac{\partial x_i}{\partial X_j}, \quad L_{ij} = \frac{\partial v_i}{\partial x_j} \quad (i, j = 1, 2, 3). \tag{6.73}$$

The two gradients, $\boldsymbol{F}$ and $\boldsymbol{L}$, describe the evolution of the deformation field with time in the form of the following kinematic relation (Chadwick 1999):

$$\dot{\boldsymbol{F}} = \boldsymbol{L}\boldsymbol{F}, \tag{6.74}$$

where the superposed dot denotes the material time derivative. The latter tensorial equation is used to calculate current strains as the deformation of the body progresses, starting from an initial unstrained configuration at $t = t_0$, when $\boldsymbol{F} = \boldsymbol{I}$. A proper description of the deformation field requires a strain measure be frame-indifferent (objective), see Appendix B. For this purpose, the Euler-Almansi finite strain tensor, $\boldsymbol{E}$, defined by (Spencer 1980)

$$\boldsymbol{E} = \tfrac{1}{2}\left(\boldsymbol{I} - (\boldsymbol{F}\boldsymbol{F}^T)^{-1}\right), \tag{6.75}$$

is chosen, with components given by

$$E_{ij} = \frac{1}{2}\left(\delta_{ij} - \frac{\partial X_k}{\partial x_i}\frac{\partial X_k}{\partial x_j}\right) \quad (i, j, k = 1, 2, 3), \tag{6.76}$$

and the symmetry property $E_{ij} = E_{ji}$. Note that $\boldsymbol{F}$ itself is not an objective tensor field.

By employing the deformation measure $\boldsymbol{E}$, the micro-macroscopic strain difference for a k-th crystal in the aggregate, denoted by $\hat{\boldsymbol{E}}_k$, can be expressed by the relation

$$\hat{\boldsymbol{E}}_k = \boldsymbol{E}_k - \bar{\boldsymbol{E}} \quad (k = 1, \ldots, N_c), \tag{6.77}$$

where $\bar{\boldsymbol{E}}$ is the macroscopic strain tensor describing the bulk deformation of the polycrystal. To establish a criterion for the onset of the dynamic recrystallization process, an invariant form of the strain-difference measure is required. This is adopted as the second moment (Truesdell and Noll 2004) of the strain-difference tensor:

$$I_2^{(k)} = \operatorname{tr}\hat{\boldsymbol{E}}_k^2 = \hat{E}_{ij}^{(k)}\hat{E}_{ji}^{(k)} \quad (i, j = 1, 2, 3). \tag{6.78}$$

To define the migration recrystallization condition, a critical level of the latter invariant is adopted, I_2^{cr}, and it is assumed that a given crystal k is subjected to recrystallization as soon as the strain-difference invariant $I_2^{(k)}$ in that crystal is greater than, or equal to, the critical value of the strain difference invariant I_2^{cr}. That is, a k-th crystal recrystallizes when the following condition is fulfilled

$$I_2^{(k)} \geq I_2^{cr}. \tag{6.79}$$

It is supposed that the value of the critical strain-difference parameter I_2^{cr} is related to an invariant of the macroscopic strain $\bar{\boldsymbol{E}}$ for the whole polycrystalline aggregate. Hence, a parameter $\bar{I}_2$ is introduced which, by analogy to (6.78), is defined by

$$\bar{I}_2 = \operatorname{tr} \bar{\boldsymbol{E}}^2 = \bar{E}_{ij}\bar{E}_{ji} \quad (i, j = 1, 2, 3). \tag{6.80}$$

Accordingly, the critical strain-difference invariant is expressed in the form

$$I_2^{cr} = \psi^{cr} \bar{I}_2, \tag{6.81}$$

so that the recrystallization criterion (6.79) becomes

$$I_2^{(k)} \geq \psi^{cr} \bar{I}_2. \tag{6.82}$$

The dimensionless parameter ψ^{cr} entering (6.82) defines the relative threshold strain-difference at which dynamic recrystallization occurs, and is a free parameter of the model. Unfortunately, the magnitude of this parameter cannot be determined directly by correlation with empirical data, as these are not available. Therefore, the value of this parameter has been evaluated on the basis of the results of numerical simulations described in the next section. To do this, laboratory observations reported by Jacka and Maccagnan (1984) and Budd and Jacka (1989) were used, showing girdle fabrics of recrystallized ice formed in uniaxial compression conditions. For very large axial strains, these fabrics have steady-size girdles, with all the crystal c-axes distributed at angles ranging between 25° and 45° to the principal axis of compression. By the method of trial and error, it was found that the best fit of the model predictions to the observations is achieved for $\psi^{cr} = 0.39$. Hence, this particular value of the invariant was used in all the simulations, the results of which are presented in the following Sect. 6.5.

Once the dynamic recrystallization of a given grain has been initiated, a new crystal starts to develop from the old one undergoing recrystallization. The spatial orientations of the new crystals are determined in the same way as described in the previous Sect. 6.4.2. Hence, it is assumed that the initial c-axes orientations of the new crystals favour their viscous deformation in the current macroscopic stress configurations; that is, the strain-rates of these crystals are maximized in the current stress field. This means that the new crystal c-axes are distributed on a conical surface around the current principal axis of maximum compression, at the zenith angles θ_{max} determined from Eq. (6.72) on p. 202.

6.5 Numerical Simulations

The three multi-grain dynamic recrystallization models described in Sect. 6.4, based on three different recrystallization criteria, have been used to simulate the behaviour of polycrystalline ice in sustained uniaxial compression and simple shear flows. The results of the simulations illustrate the process of the ice fabric evolution with increasing deformation and time, and show the variation of macroscopic viscosities with strains. The principal objective of all the simulations was to investigate the effect of the recrystallization mechanism on the characteristic macroscopic features of ice. For this purpose, the results obtained for ice undergoing recrystallization are compared with the predictions of the models described in Sect. 6.3, in which the dynamic recrystallization process is not incorporated. Also, inter-comparisons of the results given by the three proposed recrystallization models are presented.

The simulations were conducted for the rheological parameters pertaining to polar ice, and for the stress and strain-rate magnitudes typically occurring in flows of large polar ice sheets. Hence, the macroscopic viscosity of the isotropic ice was adopted as $\mu_0 = 25$ MPa · a (with the unit 'a' denoting the year, alternatively, also the unit 'yr' is used in the plots to denote the same). The enhancement factors for compression and shear were assumed as those for cold ice, that is, $E_a = 1/3$ and $E_s = 5$, respectively. The latter determine the microscopic rheological parameter values $A = 15$ and $B = 4$. The calculations were carried out by adopting in the initial configuration $N_c = 1000$ discrete grains of equal volumes, with a random distribution of the grain c-axes to represent a macroscopically isotropic polycrystalline aggregate. It was assumed that the process of nucleation of new grains from the old ones takes 1000 years (so that there are simultaneously two discrete grains in the model, of varying volumes, during a crystal recrystallization event). Due to the lack of empirical data, the adopted recrystallization time was chosen arbitrarily, though the values of similar magnitudes were suggested in the literature (De La Chapelle et al. 1998).

The following four figures illustrate the behaviour of ice predicted by the uniform strain model described in Sect. 6.4.1, based on the assumption that the dynamic recrystallization process is controlled by deviatoric stresses. The results were obtained by assuming that the uniaxial flow (along the x_3-axis) occurred at the macroscopic strain-rate $\bar{D}_{33} = 10^{-4}\ \mathrm{a}^{-1}$, and the shear strain-rate in the plane Ox_1x_3 was $\bar{D}_{13} = 10^{-4}\ \mathrm{a}^{-1}$, both kept constant throughout the flow. The model predictions for unconfined uniaxial flow are shown in Figs. 6.16 and 6.17. The plots in the first of these figures illustrate the evolution of the anisotropic fabric, by showing the distributions of crystal c-axes on equal area Schmidt diagrams. These diagrams demonstrate how the ice fabric changes with an increasing axial strain $\varepsilon = 1 - \lambda_3$, shown in the figure, together with the corresponding flow times t (given in thousands of years). The upper row of diagrams, Fig. 6.16a, displays the development of a typical single-maximum fabric at large compressive strains when no recrystallization occurs, similar fabrics were already presented in Fig. 6.5 on p. 186. For comparison, the lower row of diagrams, Fig. 6.16b, shows the evolution of the ice fabric when the migration recrystallization process takes place, acting alongside the crystal lattice rotation mechanism. The presented plots were obtained by assuming that the crystal recrystallization starts when the equivalent stress ratio $\zeta_{eq}^{cr} = S_{eq}^{cr}/\bar{S}_{eq}$ is equal to 2.2, see Eq. (6.67) and Fig. 6.13. The latter particular threshold stress level was chosen by comparing the model predictions with experimental observations by Budd and Jacka (1989), indicating that typical fabrics formed in recrystallizing ice under uniaxial compression have girdles at the angle of about 25° to the principal axis of maximum compression. This means that all the grains with the orientation angles $\theta \lesssim 25°$ are subject to the dynamic recrystallization. The solid curve in Fig. 6.13 (that for $A = 15$ and $B = 4$) shows that the equivalent stress level which corresponds to the limit angle $\theta = 25°$ is equal to about 2.2; hence the choice of the latter value of $\zeta_{eq}^{cr} = 2.2$ for the simulations. It is seen in Fig. 6.16b that the characteristic girdle fabrics start to

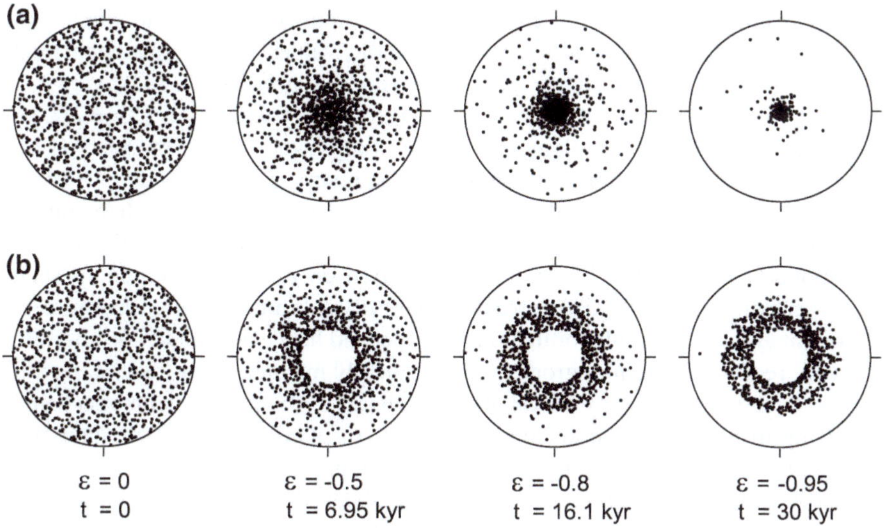

Fig. 6.16 Evolution of fabric in uniaxial compression along the x_3-axis as a function of the macroscopic axial strain $\varepsilon = \lambda_3 - 1$ and time t (given in thousands of years, kyr) for (**a**) non-recrystallizing and (**b**) recrystallizing ice. Predictions of the uniform strain (Taylor-Voigt) model. Reprinted from Staroszczyk (2009), Fig. 4. Copyright 2009 by the Institute of Fundamental Technological Research of the Polish Academy of Sciences

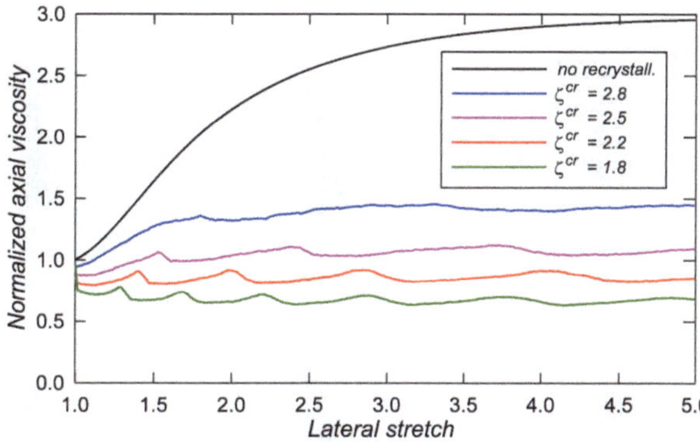

Fig. 6.17 Variation of the normalized axial viscosity μ_{33}/μ_0 with the lateral stretch λ_1 in uniaxial compression for non-recrystallizing ice, and for ice recrystallizing at different critical equivalent stresses ζ_{eq}^{cr}. Adapted from Staroszczyk (2009), Fig. 6. Copyright 2009 by the Institute of Fundamental Technological Research of the Polish Academy of Sciences

appear at the axial strains as small as $\varepsilon = -0.5$ (two-fold axial compression of ice), and subsequently fully develop with increasing deformation, which is illustrated by the strong girdle fabric corresponding to the strain $\varepsilon = -0.95$ (twenty-fold compression of ice, which is a typical magnitude for bottom layers in polar ice masses).

In the context of the polar ice sheet flow modelling, more important than the differences in qualitative properties of recrystallizing and non-recrystallizing ice fabrics are the respective differences between the macroscopic viscosities of ice. Figure 6.17 illustrates the variation of the dimensionless macroscopic axial viscosities μ_{33}/μ_0 with increasing lateral stretch λ_1. Compared are the model results for non-recrystallizing ice with those for ice that recrystallizes at different critical stress ratios $\zeta_{eq}^{cr} = S_{eq}^{cr}/\bar{S}_{eq}$. Depending on the specific values of ζ_{eq}^{cr} illustrated in the figure, the average macroscopic viscosities (for the stretches $\lambda_1 \geq 2$) vary from $\mu_{33}/\mu_0 \sim 1.36$ for $\zeta_{eq}^{cr} = 2.8$, to the value ~ 0.57 for $\zeta_{eq}^{cr} = 1.5$. For $\zeta_{eq}^{cr} = 2.2$, best correlating with the observations (see the red line in the figure), the macroscopic axial viscosities are, on average, slightly less than unity. This means that the recrystallized anisotropic ice is slightly less viscous than the isotropic ice with the random distribution of the crystal c-axes. A characteristic feature that is seen in Fig. 6.17 is a non-monotonic, wave-like variation of the axial viscosity with deformation. Such a quasi-periodic pattern is due to the fact that new crystals are nucleated at similar spatial orientations, favourable for their basal glide, and hence they undergo subsequent dynamic recrystallization in groups, not in a steady manner. Such 'recrystallization waves' have been hinted by Duval and Castelnau (1995) as a possible mechanism occurring in natural ice, producing a sequence of alternating softening/hardening phases experienced by the material during its creep.

The results of simulations for the simple shear flow regime are presented in Figs. 6.18 and 6.19. The evolution of the anisotropic microstructure with increasing shear strain κ and time t is illustrated by the diagrams in Fig. 6.18. Again, the crystal c-axis distributions for non-recrystallizing, (a), and recrystallizing ice, (b), are shown. The results were obtained for the same value of the critical equivalent stress as for the axial compression case; that is, for $\zeta_{eq}^{cr} = S_{eq}/\bar{S}_{eq} = 2.2$. Comparison of

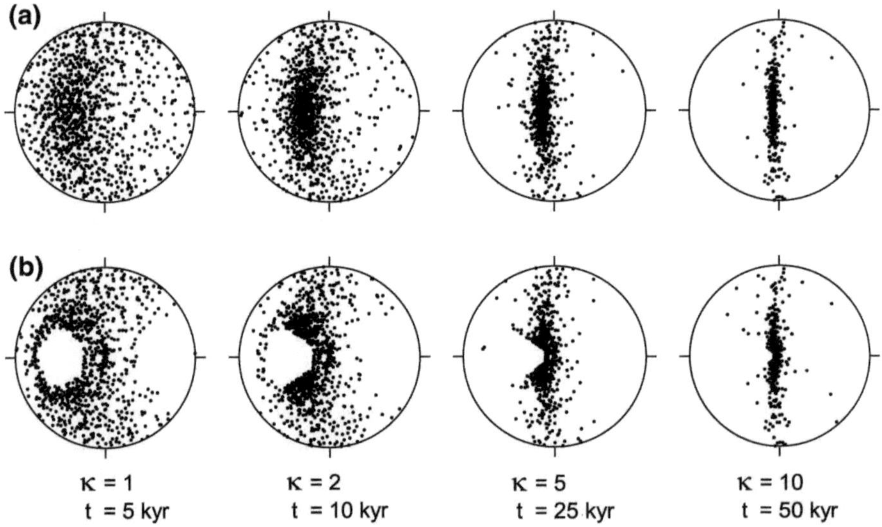

Fig. 6.18 Evolution of fabric in simple shear in the Ox_1x_3 plane as a function of the macroscopic sher strain κ and time t (in thousands of years) for (**a**) non-recrystallizing and (**b**) recrystallizing ice. Predictions of the uniform strain (Taylor-Voigt) model. Reprinted from Staroszczyk (2009), Fig. 7. Copyright 2009 by the Institute of Fundamental Technological Research of the Polish Academy of Sciences

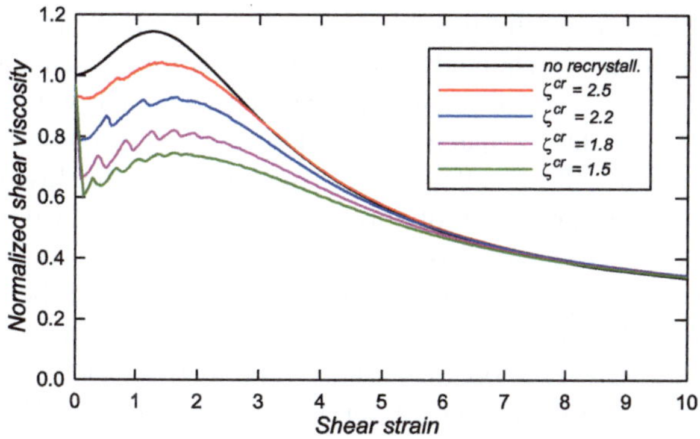

Fig. 6.19 Variation of the normalized shear viscosity μ_{13}/μ_0 with the shear strain κ in simple shearing for non-recrystallizing ice, and for ice recrystallizing at different critical equivalent stresses ζ_{eq}^{cr}. Adapted from Staroszczyk (2009), Fig. 9. Copyright 2009 by the Institute of Fundamental Technological Research of the Polish Academy of Sciences

the diagrams in the upper and lower rows shows that the most significant differences between the fabrics developing in non-recrystallizing and recrystallizing ice occur at small shear strains κ. When only the lattice rotation mechanism operates, see diagrams in Fig. 6.18a, the crystal c-axes first rotate towards the plane Ox_2x_3 (normal to the plane of the plot), and then, for very large shear strains ($\kappa \gg 10$), they start to cluster around the x_3-axis, the latter becoming the principal axis of compression at $\kappa \to \infty$. In the case of the migration recrystallization mechanism being active, see diagrams in Fig. 6.18b, the unfavourably oriented grains (those with the c-axes near the plane Ox_1x_3 and at the angles of about 45° to the x_3-axis) are replaced by new grains, which are much better oriented for basal glide. These new grains, with the c-axes once getting around the x_3 axis, remain at these stable orientations, at which they do not undergo any further recrystallization. It can be seen in Fig. 6.18b that by the time the shear strain κ exceeds a magnitude of about 5, most of the crystals have already reached stable orientations near the plane Ox_2x_3, so that further shearing does not cause any significant changes in the fabrics, which eventually become very close to those for non-recrystallizing ice at the same strain levels $\kappa \gtrsim 10$.

Finally, Fig. 6.19 demonstrates how the macroscopic shear viscosities μ_{13}/μ_0 are affected by the magnitude of the critical equivalent stress ζ_{eq}^{cr}. It can be observed that, irrespective of the critical stress level ζ_{eq}^{cr}, the irregular changes in the viscosity values, reflecting the activity of recrystallization processes taking place on the microscopic level, are essentially confined to the range of small shear strains, $\kappa \lesssim 2.5$. It can be also noted that the softening effect of the dynamic recrystallization on the creep behaviour of ice in simple shear is limited to the range of small strains. For larger strains, $\kappa \gtrsim 10$, the behaviour of the polycrystalline aggregate is practically insensitive to the magnitude of the stress controlling the process of recrystallization, and the macroscopic shear viscosities have practically the same values, whether or not the material has previously recrystallized.

Now the predictions of the two versions of the uniform stress (Sachs-Reuss) model are presented, in which the dynamic recrystallization mechanism is assumed to be either strain-rate or strain – controlled. These two models will be denoted here as SR-D and SR-F models, respectively, while the above discussed uniform strain (Taylor-Voigt) model will be called the TV model. As in the Taylor-Voigt model simulations, the viscous properties of polycrystalline ice are defined by the axial and shear enhancement factors equal to $E_a = 1/3$ and $E_s = 5$, respectively. Since the predictions of the two Sachs-Reuss recrystallization models will be compared with the results given by the Taylor-Voigt model, the microscopic rheological parameters $\alpha = 1/15$ and $\beta = 1/4$ are adopted, as these values yield exactly the desired limit macroscopic behaviour of ice in the Taylor-Voigt approximation. The adopted combination of α and β gives the favourable crystal c-axis orientation angle $\theta_{max} \approx 45.4°$ for new grains nucleated during the dynamic recrystallization process (see the solid line in Fig. 6.15 on p. 202).

The calculations were carried out with the same input parameters as those used in the uniform strain recrystallization model, the results of which are presented earlier in this section. Thus, the macroscopic isotropic ice viscosity was $\mu_0 = 25$ MPa· a, and the macroscopic deviatoric stresses (changing in time) applied to the ice polycrystal were such that the strain-rates were constant and equal to $\bar{D}_{33} = 10^{-4}$ a^{-1} in the uniaxial compression flow along the x_3-axis, and $\bar{D}_{13} = 10^{-4}$ a^{-1} in the simple shear flow in the plane Ox_1x_3. The diagrams in Fig. 6.20 illustrate the evolution of the anisotropic ice fabric with axial deformation $\varepsilon_{33} = \lambda_3 - 1$ and time t, predicted by the Sachs-Reuss model based on the strain-rate recrystallization criterion. One can see that the girdle-like fabrics shown in the figure are very similar to those presented in Fig. 6.16b, predicted by the Taylor-Voigt model with the deviatoric stress recrystallization condition. The only exception is that the SR-D formulation predicts much faster removal of grains oriented at high zenith angles ($\theta \gtrsim 75°$) than the TV approximation. The results presented in Fig. 6.20 were obtained by assuming that the recrystallization is controlled by the critical strain-rate invariant level $\xi_{eq}^{cr} = 1$, as this particular value ensures that the girdles are formed at the angle of about 25° to the axis of compression, as observed in polar ice samples (Budd and Jacka 1989; Gow et al. 1997).

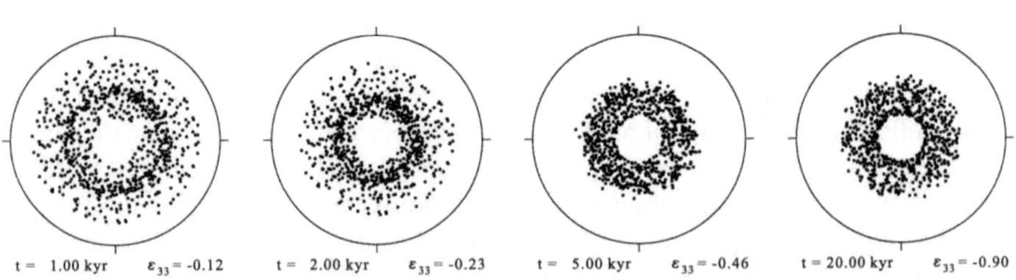

Fig. 6.20 Evolution of recrystallizing ice fabric in uniaxial compression along the x_3-axis as a function of the macroscopic axial strain $\varepsilon_{33} = \lambda_3 - 1$ and time t (given in thousands of years). Predictions of the SR-D model. Reprinted with permission from Staroszczyk (2011), Fig. 3b. Copyright 2011 by Springer Nature

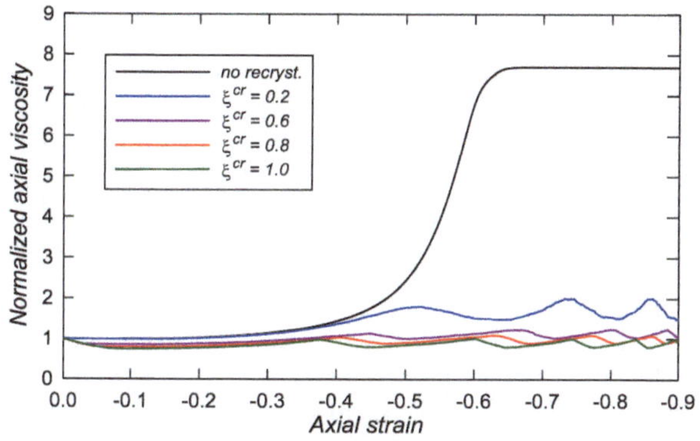

Fig. 6.21 Variation of the normalized macroscopic axial viscosity μ_{33}/μ_0 with the axial strain $\varepsilon_{33} = \lambda_3 - 1$ in uniaxial compression of recrystallizing ice, for different levels of the critical strain-rate invariant $\xi^{cr} = D_{eq}/\bar{D}_{eq}$. Predictions of the SR-D model

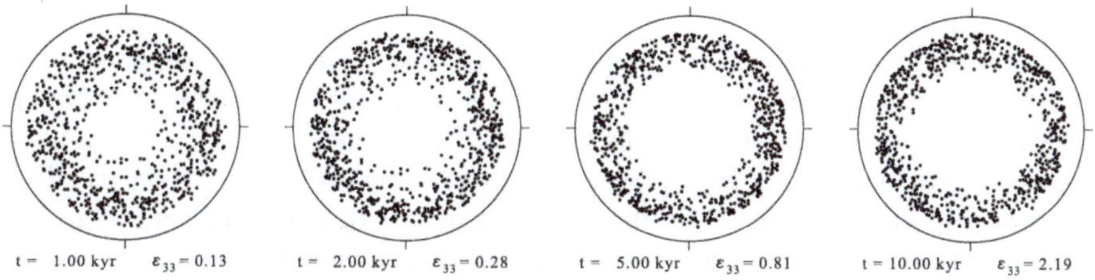

Fig. 6.22 Evolution of recrystallizing ice fabric in uniaxial tension along the x_3-axis as a function of the macroscopic axial strain $\varepsilon_{33} = \lambda_3 - 1$ and time t (given in thousands of years). Predictions of the SR-D model. Reprinted with permission from Staroszczyk (2011), Fig. 6b. Copyright 2011 by Springer Nature

Figure 6.21 presents the evolution of the normalized axial viscosity μ_{33}/μ_0 with increasing deformation measured by the axial strain $\varepsilon_{33} = \lambda_3 - 1$, $\varepsilon_{33} \leq 0$, starting from an initial isotropic state of the polycrystal. The plots demonstrate the effect of the critical level of the strain-rate invariant ξ_{eq}^{cr} on the magnitude of the macroscopic viscosity of recrystallizing ice; the variation of μ_{33}/μ_0 for non-recrystallizing ice is also shown for reference. It is seen that the softening effect of the recrystallization process is significant. For instance, in the case of $\xi_{eq}^{cr} = 1$, best correlating with the observations, an average value of μ_{33}/μ_0 at large deformations is equal to about 0.87. It is worth noting that a corresponding result predicted by the Taylor-Voigt model is 0.85 (Staroszczyk 2009), see also Fig. 6.17, so the two approximations yield the results that agree very well.

The next two figures illustrate the behaviour of recrystallizing ice under uniaxial tension. The pole diagrams in Fig. 6.22 display the evolution of the oriented structure of ice with increasing axial deformation, measured by the strain ε_{33} along the axis of tension, x_3. It can be seen that the migration recrystallization process leads to the removal of the grains at high (that is close to 90°) zenith angles, which gives

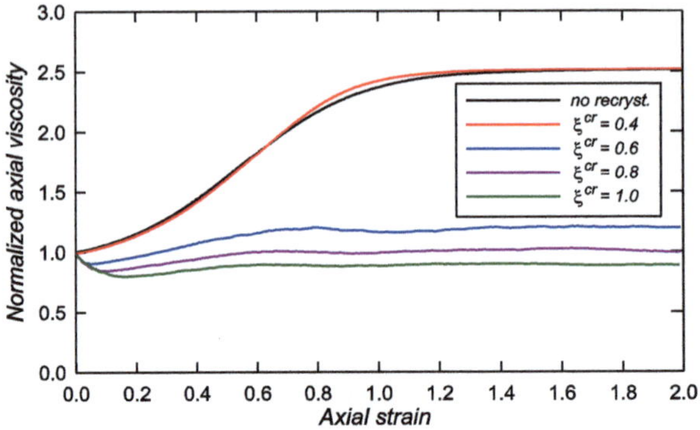

Fig. 6.23 Variation of the normalized macroscopic axial viscosity μ_{33}/μ_0 with the axial strain $\varepsilon_{33} = \lambda_3 - 1$ in uniaxial tension of recrystallizing ice, for different levels of the critical strain-rate invariant $\xi^{cr} = D_{eq}/\bar{D}_{eq}$. Predictions of the SR-D model

rise to girdle fabrics. However, compared to the case of uniaxial compression, see Fig. 6.20, the girdles are now broader, with the majority of crystal c-axes having, for strains $\varepsilon_{33} \gtrsim 0.5$, the orientations within the range $55° \lesssim \theta \lesssim 75°$. Corresponding to the previous plots is Fig. 6.23, demonstrating the evolution of the normalized macroscopic viscosity μ_{33}/μ_0 with increasing axial extension, for various magnitudes of the critical strain-rate invariant ξ^{cr}_{eq}. For reference, the viscosity variation of non-recrystallizing ice is also shown. We observe that for $\xi^{cr}_{eq} = 1$, for which the fabrics plotted in Fig. 6.22 were obtained, an average viscosity of $\mu_{33}/\mu_0 = 0.89$ is not much different from that in compression (0.87). Further, we see in Fig. 6.23 that the characteristic wave-like behaviour of a polycrystal under tensile stress is much less pronounced than in the case of compression. The latter is due to the fact (observed in the simulations) that, at given time and stress, an average number of grains undergoing recrystallization is smaller in uniaxial tension than it is in the case of uniaxial compression. More results obtained by applying the multi-grain TV and SR-D dynamic recrystallization models can be found in the papers by Staroszczyk (2009, 2011).

Now consider the predictions of the uniform stress model incorporating the strain-incompatibility recrystallization criterion (the SR-F model). Figure 6.24 illustrates the evolution of the anisotropic fabric with increasing axial strain ε_{33} in unconfined uniaxial compression, by showing the distribution of the crystal c-axes in pole diagrams plotted on the plane normal to the principal axis of compression. It is seen that the presented girdle-like fabrics are very similar to those predicted by the other two approaches (TV and SR-D), shown in Figs. 6.16b and 6.20, though the strains ε_{33} at which similar fabrics occur are slightly different, with larger discrepancies occurring at the initial stages of deformation. It appears from the diagrams in Fig. 6.24 that the effects of the two counteracting mechanisms, the crystal lattice rotation and the migration recrystallization, start to balance each other at axial strains $\varepsilon_{33} \sim -0.5$

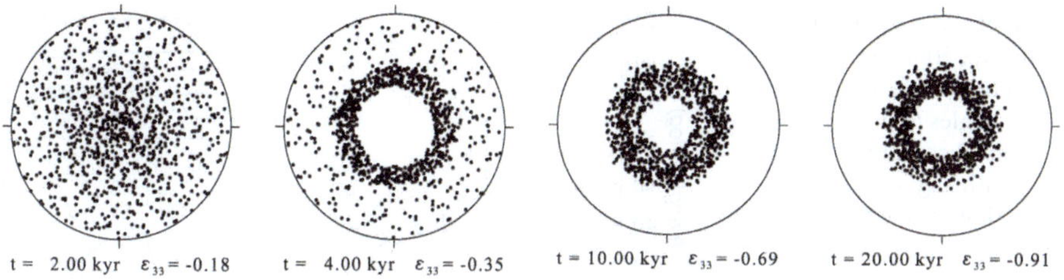

Fig. 6.24 Evolution of recrystallizing ice fabric in uniaxial compression along the x_3-axis as a function of the macroscopic axial strain $\varepsilon_{33} = \lambda_3 - 1$ and time t (given in thousands of years). Predictions of the SR-F model

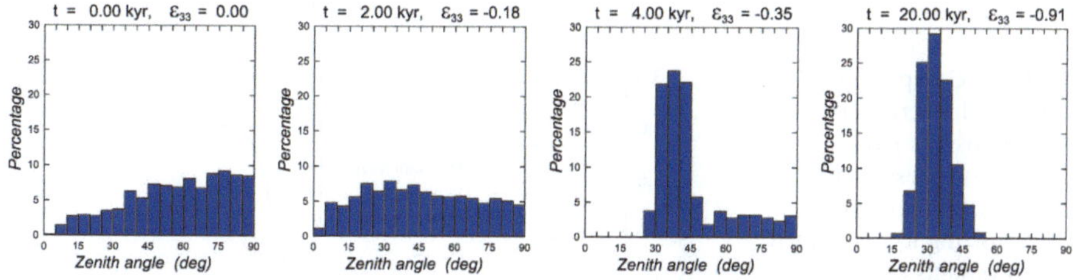

Fig. 6.25 Changes in the distribution of the crystal c-axis zenith angles θ with increasing time and axial strain in uniaxial compression of recrystallizing ice. Predictions of the SR-F model

(twofold axial compression of the material), and at the later stages of the deformation the characteristic girdles have approximately the same width.

In order to gain some quantitative insight into the development of the girdle fabrics, Fig. 6.25 shows histograms of the crystal c-axis distributions for different axial strains ε_{33}. In the diagrams, the zenith angle $\theta = 0$ corresponds to a crystal with its c-axis directed along the principal axis of compression x_3, while $\theta = 90°$ indicates a crystal with its c-axis being normal to x_3-axis (refer to Fig. 6.3 on p. 171). The plots illustrate the process of gradual formation of the anisotropic microstructure starting from the initially isotropic fabric, up to the well developed girdle fabric (the rightmost diagram). In the latter plot, for $\varepsilon_{33} = -0.91$ (about an eleven-fold compression of a sample of ice), one can see that the majority of crystal c-axes are between the zenith angles 20° and 50°, forming a roughly symmetric distribution centred at $\theta \sim 35°$. This prediction agrees well with the observations (Jacka and Maccagnan 1984), proving thus that the threshold strain-difference parameter $\psi^{cr} = 0.39$ was properly selected.

More quantitative information on the process of girdle fabric evolution and the corresponding changes in the c-axis distribution can be inferred from the plots in Fig. 6.26. The plots illustrate the variations in time of the 5, 50 and 95th percentiles of the c-axis zenith angle distributions in fabrics formed in ice which recrystallizes under uniaxial compressive stress. A very characteristic feature is observed, consisting of periodic, and regular, changes in the microstructure of ice, beginning from the second cycle of the dynamic recrystallization process. This feature seems to support the

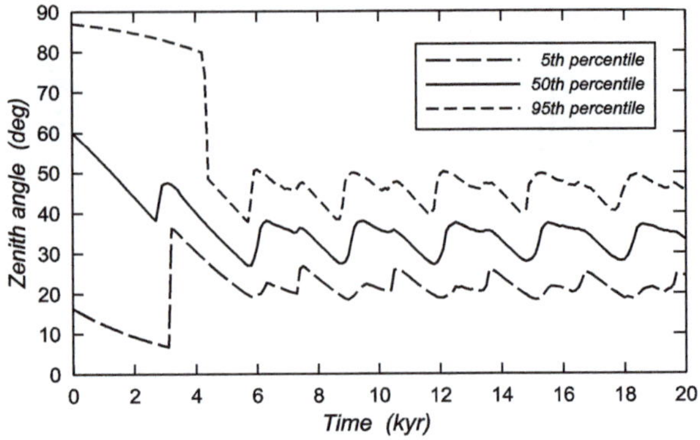

Fig. 6.26 Variation in time of the 5th, 50th and 95th percentiles of the c-axis zenith angles θ distribution in uniaxial compression of recrystallizing ice. Predictions of the SR-F model

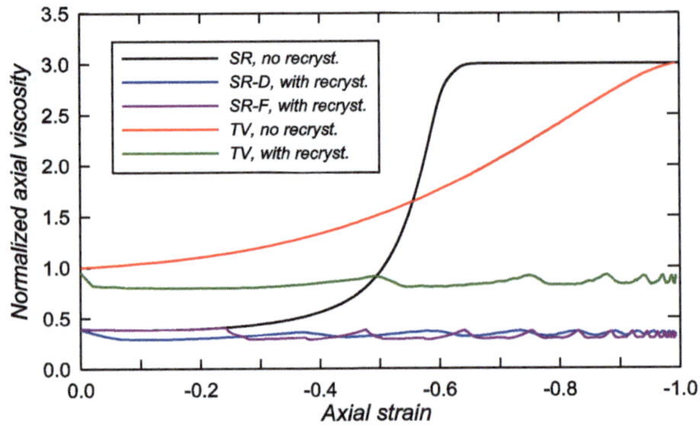

Fig. 6.27 Comparison of the normalized macroscopic axial viscosities μ_{33}/μ_0^{TV} in uniaxial compression for non-recrystallizing and recrystallizing ice, predicted by two variants of the Sachs-Reuss models (SR-D and SR-F) and the Taylor-Voigt model(TV)

concept of the existence of 'recrystallization waves' in polycrystalline ice subjected to sustained compressive stresses.

The following Fig. 6.27 compares the predictions of the three recrystallization models described in Sect. 6.4, showing the variation of the macroscopic viscosities μ_{33}/μ_0^{TV} with axial strain ε_{33} (note that the viscosities are normalized by the magnitude of the isotropic ice viscosity μ_0^{TV} predicted by the uniform strain approximation). The axial viscosities for non-recrystallizing ice, given by the two limit, uniform stress (SR) and uniform strain (TV), approximations are plotted in the figure as well. It is seen that both the strain-rate – induced (SR-D) and the strain–induced (SR-F) versions of the uniform stress (Sachs-Reuss) model yield very similar magnitudes of the axial viscosity. It is worth noting that the latter predictions for the recrystallizing cold ice axial viscosity are very close to the viscosities measured in the laboratory for warm ice at relatively high strain-rates by Budd and Jacka (1989), in which case they are equal to $1/E_a = 1/3$ (the reciprocal of the axial enhancement factor $E_a = 3$ for the cold ice). Some noticeable differences between the viscosities predicted by the SR-D and SR-F models occur only at small strains, which can be explained by the fact that some period of time (that is, some macroscopic deformation) is required for strain differences between crystals and the matrix to accumulate before the recrystallization process is initiated. Therefore, the process starts earlier

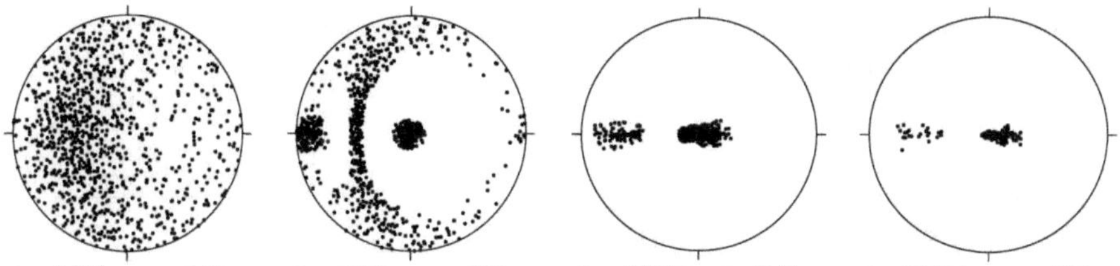

Fig. 6.28 Evolution of recrystallizing ice fabric in simple shear in the Ox_1x_3 plane as a function of the macroscopic shear strain κ and time t (given in thousands of years). Predictions of the SR-F model

when it is controlled by strain-rates (compare the curves labelled SR-D and SR-F). Obviously, the viscosities of recrystallized ice predicted by the uniform strain (Taylor-Voigt) theory, representing the upper bound approximation, are larger than those given by the lower bound Sachs-Reuss approximation. Quantitatively, they differ roughly by a factor μ_0^{TV}/μ_0^{SR} defined by (6.62) on p. 191 (equal to 2.5667 for the microscopic rheological parameters $A = 1/\alpha = 15$ and $B = 1/\beta = 4$ used in the simulations to describe the properties of cold ice).

Finally, the results of the simulations of the simple shear flow, assumed to occur in the Ox_1x_3 plane, are presented. Figure 6.28 illustrates the fabric evolution with increasing strain κ and time t for ice undergoing recrystallization, predicted by the strain–induced Sachs-Reuss (SR-F) model. The crystal c-axes distributions are plotted on the plane Ox_1x_2, being the macroscopic glide plane. The model predicts, for sufficiently large shear strains ($\kappa \gtrsim 2$ say), the development of characteristic two-maxima fabrics, with a stronger maximum occurring in the direction of the principal axis of compression (in the central part of the diagrams). These two-maxima fabrics, predicted by the SR-F model, are in a very good agreement with the fabrics found in polar ice and shown by Duval (1981), Alley (1992) and Paterson (1994). Note that these fabrics qualitatively differ from those illustrated in Fig. 6.18b, obtained by employing the Taylor-Voigt model, with the deviatoric stress controlling the recrystallization process. As the shear deformation continues, then, due to repeating recrystallization events, the number of crystals forming the stronger c-axes concentration gradually increases, while the number of crystals in the other, weaker, concentration decreases. The prediction of such a mechanism supports a hypothesis of Paterson (1994), based on energy considerations. Eventually, at very large strains, $\kappa \gtrsim 10$, the weaker maximum disappears, and a very strong single-maximum fabric is formed, resembling that developing under uniaxial compression of non-recrystallizing ice, see Fig. 6.16a on p. 207.

The last Fig. 6.29 illustrates the variation of the macroscopic shear viscosities μ_{13}/μ_0^{TV} with shear strain κ, and compares the predictions of the three migration recrystallization models. As previously in Fig. 6.27, the viscosity variation of non-recrystallizing ice (the black line) is plotted for reference. One can observe that the results given by the two variants of the uniform-stress formulations (SR-D and SR-F) differ noticeably only within the range of small strains, $\kappa \lesssim 0.8$. With increasing shear deformation, the discrepancies between the results of all the three proposed models steadily decrease, so that at large strains, $\kappa \gtrsim 7$, they become practically neg-

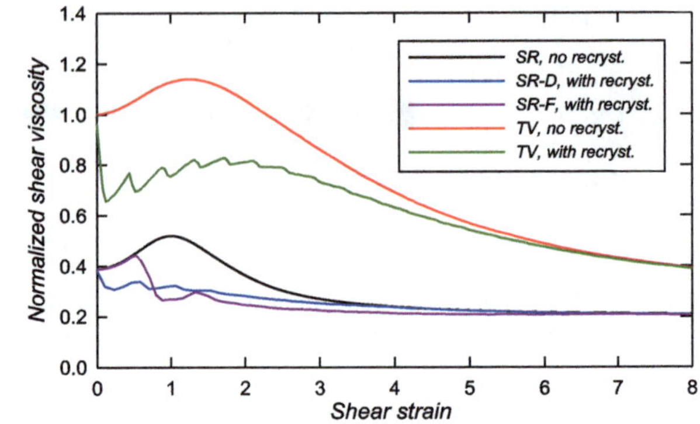

Fig. 6.29 Comparison of the normalized macroscopic shear viscosities μ_{13}/μ_0^{TV} in simple shear for non-recrystallizing and recrystallizing ice, predicted by two variants of the Sachs-Reuss models (SR-D and SR-F) and the Taylor-Voigt model (TV)

ligible, and the SR-F, SR-D and TV approaches yield the macroscopic shear viscosities prescribed by the limit $1/E_s = 0.2$ (recall that E_s denotes the shear enhancement factor, a macroscopic quantity used to calibrate the models).

References

Alley RB (1992) Flow-law hypotheses for ice-sheet modelling. J Glaciol 38(129):245–256

Arminjon M (1991) Limit distributions of the states and homogenization in random media. Acta Mech 88:27–59

Azuma N (1994) A flow law for anisotropic ice and its application to ice sheets. Earth Planet Sci Lett 128(3–4):601–614

Azuma N (1995) A flow law for anisotropic polycrystalline ice under uniaxial compressive deformation. Cold Reg Sci Technol 23:137–147

Bishop JFW, Hill R (1951) A theory of plastic distortion of a polycrystalline aggregate under combined stresses. Phil Mag (7th Ser) 42(327):414–427

Boehler JP (1987) Representations for isotropic and anisotropic non-polynomial tensor functions. In: Boehler JP (ed) Applications of tensor functions in solid mechanics. Springer, Wien, pp 31–53

Budd WF, Jacka TH (1989) A review of ice rheology for ice sheet modelling. Cold Reg Sci Technol 16(2):107–144. https://doi.org/10.1016/0165-232X(89)90014-1

Castelnau O, Duval P, Lebensohn RA, Canova GR (1996) Viscoplastic modeling of texture development in polycrystalline ice with a self-consistent approach: comparison with bound estimates. J Geophys Res 101(B6):13851–13868. https://doi.org/10.1029/96JB00412

Chadwick P (1999) Continuum mechanics: concise theory and problems, 2nd edn. Dover, Mineola, New York

De La Chapelle S, Castelnau O, Lipenkov V, Duval P (1998) Dynamic recrystallization and texture development in ice as revealed by the study of deep ice cores in Antarctica and Greenland. J Geophys Res 103(B3):5091–5105. https://doi.org/10.1029/97JB02621

Durand G, Svensson A, Persson A, Gagliardini O, Gillet-Chaulet F, Sjolte J, Montagnat M, Dahl-Jensen D (2009) Evolution of the texture along the EPICA Dome C ice core. In: Hondoh T (ed) Physics of ice core records II. Hokkaido University, Hokkaido, pp 91–105

Duval P (1981) Creep and fabric of polycrystalline ice under shear and compression. J Glaciol 27(95):129–140

Duval P, Arnaud L, Brissaud O, Montagnat M, De La Chapelle S (2000) Deformation and recrystallization processes of ice from polar ice sheets. Ann Glaciol 30:83–87

Duval P, Castelnau O (1995) Dynamic recrystallization of ice in polar ice sheets. J Phys IV 5(C3):197–205

Elvin AA (1996) Number of grains required to homogenize elastic properties of polycrystalline ice. Mech Mat 22(1):51–64

Faria SH (2006) Creep and recrystallization of large polycrystalline masses. III. Continuum theory of ice sheets. Proc R Soc Lond A 462(2073):2797–2816. https://doi.org/10.1098/rspa.2006.1698

Faria SH, Kremer GM, Hutter K (2003) On the inclusion of recrystallization processes in the modeling of induced anisotropy in ice sheeets: a thermodynamicist's point of view. Ann Glaciol 37:29–34

Faria SH, Ktitarev D, Hutter K (2002) Modelling evolution of anisotropy in fabric and texture of polar ice. Ann Glaciol 35:545–551

Faria SH, Weikusat I, Azuma N (2014) The microstructure of polar ice. Part I: highlights from ice core research. J Struct Geol 61:2–20. https://doi.org/10.1016/j.jsg.2013.09.010

Gagliardini O, Arminjon M, Imbault D (2001) An inhomogeneous variational model applied to predict the behaviour of isotropic polycrystalline ice. Arch Mech 53(1):3–21

Gagliardini O, Meyssonnier J (1999) Analytical derivations for the behavior and fabric evolution of a linear orthotropic ice polycrystal. J Geophys Res 104(B8):17797–17809. https://doi.org/10.1029/1999JB900146

Gödert G, Hutter K (1998) Induced anisotropy in large ice shields: theory and its homogenization. Contin Mech Thermodyn 10(5):293–318

Gow AJ, Meese DA (2007) Physical properties, crystalline textures and c-axis fabrics of the Siple Dome (Antarctica) ice core. J Glaciol 53(183):573–584. https://doi.org/10.3189/002214307784409252

Gow AJ, Meese DA, Alley RB, Fitzpatrick JJ, Anandakrishnan S, Woods GA, Elder BC (1997) Physical and structural properties of the Greenland Ice Sheet Project 2 ice core: a review. J Geophys Res 102(C12):26559–26575. https://doi.org/10.1029/97JC00165

Hashin Z, Shtrikman S (1963) A variational approach to the theory of the elastic behaviour of multiphase materials. J. Mech. Phys. Solids 11(2):127–140

Hill R (1952) The elastic behaviour of a crystalline aggregate. Proc Phys Soc A 65(389):349–354

Hutchinson JW (1976) Bounds and self-consistent estimates for creep of polycrystalline materials. Proc R Soc Lond A 348(1652):101–127

Jacka TH, Maccagnan M (1984) Ice crystallographic and strain rate changes with strain in compression and extension. Cold Reg Sci Technol 8(3):269–286. https://doi.org/10.1016/0165-232X(84)90058-2

Jones SJ (1982) The confined compressive strength of polycrystalline ice. J Glaciol 28(98):171–177

Kamb WB (1961) The glide direction in ice. J Glaciol 3(30):1097–1106

Kennedy JH, Pettit EC, Di Prinzio CL (2013) The evolution of crystal fabric in ice sheets and its link to climate history. J Glaciol 59(214):357–373. https://doi.org/10.3189/2013JoG12J159

Ktitarev D, Gödert G, Hutter K (2002) Cellular automaton model for recrystallization, fabric, and texture development in polar ice. J Geophys Res 107(B8):2165. https://doi.org/10.1029/2001JB000621

Lile RC (1978) The effect of anisotropy on the creep of polycrystalline ice. J Glaciol 21(85):475–483

Liu IS (2002) Continuum mechanics. Springer, Berlin

Lliboutry L (1993) Anisotropic, transversely isotropic nonlinear viscosity of rock ice and rheological parameters inferred from homogenization. Int J Plast 9(5):619–632

Ma Y, Gagliardini O, Ritz C, Gillet-Chaulet F, Durand G, Montagnat M (2010) Enhancement factors for grounded ice and ice shelves inferred from an anisotropic ice-flow model. J Glaciol 56(199):805–812

Mangeney A, Califano F, Castelnau O (1996) Isothermal flow of an anisotropic ice sheet in the vicinity of an ice divide. J Geophys Res 101(B12):28189–28204

Mellor M, Cole DM (1982) Deformation and failure of ice under constant stress or constant strain-rate. Cold Reg Sci Technol 5(3):201–219

Meyssonnier J, Philip A (1996) A model for tangent viscous behaviour of anisotropic polar ice. Ann Glaciol 23:253–261

Meyssonnier J, Philip A (1999) Remarks on self-consistent modelling of polycrystalline ice. In: Hutter K, Wang Y, Beer H (eds) Advances in cold-region thermal engineering and sciences. Springer, Berlin, pp 225–236

Molinari A, Canova GR, Ahzy S (1987) A self-consistent approach of the large deformation polycrystal viscoplasticity. Acta Metallurgica 35(12):2983–2994
Mori T, Tanaka K (1973) Average stress in matrix and average elastic energy of materials with misfitting inclusions. Acta Metallurgica 21(5):571–574
Morland LW (2002) Influence of lattice distortion on fabric evolution in polar ice. Continuum Mech Thermodyn 14(1):9–24. https://doi.org/10.1007/s001610100068
Morland LW, Staroszczyk R (2009) Ice viscosity enhancement in simple shear and uni-axial compression due to crystal rotation. Int J Eng Sci 47(11–12):1297–1304. https://doi.org/10.1016/j.ijengsci.2008.09.011
Paterson WSB (1994) The physics of glaciers, 3rd edn. Butterworth-Heinemann, Oxford
Pimienta P, Duval P, Lipenkov VY (1987) Mechanical behavior of anisotropic polar ice. In: International association of hydrological sciences publication, no 170, pp 57–66. (Symp. Physical Basis of Ice Sheet Modelling, Vancouver 1987)
Placidi L, Greve R, Seddik H, Faria SH (2010) Continuum-mechanical, anisotropic flow model for polar ice masses, based on an anisotropic flow enhancement factor. Continuum Mech Thermodyn 22(3):221–237. https://doi.org/10.1007/s00161-009-0126-0
Placidi L, Hutter K, Faria SH (2006) A critical review of the mechanics of polycrystalline polar ice. GAMM-Mitt 29(1):80–117
Rigsby GP (1958) Effect of hydrostatic pressure on velocity of shear deformation of single ice crystals. J Glaciol 3(24):273–278
Spencer AJM (1980) Continuum mechanics. Longman, Harlow
Staroszczyk R (2001) A uniform stress, discrete-grain model for induced anisotropy of ice. In: Szmidt K (ed) Applications of mechanics in civil- and hydro-engineering. IBW PAN Publishing House, Gdańsk, pp 295–314
Staroszczyk R (2002) A uniform strain, discrete-grain model for evolving anisotropy of polycrystalline ice. Arch Mech 54(2):103–126
Staroszczyk R (2004) Constitutive modelling of creep induced anisotropy of ice. IBW PAN Publishing House, Gdańsk
Staroszczyk R (2009) A multi-grain model for migration recrystallization in polar ice. Arch Mech 61(3–4):259–282
Staroszczyk R (2011) A uniform stress, multi-grain model for migration recrystallization in polar ice. Acta Geophys 59(5):833–857. https://doi.org/10.2478/s11600-011-0026-0
Staroszczyk R, Morland LW (2000) Plane ice-sheet flow with evolving orthotropic fabric. Ann Glaciol 30:93–101
Staroszczyk R, Morland LW (2001) Strengthening and weakening of induced anisotropy in polar ice. Proc R Soc Lond A 457(2014):2419–2440. https://doi.org/10.1098/rspa.2001.0817
Svendsen B, Hutter K (1996) A continuum approach for modelling induced anisotropy in glaciers and ice sheets. Ann Glaciol 23:262–269
Thorsteinsson T (2002) Fabric development with nearest-neighbor interaction and dynamic recrystallization. J Geophys Res 107(B1). https://doi.org/10.1029/2001JB000244
Thorsteinsson T, Kipfstuhl J, Miller H (1997) Textures and fabrics in the GRIP ice core. J Geophys Res 102(C12):26583–26599. https://doi.org/10.1029/97JC00161
Thorsteinsson T, Waddington ED, Taylor KC, Alley RB, Blankenship DD (1999) Strain-rate enhancement at Dye 3, Greenland. J Glaciol 45(150):338–345
Treverrow A, Budd WF, Jacka TH, Warner RC (2012) The tertiary creep of polycrystalline ice: experimental evidence for stress-dependent levels of strain-rate enhancement. J Glaciol 58(208):301–314. https://doi.org/10.3189/2012JoG11J149
Truesdell C, Noll W (2004) The non-linear field theories of mechanics, 3rd edn. Springer, Berlin
Van der Veen CJ, Whillans IM (1994) Development of fabric in ice. Cold Reg Sci Technol 22(2):171–195. https://doi.org/10.1016/0165-232X(94)90027-2
Wenk HR, Canova GR, Molinari A, Kocks UF (1989) Viscoplastic modelling of texture development in quartzite. J Geophys Res 94(B12):17895–17906
Zhang Y, Jenkins JT (1993) The evolution of the anisotropy of a polycrystalline aggregate. J Mech Phys Solids 41(7):1213–1243

Chapter 7
Phenomenological Constitutive Models for Polar Ice

In Chap. 6, by following the micro-mechanical approach, several constitutive models for the description of the macroscopic creep behaviour of polar ice are formulated. These models have been derived from the properties of individual ice crystals, for which microscopic constitutive laws are first developed, and then used to determine the macroscopic behaviour of a polycrystalline aggregate by applying either the uniform stress or uniform strain homogenization techniques.

In this chapter a fundamentally different method is applied, in which the macroscopic properties of polar ice are described entirely in terms of the macroscopic quantities that can be measured empirically; that is, no variables which define the underlying microstructure of the material are involved in the constitutive description. Hence, a phenomenological theory is formulated which expresses macroscopic stresses in terms of macroscopic strains and strain-rates, and which also enables the description of the evolving macroscopic anisotropy of polycrystalline ice. However, unlike the common phenomenological models, the proposed constitutive theory does account, although indirectly, for some important microscopic mechanisms which induce the development of the macroscopic anisotropy in the material, such as the mechanism of the crystal lattice rotation described in detail in Chap. 6. In the constitutive models discussed below it is deduced that the ice fabric which develops from an initially isotropic state at the free surface of an ice sheet is macroscopically orthotropic, and retains orthotropic symmetries during its further evolution as ice descends downwards through an ice sheet; the motivation for such an approximation is given in Sect. 7.1. A general frame-indifferent orthotropic viscous flow law expressing deviatoric stress in terms of strain-rate, strain, and three structure tensors defining orthotropic symmetries in the material is formulated in Sect. 7.2. This general form of the constitutive equation is subsequently reduced in Sect. 7.3 to enable the model correlation with the results of laboratory tests. The reduced model is then applied to illustrate the behaviour of ice in continued uniaxial compression and simple shear configurations. Alternative forms of the orthotropic flow laws are presented in Sects. 7.4 and 7.5. The last section of the chapter deals with the formulation of phenomenological constitutive models describing the process of dynamic recrystallization of polar ice.

R. Staroszczyk, *Ice Mechanics for Geophysical and Civil Engineering Applications*, GeoPlanet: Earth and Planetary Sciences, https://doi.org/10.1007/978-3-030-03038-4_7

7.1 Continuum Mechanics Preliminaries

The macroscopic constitutive laws considered in this chapter are motivated by a simple picture of individual ice crystal easy glide planes (crystal basal planes) being rotated towards planes normal to principal axes of compression, and away from planes normal to principal axes of extension (see Fig. 6.2 on p. 169). Such planes, the normals of which are principal stretch axes, are henceforth called principal stretch planes. Given that the initial macroscopic isotropy of ice implies a random distribution of crystal easy glide planes symmetrically distributed about all planes, it is assumed that the new orientations of the glide planes will then be distributed symmetrically about the principal stretch planes, and so the new instantaneous creep response of ice will have reflectional symmetries in these planes. Thus, the instantaneous response of the material is orthotropic with respect to current principal stretch planes. This idealized view of rotating crystal basal planes, retaining reflectional material symmetries in a current orthogonal system of three principal stretch planes, asserts that the instantaneous creep response due to glide on all crystal basal planes must satisfy the orthotropic symmetry. That is, orthotropic symmetry is maintained, but the three planes of reflectional symmetry evolve as the deformation proceeds, so while each deformed configuration has an orthotropic symmetry, this is with respect to changing symmetry planes (Morland and Staroszczyk 1998).

The above overview ignores the local behaviour of individual crystals, the deformations of which, in general, will induce a fabric with no material symmetries, rather than a fabric with orthotropic symmetries. Nonetheless, it is believed that the actual fabric will not considerably differ from the orthotropic fabric assumed to occur. Further, it is also supposed that the induced anisotropy of the material depends only on the evolving current deformation, and not on the deformation history, though it is realized that the effects of individual crystal interactions may depend on the nature of the deformation process, and therefore induce different fabrics for different deformation paths.

The local deformation field is described by means of the material (referential) deformation gradient tensor $\boldsymbol{F}$, with the components defined by

$$F_{ij} = \frac{\partial x_i}{\partial X_j} \quad (i, j = 1, 2, 3), \tag{7.1}$$

where x_i and X_j denote the spatial and material Cartesian coordinates, respectively. A plane view of the macroscopic deformation of a polycrystalline aggregate, described by the tensor $\boldsymbol{F}$, and the rotation of a symmetric quadruple of crystal easy glide planes (represented by four solid lines inside each rectangle), is illustrated in Fig. 7.1 (Morland and Staroszczyk 1998).

The deformation gradient $\boldsymbol{F}$, like any second-order tensor, can be decomposed, applying the polar decomposition theorem, into a product of two second-order tensors (Spencer 1980; Chadwick 1999; Truesdell and Noll 2004): an orthogonal tensor and a positive definite symmetric tensor; that is,

Fig. 7.1 Plane view of deformation of a material element, illustrating the polar decomposition of the deformation gradient $\boldsymbol{F}$, and showing the rotation of a symmetric quadruple of crystal easy glide planes (represented by four solid lines inside each rectangle)

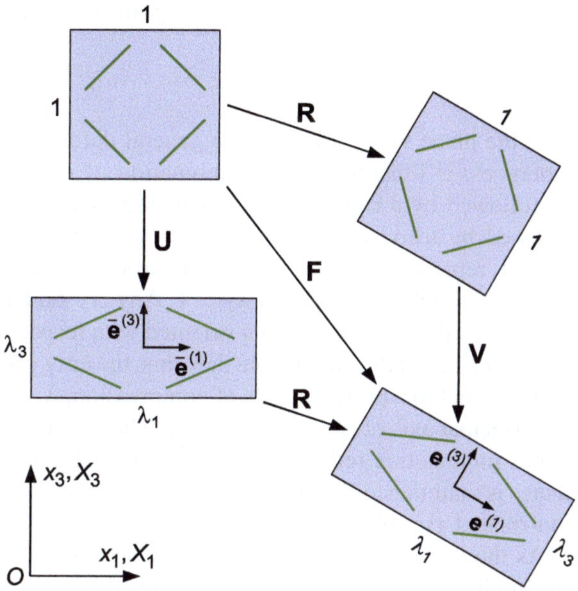

$$\boldsymbol{F} = \boldsymbol{R}\,\boldsymbol{U} = \boldsymbol{V}\boldsymbol{R}. \tag{7.2}$$

In Eq. (7.2), $\boldsymbol{R}$, the rotation tensor, is a proper orthogonal tensor, with the properties $\boldsymbol{R}^{-1} = \boldsymbol{R}^T$ and $\det \boldsymbol{R} = +1$. The tensors $\boldsymbol{U}$ and $\boldsymbol{V}$ are both symmetric and positive definite, and are termed, respectively, the right and the left *stretch tensors*. The principal stretches λ_r ($r = 1, 2, 3$) are along the principal axes $\bar{\boldsymbol{e}}^{(r)}$ (unit vectors) of $\boldsymbol{U}$ in the first decomposition, and along the principal axes $\boldsymbol{e}^{(r)}$ of $\boldsymbol{V}$ in the second decomposition, and the principal unit vectors are related by

$$\bar{\boldsymbol{e}}^{(r)} = \boldsymbol{R}\boldsymbol{e}^{(r)}. \tag{7.3}$$

The principal stretches λ_r along the principal stretch axes $\boldsymbol{e}^{(r)}$ or $\bar{\boldsymbol{e}}^{(r)}$ are the three eigenvalues (necessarily real and positive) of the equations

$$\det(\boldsymbol{U} - \lambda\boldsymbol{I}) = 0 \quad \text{or} \quad \det(\boldsymbol{V} - \lambda\boldsymbol{I}) = 0, \tag{7.4}$$

and

$$\boldsymbol{U}\bar{\boldsymbol{e}}^{(r)} = \lambda_r\bar{\boldsymbol{e}}^{(r)}, \quad \boldsymbol{V}\boldsymbol{e}^{(r)} = \lambda_r\boldsymbol{e}^{(r)}. \tag{7.5}$$

Ice is assumed to be incompressible, so that

$$\det \boldsymbol{F} = \det \boldsymbol{U} = \det \boldsymbol{V} = \lambda_1\lambda_2\lambda_3 = 1, \tag{7.6}$$

and, without loss of generality, we adopt the ordering

$$\lambda_1 \geq \lambda_2 \geq \lambda_3, \quad \lambda_1 \geq 1, \quad \lambda_3 \leq 1. \tag{7.7}$$

In the plane view in Fig. 7.1, a general deformation $\boldsymbol{F}$ which an element of ice undergoes, is illustrated as a combination of either triaxial stretch (described by $\boldsymbol{U}$) followed by a rigid body rotation (represented by $\boldsymbol{R}$), or a rigid body rotation followed by triaxial stretch (given by $\boldsymbol{V}$). The square with unit sides, shown in Fig. 7.1, represents macroscopically isotropic ice that has not deformed yet, with the principal stretches $\lambda_1 = \lambda_2 = \lambda_3 = 1$. Any crystal basal plane in the aggregate will have three others symmetrically oriented with respect to the chosen coordinate axes. As the polycrystalline aggregate deforms, the easy glide planes are rotated towards a plane normal to a principal axis of compression, $\lambda_r < 1$, and away from that normal to a principal axis of extension, $\lambda_r > 1$. The symmetric distribution of the easy glide planes implies that reflectional symmetry in the three orthogonal principal stretch planes is maintained throughout the whole deformation process, either viewed in the non-rotated axes $\bar{\boldsymbol{e}}^{(r)}$, or in the rotated axes $\boldsymbol{e}^{(r)}$.

As the crystal basal planes are those over which the ice can shear most easily, this implies that macroscopic shearing over the principal stretch planes should have ease of shearing, with fluidities (reciprocal viscosities) ordered by respective normal compressions, or inverse stretches, λ_r^{-1}. Furthermore, the relative magnitudes of such shear viscosities should depend on the mean rotations, and hence on, at least, the inverse stretches λ_r^{-1}. An instantaneous viscous creep response of the material must therefore include dependence on at least the principal stretches as arguments of response coefficients, but possibly more generally on the deformation. The most simple approach to a constitutive law which captures an evolving orthotropic fabric suggested by the above picture is to relate the Cauchy stress to strain-rate, current deformation and a set of tensors describing reflectional symmetries in the material.

7.2 General Orthotropic Constitutive Law

In order to account for the orthotropic symmetries at each deformed configuration, the material response of the body is described in terms of three orthogonal unit vectors along the current principal stretches λ_r (r = 1, 2, 3). These orthogonal vectors, either $\bar{\boldsymbol{e}}^{(r)}$ in the non-rotated configuration, or $\boldsymbol{e}^{(r)}$ in the rotated configuration, define the axes of the orthotropic material symmetry, and determine three *structure tensors* given by their respective outer (dyadic) products as

$$\bar{\boldsymbol{M}}^{(r)} = \bar{\boldsymbol{e}}^{(r)} \otimes \bar{\boldsymbol{e}}^{(r)} \quad \text{or} \quad \boldsymbol{M}^{(r)} = \boldsymbol{e}^{(r)} \otimes \boldsymbol{e}^{(r)} \quad (r = 1, 2, 3). \tag{7.8}$$

The above tensors are used in the constitutive formulation to describe the three planes of the orthotropic symmetry in the material.

Since the directional strengths of the creep response, that is the degrees of the crystal basal planes alignment towards the principal stretch planes, depend on the current deformation, the orthotropic constitutive law needs to involve at least the three principal stretches, as the latter, according to the rotation picture described in Sect. 7.1, govern the rotation of easy glide planes. Necessarily, the dependence of the creep response on the principal stretches must be expressed in a symmetric manner, since there can be no material distinction between the principal stretch axes.

Any valid constitutive law has to satisfy the principle of material *frame-indifference* (or *objectivity*), which requires material properties to be independent of an observer (see Appendix B). In mathematical terms, the principle of objectivity requires the directions associated with vector and tensor fields be unaltered by a frame transformation preserving the essential properties of space and time (Truesdell and Noll 2004; Chadwick 1999). The theory of the objective constitutive relations was pioneered by Rivlin and his associates (see, for instance, Ericksen and Rivlin 1954; Rivlin and Ericksen 1955; Rivlin 1955; Smith and Rivlin 1957; Smith 1994), and has been extensively developed since then (see reviews by Spencer 1987a, b).

We are concerned with a symmetric tensor relation for deviatoric stress in terms of strain-rate, strain, and three structure tensors defining orthotropic response with respect to the current principal stretch planes, or alternatively, for strain-rate in terms of the other variables. The Cauchy stress, strain-rate, and structure tensors (7.8) are all frame-indifferent. As regards the deformation measures, from among the three tensors $\boldsymbol{F}$, $\boldsymbol{U}$ and $\boldsymbol{V}$, only the left stretch tensor $\boldsymbol{V}$ is frame-indifferent, whereas the other two are not (Hunter 1983; Liu 2002). For this reason, the tensor $\boldsymbol{V}$ must be used in a constitutive relation as the deformation measure, together with the stress, strain-rate and structure tensors. However, for a given deformation field described by $\boldsymbol{F}$, the direct calculation of $\boldsymbol{V}$ from (7.2) is inconvenient. Instead, it is much simpler to use a frame-indifferent tensor which is closely related to $\boldsymbol{V}$, namely the *left Cauchy-Green deformation tensor* $\boldsymbol{B}$, defined by

$$\boldsymbol{B} = \boldsymbol{V}^2 = \boldsymbol{F}\boldsymbol{F}^T. \tag{7.9}$$

It can be easily proved that the principal directions of $\boldsymbol{B}$ coincide with the current stretch axes of $\boldsymbol{V}$, and the principal values b_r ($r = 1, 2, 3$) of $\boldsymbol{B}$ are the squares of the corresponding principal stretches λ_r. Thus,

$$\det(\boldsymbol{B} - b_r\boldsymbol{I}) = 0, \quad \boldsymbol{B}\boldsymbol{e}^{(r)} = b_r\boldsymbol{e}^{(r)}, \quad b_r = \lambda_r^2 \quad (r = 1, 2, 3), \tag{7.10}$$

and, by ice incompressibility,

$$\det \boldsymbol{B} = b_1 b_2 b_3 = 1. \tag{7.11}$$

Hence, the deformation measure $\boldsymbol{B}$ which will be used in the constitutive law is expressed directly, by (7.9), in terms of the deformation gradient $\boldsymbol{F}$. The evolution of the deformation field with time is governed by the kinematic relation

$$\dot{\boldsymbol{F}} = \boldsymbol{L}\boldsymbol{F}, \tag{7.12}$$

where the superposed dot denotes the material time derivative, and $\boldsymbol{L}$ is the spatial velocity gradient tensor, with the components given by

$$L_{ij} = \frac{\partial v_i}{\partial x_j}. \tag{7.13}$$

With the above definition, the deformation evolution equation (7.12) becomes

$$\dot{F}_{ij} = \frac{\partial F_{ij}}{\partial t} + v_k \frac{\partial F_{ij}}{\partial x_k} = L_{ik} F_{kj} \quad (i, j, k = 1, 2, 3), \tag{7.14}$$

where the summation convention for repeated subscripts applies.

We focus on a viscous constitutive law which expresses stress in terms of strain-rate and strain, which is the most convenient form to use with the momentum balance equations. An inverse law, expressing strain-rate in terms of stress and deformation, is considered in Sect. 7.4. Due to the ice incompressibility assumption, the hydrostatic pressure is not prescribed by the constitutive equation. Therefore, the viscous behaviour of the material is entirely governed by the deviatoric stress, defined by (6.12) on p. 174. Accordingly, a general frame-indifferent orthotropic representation, relating one symmetric tensor (in our case the deviatoric Cauchy stress $\boldsymbol{S}$) to two other symmetric tensors (the strain-rate $\boldsymbol{D}$ and the left Cauchy-Green deformation $\boldsymbol{B}$) is expressed in the following form (see Appendix B, Eqs. (B.10) and (B.11) on p. 325):

$$\begin{aligned} \boldsymbol{S} = & \sum_{r=1}^{3} \left[\phi_r \boldsymbol{M}^{(r)} + \phi_{r+3} \left(\boldsymbol{M}^{(r)} \boldsymbol{D} + \boldsymbol{D} \boldsymbol{M}^{(r)} \right) + \phi_{r+6} \left(\boldsymbol{M}^{(r)} \boldsymbol{B} + \boldsymbol{B} \boldsymbol{M}^{(r)} \right) \right] + \\ & + \ \phi_{10} \boldsymbol{D}^2 + \phi_{11} \boldsymbol{B}^2 + \phi_{12} \left(\boldsymbol{B}\boldsymbol{D} + \boldsymbol{D}\boldsymbol{B} \right), \end{aligned} \tag{7.15}$$

where the 12 response coefficients ϕ_i $(i = 1, \ldots, 12)$ are the functions of 19 invariants I_k $(k = 1, \ldots, 19)$ formed from the tensors $\boldsymbol{M}^{(r)}$, $\boldsymbol{D}$ and $\boldsymbol{B}$:

$$\begin{aligned} & I_r = \operatorname{tr} \boldsymbol{M}^{(r)} \boldsymbol{D}, \quad I_{r+3} = \operatorname{tr} \boldsymbol{M}^{(r)} \boldsymbol{B}, \quad I_{r+6} = \operatorname{tr} \boldsymbol{M}^{(r)} \boldsymbol{D}^2, \\ & I_{r+9} = \operatorname{tr} \boldsymbol{M}^{(r)} \boldsymbol{B}^2, \quad I_{r+12} = \operatorname{tr} \boldsymbol{M}^{(r)} \boldsymbol{B}\boldsymbol{D} \quad (r = 1, 2, 3), \\ & I_{16} = \operatorname{tr} \boldsymbol{B}\boldsymbol{D}^2, \quad I_{17} = \operatorname{tr} \boldsymbol{B}^2 \boldsymbol{D}, \quad I_{18} = \det \boldsymbol{D}, \quad I_{19} = \det \boldsymbol{B}. \end{aligned} \tag{7.16}$$

Due to the ice incompressibility condition (7.11), only 18 invariants are nontrivial, since $I_{19} = \det \boldsymbol{B} = 1$. Further constraints on the material response functions ϕ_i and the invariants I_k are imposed by a viscous response in which the deviatoric stress vanishes when there is no motion; that is, when the strain-rate vanishes.

The general flow law defined by Eq. (7.15), with 12 response functions and 18 invariants as their possible arguments, is far beyond a theory that can be correlated with available experimental data for polar ice. For this reason, relation (7.15) has to be significantly simplified by reducing the number of the functions ϕ_i and invariants I_k. The simplification should be carried out in such a way that the main features of the observed creep behaviour of ice are captured, and all the coefficients in the reduced constitutive model can be determined in feasible laboratory tests. The method based on the concept of so-called *instantaneous directional viscosities* (Morland and Staroszczyk 1998; Staroszczyk and Morland 2000a; Staroszczyk 2004), which can be measured in a series of simple shear tests, is applied here to simplify the general form (7.15).

Accordingly, let us consider an ice element, the current deformation of which is described by distinct principal stretches λ_1, λ_2, λ_3 along the fixed coordinate axes x_1, x_2, x_3 as follows:

$$x_1 = \lambda_1 X_1, \quad x_2 = \lambda_2 X_2, \quad x_3 = \lambda_3 X_3, \quad \lambda_1 \lambda_2 \lambda_3 = 1, \tag{7.17}$$

where X_1, X_2, X_3 are ice particle coordinates in the initial (undeformed) isotropic reference state. The left stretch tensor $\boldsymbol{V}$, the deformation gradient tensor $\boldsymbol{F}$, the rotation tensor $\boldsymbol{R}$, and the left Cauchy-Green deformation tensor $\boldsymbol{B}$ are given then by

$$\boldsymbol{V} = \boldsymbol{F} = \begin{pmatrix} \lambda_1 & 0 & 0 \\ 0 & \lambda_2 & 0 \\ 0 & 0 & \lambda_3 \end{pmatrix}, \quad \boldsymbol{R} = \boldsymbol{I}, \quad \boldsymbol{B} = \begin{pmatrix} \lambda_1^2 & 0 & 0 \\ 0 & \lambda_2^2 & 0 \\ 0 & 0 & \lambda_3^2 \end{pmatrix}. \tag{7.18}$$

Since the principal stretch axes $\boldsymbol{e}^{(r)}$ coincide in the above configuration with the coordinate axes, therefore the three structure tensors $\boldsymbol{M}^{(r)}$ (7.8) are defined by the single diagonal element matrices

$$\boldsymbol{M}^{(1)} = \begin{pmatrix} 1 & 0 & 0 \\ 0 & 0 & 0 \\ 0 & 0 & 0 \end{pmatrix}, \quad \boldsymbol{M}^{(2)} = \begin{pmatrix} 0 & 0 & 0 \\ 0 & 1 & 0 \\ 0 & 0 & 0 \end{pmatrix}, \quad \boldsymbol{M}^{(3)} = \begin{pmatrix} 0 & 0 & 0 \\ 0 & 0 & 0 \\ 0 & 0 & 1 \end{pmatrix}. \tag{7.19}$$

Let remove now the stress and strain-rate, so the fabric defined by the current axial stretches λ_1, λ_2 and λ_3 is 'frozen', and consider instantaneous responses to shearing performed in different directions on different coordinate planes. For simple shear performed in the x_i direction on a glide plane which is normal to the x_j direction $(i \neq j)$, with no summation implied by a repeated index, the new deformation field is defined by

$$x_i = \lambda_i X_i + \kappa_{ij} X_j \,, \quad x_j = \lambda_j X_j \,, \quad x_k = \lambda_k X_k \,, \tag{7.20}$$

where i, j, k are distinct permutations of 1, 2, 3, and κ_{ij} is a shear strain in the Ox_ix_j plane. For the shearing occurring in the latter plane, all the components of the strain-rate tensor $\boldsymbol{D}$ are zero except the two symmetric entries $D_{ij} = D_{ji}$. Such a flow configuration induces a viscous response, described by the law (7.15), in which the deviatoric stress tensor has, in general, three non-zero diagonal components and two non-zero off-diagonal symmetric components $S_{ij} = S_{ji}$. Instantaneously, when the shearing has only started and the strain κ_{ij} is very small, the deformation field is still described be the tensor $\boldsymbol{B}$ given by (7.18)$_3$, and the related three structure tensors are expressed by (7.19). The symmetric tensor generators in Eq. (7.15) have then, for $i \neq j$, the following instantaneous (ij) components, equal to the (ji) components:

$$\left(\boldsymbol{M}^{(r)}\boldsymbol{D} + \boldsymbol{D}\boldsymbol{M}^{(r)}\right)_{ij} = \begin{cases} D_{ij} & (r = i \ \text{ or } \ r = j) \\ 0 & (r \neq i \ \text{ and } \ r \neq j) \end{cases}, \tag{7.21}$$

$$\left(\boldsymbol{M}^{(r)}\boldsymbol{B} + \boldsymbol{B}\boldsymbol{M}^{(r)}\right)_{ij} = 0, \quad \left(\boldsymbol{D}^2\right)_{ij} = 0, \quad \left(\boldsymbol{B}^2\right)_{ij} = 0, \tag{7.22}$$

$$(\boldsymbol{B}\boldsymbol{D} + \boldsymbol{D}\boldsymbol{B})_{ij} = (b_i + b_j)D_{ij}\,, \tag{7.23}$$

where $b_i = \lambda_i^2$ and $b_j = \lambda_j^2$ are the principal values of $\boldsymbol{B}$. The diagonal components of the instantaneous deviatoric stress $\boldsymbol{S}$ are of no interest at this point, since they do not contribute to the instantaneous shear response of the material. Expressions (7.21)–(7.23) show that only four response functions entering Eq. (7.15), namely ϕ_4, ϕ_5, ϕ_6 and ϕ_{12}, can be detected in simple shear tests. Therefore, only these four functions will be retained in the reduced model considered further.

7.3 Reduced Orthotropic Flow Law

In view of the results presented in the previous section, the general orthotropic flow law (7.15) is reduced to the form that retains only the terms which contribute to the instantaneous shear response of the material. Thus, the reduced constitutive equation is considered, which includes the fabric response functions ϕ_4, ϕ_5, ϕ_6 and ϕ_{12} and the associated tensor generators. Accordingly, the reduced orthotropic flow law has the following form (Staroszczyk and Morland 1999, 2000a):

$$\boldsymbol{S} = \sum_{r=1}^{3} \phi_{r+3}\left(\boldsymbol{M}^{(r)}\boldsymbol{D} + \boldsymbol{D}\boldsymbol{M}^{(r)}\right) + \phi_{12}\,(\boldsymbol{B}\boldsymbol{D} + \boldsymbol{D}\boldsymbol{B})\,. \tag{7.24}$$

An equivalent form is obtained by equating the deviatoric parts of both sides of (7.24), which yields

$$\begin{aligned} \boldsymbol{S} = {} & \sum_{r=1}^{3} \phi_{r+3}\left[\boldsymbol{M}^{(r)}\boldsymbol{D} + \boldsymbol{D}\boldsymbol{M}^{(r)} - \tfrac{2}{3}\mathrm{tr}\,(\boldsymbol{M}^{(r)}\boldsymbol{D})\boldsymbol{I}\right] + \\ & + \phi_{12}\left[\boldsymbol{B}\boldsymbol{D} + \boldsymbol{D}\boldsymbol{B} - \tfrac{2}{3}\mathrm{tr}\,(\boldsymbol{B}\boldsymbol{D})\boldsymbol{I}\right]. \end{aligned} \tag{7.25}$$

The invariants appearing in the latter equation are: tr $(\boldsymbol{M}^{(r)}\boldsymbol{D}) = I_r$ $(r = 1, 2, 3)$, and tr $(\boldsymbol{B}\boldsymbol{D}) = I_{13} + I_{14} + I_{15}$, see definitions (7.16).

The flow law (7.25) determines the (ij) component $(i \neq j)$ of the deviatoric stress tensor $\boldsymbol{S}$ as

$$S_{ij} = \left[\phi_{i+3} + \phi_{j+3} + (b_i + b_j)\phi_{12}\right] D_{ij} \quad (i \neq j), \tag{7.26}$$

which defines an instantaneous viscosity μ_{ij} for shearing in the x_i direction on a glide plane normal to the x_j direction by

$$\mu_{ij} = \frac{S_{ij}}{2D_{ij}} = \tfrac{1}{2}\left[\phi_{i+3} + \phi_{j+3} + (b_i + b_j)\phi_{12}\right] \quad (i \neq j). \tag{7.27}$$

Similarly, for unconfined compressions carried out along the principal stretch axes x_i $(i = 1, 2, 3)$, with the current deformation tensor $\boldsymbol{B}$ and the three structure tensors $\boldsymbol{M}^{(r)}$ given by (7.18) and (7.19), respectively, the axial components of the deviatoric stress tensor are expressed by

$$S_{ii} = \tfrac{1}{3}\left[4\,\phi_{i+3} + \phi_{j+3} + \phi_{k+3} + (4b_i + b_j + b_k)\,\phi_{12}\right] D_{ii}\,, \tag{7.28}$$

where i, j, k are distinct permutations of $1, 2, 3$ and no summation over repeated indices is invoked. Equation (7.28) defines an instantaneous axial viscosity for compression in the x_i direction by

$$\mu_{ii} = \frac{S_{ii}}{2D_{ii}} = \tfrac{1}{6}\left[4\,\phi_{i+3} + \phi_{j+3} + \phi_{k+3} + (4b_i + b_j + b_k)\,\phi_{12}\right]. \tag{7.29}$$

The shear and axial viscosity relations (7.27) and (7.29), due to their symmetry properties (the viscosities cannot change when the indices $1, 2, 3$ are interchanged) can be used to infer the properties of the response coefficients ϕ_4, ϕ_5, ϕ_6 and ϕ_{12}, and to determine on which scalar invariants I_k defined by (7.16) the viscous response of the material depends, and on which invariants it does not depend (so that such invariants can be discarded). Further information on the properties of the four response functions and the scalars I_k involved in the constitutive description can be obtained by requiring that the orthotropic flow law (7.25) reduces to the law for an isotropic viscous fluid when the ice is undeformed and there is no anisotropic fabric (that is, when $\boldsymbol{F} = \boldsymbol{I}$, or $\boldsymbol{B} = \boldsymbol{I}$). Hence, it is assumed that for $\boldsymbol{F} = \boldsymbol{I}$, when $\lambda_1 = \lambda_2 = \lambda_3$, the viscous response of ice obeys the *Reiner-Rivlin fluid* flow law defined by (Spencer 1980):

$$\boldsymbol{S} = \Phi_1 \boldsymbol{D} + \Phi_2\left[\boldsymbol{D}^2 - \tfrac{1}{3}\mathrm{tr}\,(\boldsymbol{D}^2)\boldsymbol{I}\right], \tag{7.30}$$

where Φ_1 and Φ_2 are, in general, the functions of the density ϱ, temperature T, and the three strain-rate invariants: tr $\boldsymbol{D}$, tr $\boldsymbol{D}^2$ and det $\boldsymbol{D}$. The conventional approach in glaciology is that $\Phi_2 = 0$ and Φ_1 is independent of det $\boldsymbol{D} = I_{18}$. Since, due to ice incompressibility, tr $\boldsymbol{D} = 0$, the function Φ_1 depends only on one invariant, tr $\boldsymbol{D}^2 = I_7 + I_8 + I_9$ (and on T as well, but the effects of temperature are ignored at this stage).

The detailed analysis conducted by Staroszczyk and Morland (2000a) showed that the response function ϕ_{12} can depend only on the combinations of two invariants

$$I_{20} = \sum_{r=1}^{3} I_{r+3} = \operatorname{tr} \boldsymbol{B}, \quad I_{21} = \sum_{r=1}^{3} I_{r+9} = \operatorname{tr} \boldsymbol{B}^2, \tag{7.31}$$

while the functions ϕ_4, ϕ_5 and ϕ_6 can have common dependence on both I_{20} and I_{21}, as well as dependence on $I_4 = b_1$, $I_5 = b_2$ and $I_6 = b_3$, respectively. Moreover, it was found that, in the initially isotropic state with $\boldsymbol{B} = \boldsymbol{I}$, the following relations hold

$$\phi_4 = \phi_5 = \phi_6 \quad \text{and} \quad \phi_4 + \phi_{12} = \tfrac{1}{2}\Phi_1, \tag{7.32}$$

with the invariants (7.16) becoming, at $\boldsymbol{B} = \boldsymbol{I}$,

$$\begin{aligned} &I_r = \operatorname{tr} \boldsymbol{M}^{(r)}\boldsymbol{D}, \quad I_{r+3} = \operatorname{tr} \boldsymbol{M}^{(r)} = 1, \quad I_{r+6} = \operatorname{tr} \boldsymbol{M}^{(r)}\boldsymbol{D}^2, \\ &I_{r+9} = \operatorname{tr} \boldsymbol{M}^{(r)} = 1, \quad I_{r+12} = I_r \quad (r = 1, 2, 3), \\ &I_{16} = \operatorname{tr} \boldsymbol{D}^2, \quad I_{17} = \operatorname{tr} \boldsymbol{D} = 0, \quad I_{18} = \det \boldsymbol{D}. \end{aligned} \tag{7.33}$$

Since for an isotropic material the response coefficients should be symmetric functions of the invariants (meaning that they do not change their values when the indices $r = 1, 2, 3$ are permuted), (7.33) implies, at $\boldsymbol{B} = \boldsymbol{I}$, the dependence of the response functions on the invariants

$$\begin{aligned} &I_{22} = I_1 + I_2 + I_3 = 0, \qquad I_{23} = I_7 + I_8 + I_9 = \operatorname{tr} \boldsymbol{D}^2, \\ &I_{24} = I_{13} + I_{14} + I_{15} = 0, \quad I_{16} = \operatorname{tr} \boldsymbol{D}^2. \end{aligned} \tag{7.34}$$

Thus, the four response coefficients ϕ_4, ϕ_5, ϕ_6 and ϕ_{12} can involve three invariants depending on the strain-rate $\boldsymbol{D}$, namely $I_{16} = \operatorname{tr} \boldsymbol{B}\boldsymbol{D}^2$, $I_{23} = \operatorname{tr} \boldsymbol{D}^2$ and $I_{24} = \operatorname{tr} \boldsymbol{B}\boldsymbol{D}$, since $I_{22} = \operatorname{tr} \boldsymbol{D} \equiv 0$.

In spite of the restrictions imposed by relations (7.31)–(7.34) on the response functions and their scalar arguments, the constitutive model (7.25) is still too complex to allow correlation with available empirical data. For this reason, the model is further simplified by assuming that the response functions depend on only four invariants of the deformation tensor $\boldsymbol{B}$: $I_4 = b_1$, $I_5 = b_2$, $I_6 = b_3$ and $I_{20} = \operatorname{tr} \boldsymbol{B}$ (that is, the invariant $I_{21} = \operatorname{tr} \boldsymbol{B}^2$ is discarded), which constitutes a minimum set of invariants that the theory has to incorporate in order to satisfy the directional viscosity relations. Yet another simplification is to assume a separable dependence which factors out the invariants depending only on the deformation $\boldsymbol{B}$ and retains the common dependence on the strain-rate $\boldsymbol{D}$; hence, the invariants I_{16} and I_{24} involving the products of $\boldsymbol{B}$ and $\boldsymbol{D}$ are omitted. Accordingly, new material response functions f and g are introduced, by means of which the coefficients ϕ_4, ϕ_5, ϕ_6 and ϕ_{12} are expressed in the forms

$$\begin{aligned} \phi_{r+3}\,(I_{r+3}, I_{20}, I_{23}) &= \mu_0\,(\operatorname{tr} \boldsymbol{D}^2, T)\, f(b_r, \operatorname{tr} \boldsymbol{B}) \\ \phi_{12}\,(I_{r+3}, I_{20}, I_{23}) &= \mu_0\,(\operatorname{tr} \boldsymbol{D}^2, T)\, g(b_r, \operatorname{tr} \boldsymbol{B}) \end{aligned} \quad (r = 1, 2, 3), \tag{7.35}$$

where $\mu_0\,(\mathrm{tr}\,\boldsymbol{D}^2, T)$ is the viscosity of isotropic polycrystalline ice at given strain-rate invariant $\mathrm{tr}\,\boldsymbol{D}^2$ and temperature T. Since the functions (7.35) must give the isotropic viscous fluid law (7.30) with $\Phi_2 = 0$ when $\boldsymbol{B} = \boldsymbol{I}$, so $b_1 = b_2 = b_3 = 1$ and $\mathrm{tr}\,\boldsymbol{B} = 3$, one finds from the flow law (7.25) that

$$\mu_0 = \tfrac{1}{2}\,\Phi_1 \quad \text{and} \quad f(1,3) + g(1,3) = 1. \tag{7.36}$$

We focus here on a model in which the response functions f and g depend on only one invariant each, and adopt

$$f = f(b_r), \quad g = g(\mathrm{tr}\,\boldsymbol{B}), \quad f(1) + g(3) = 1. \tag{7.37}$$

With relations (7.35) and (7.37), the instantaneous directional viscosities (7.27) become

$$\mu_{ij} = \tfrac{1}{2}\mu_0\left[f(b_i) + f(b_j) + (b_i + b_j)\,g(\mathrm{tr}\,\boldsymbol{B})\right], \tag{7.38}$$

and since these viscosities must remain bounded for any axial stretch b_r increasing indefinitely, we rewrite g and the normalization condition $(7.37)_3$ as

$$g(K) = K^{-1}\,G(K), \quad f(1) + \tfrac{1}{3}\,G(3) = 1, \tag{7.39}$$

where the new function $G(K)$ is bounded, and K is the strain invariant defined by

$$K = \mathrm{tr}\,\boldsymbol{B} = b_1 + b_2 + b_3 \geq 3. \tag{7.40}$$

Accordingly, the relations for the instantaneous shear viscosities (7.27), or (7.38), and for the instantaneous axial viscosities (7.29), become

$$\mu_{ij} = \tfrac{1}{2}\mu_0\left[f(b_i) + f(b_j) + (b_i + b_j)\,K^{-1}G(K)\right], \tag{7.41}$$
$$\mu_{ii} = \tfrac{1}{6}\mu_0\left[4f(b_i) + f(b_j) + f(b_k) + (4b_i + b_j + b_k)\,K^{-1}G(K)\right], \tag{7.42}$$

where $i \neq j$ in (7.41), and $i \neq j, i \neq k$ and $j \neq k$ in (7.42). Finally, on account of relations (7.35), (7.37) and (7.39), the orthotropic constitutive equation (7.25) takes the form

$$\boldsymbol{S} = \mu_0 \Big\{ \sum_{r=1}^{3} f(b_r)\left[\boldsymbol{M}^{(r)}\boldsymbol{D} + \boldsymbol{D}\boldsymbol{M}^{(r)} - \tfrac{2}{3}\mathrm{tr}\,(\boldsymbol{M}^{(r)}\boldsymbol{D})\boldsymbol{I}\right] + $$
$$+ \; K^{-1}G(K)\left[\boldsymbol{B}\boldsymbol{D} + \boldsymbol{D}\boldsymbol{B} - \tfrac{2}{3}\mathrm{tr}\,(\boldsymbol{B}\boldsymbol{D})\boldsymbol{I}\right]\Big\}. \tag{7.43}$$

As already noted, the isotropic ice viscosity μ_0 is a function of temperature and current strain-rate, or alternatively stress. The dependence on temperature can be described by relations (3.18)–(3.20) on p. 40, derived by (Morland 1993, 2001) by correlation with experimental data. In this work, the variation of μ_0 with either strain-rate or stress is described in terms of the respective invariants, tr $\boldsymbol{D}$ or tr $\boldsymbol{S}$, being the arguments of the viscosity scaling functions $\phi(I)$ or $\psi(J)$ defined below. Hence, the relations describing the strain-rate and stress dependencies of the isotropic ice viscosity are

$$\mu_0(T, I) = \frac{\sigma_0}{2D_0}\, a^{-1}(T)\, \phi(I), \tag{7.44}$$

$$\mu_0(T, J) = \frac{\sigma_0}{2D_0}\, a^{-1}(T)\, \psi^{-1}(J), \tag{7.45}$$

where the temperature scaling factor $a(T)$ is given by (3.18), and the constants $\sigma_0 = 0.1$ MPa and $D_0 = 1\ \mathrm{yr}^{-1} = 3.17 \times 10^{-8}\ \mathrm{s}^{-1}$ are the normalizing stress and strain-rate units reflecting typical magnitudes of these quantities in natural polar ice caps. The two invariants I and J are defined by

$$I = \tfrac{1}{2}\mathrm{tr}\,(\boldsymbol{D}/D_0)^2, \quad J = \tfrac{1}{2}\mathrm{tr}\,(\boldsymbol{S}/\sigma_0)^2, \tag{7.46}$$

with the relations

$$\phi(I)\,\psi(J) = 1, \quad J = I\phi^2(I), \quad I = J\psi^2(J). \tag{7.47}$$

Specific forms of the scaling factors can be constructed by correlations with the observed creep behaviour of polar ice. One of possible forms was proposed by Smith and Morland (1981) as

$$\psi(J) = 0.3336 + 0.32J + 0.02963J^2, \tag{7.48}$$

with the corresponding function $\phi(I)$ determined by numerical inversion of the first relation (7.47). Alternatively, an algebraic representation proposed by Morland and Staroszczyk (2003b) can be used:

$$\phi(I) = \sum_{r=1}^{3} \frac{k_r}{(1+I)^{l_r}}, \tag{7.49}$$

where $k_1 = 1.4070$, $k_2 = 0.8562$, $k_3 = 0.7100$, $l_1 = 0.3679$, $l_2 = 3.2393$ and $l_3 = 21.4030$. Figure 7.2 illustrates the functions $\phi(I)$ (calculated from (7.49)) and $\psi^{-1}(J)$ (calculated from (7.48)) over the range $0 \le J \le 25$ (corresponding to shear stresses up to 0.5 MPa).

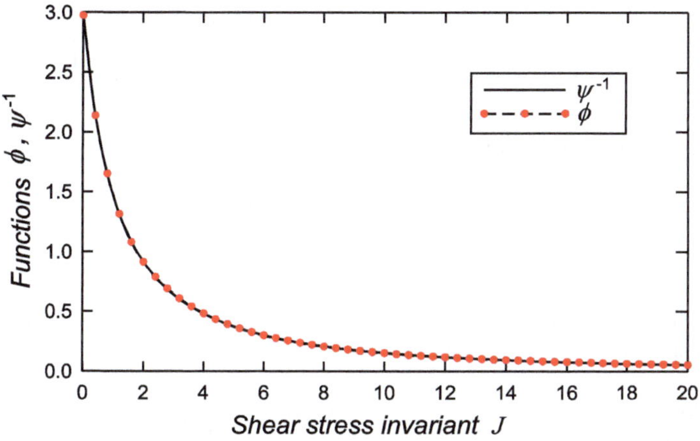

Fig. 7.2 Viscosity scaling functions $\phi(I)$ and $\psi^{-1}(J)$

7.3.1 Simple Flow Simulations

The reduced orthotropic constitutive equation (7.43) describes the viscous behaviour of ice by means of two fabric response functions f and G, the arguments of which are the current deformation measures: the principal stretches b_r $(r = 1, 2, 3)$ in f and their sum $K = b_1 + b_2 + b_3$ in G. In order to construct these two functions and correlate them with available empirical data, a methodology similar to that already applied in Chap. 6 is applied, in which the predictions of the constitutive models for viscous flows in simple configurations were correlated with the limit viscosities measured in laboratory experiments at very large strains, when the anisotropy of ice is fully developed. Accordingly, the unconfined uniaxial compression and simple shear flows illustrated in Fig. 6.4 on p. 185 are considered. For simplicity, the effects of the strain-rate/stress on the ice viscosity magnitude are here neglected.

First let us investigate the unconfined uniaxial compression flow of an initially isotropic ice sample. Referring to relations (7.17), assume that the axial compression is carried out along the x_3-axis. In this configuration, the principal axial stretch $\lambda_3 \leq 1$, and the lateral stretches $\lambda_1 = \lambda_2 = \lambda_3^{-1/2} \geq 1$. The left Cauchy-Green deformation tensor has now the components

$$\boldsymbol{B} = \begin{pmatrix} \lambda_1^2 & 0 & 0 \\ 0 & \lambda_1^2 & 0 \\ 0 & 0 & \lambda_1^{-4} \end{pmatrix}, \tag{7.50}$$

and the three structure tensors $\boldsymbol{M}^{(r)}$ are defined by (7.19). The strain-rate and deviatoric stress tensors have only diagonal components given by

$$\boldsymbol{D} = \begin{pmatrix} D_{11} & 0 & 0 \\ 0 & D_{11} & 0 \\ 0 & 0 & -2D_{11} \end{pmatrix}, \quad \boldsymbol{S} = \begin{pmatrix} S_{11} & 0 & 0 \\ 0 & S_{11} & 0 \\ 0 & 0 & -2S_{11} \end{pmatrix}, \tag{7.51}$$

and the invariants entering the flow law (7.43) are

$$\begin{aligned} &\text{tr}\,(\boldsymbol{M}^{(1)}\boldsymbol{D}) = \text{tr}\,(\boldsymbol{M}^{(2)}\boldsymbol{D}) = D_{11}, \quad &&\text{tr}\,(\boldsymbol{M}^{(3)}\boldsymbol{D}) = -2D_{11}, \\ &\text{tr}\,(\boldsymbol{B}\boldsymbol{D}) = 2D_{11}(b_1 - b_1^{-2}), &&K = \text{tr}\,\boldsymbol{B} = 2b_1 + b_1^{-2}. \end{aligned} \tag{7.52}$$

With the above definitions, the orthotropic constitutive law (7.43) gives the following viscous response of ice under uniaxial compression:

$$\frac{S_{33}}{2\mu_0 D_{33}} = \frac{1}{3}\left[f(b_1) + 2f(b_1^{-2}) + \frac{G(K)}{K}(b_1 + 2b_1^{-2}) \right] = \frac{\mu_{33}}{\mu_0}. \tag{7.53}$$

In (7.53), μ_{33}/μ_0 defines the ratio of the current fabric induced axial viscosity to the isotropic ice viscosity.

Next let us consider a simple shear flow configuration. For more generality, it is assumed that the material is not isotropic at the beginning of shear deformation; that is, it has already developed an anisotropic fabric, the strength of which is described by the principal stretches $\lambda_3 = \lambda_1^{-1}, \lambda_2 = 1$ (plane flow conditions). Now apply shearing in the plane Ox_1x_3 at a constant shear strain-rate $D_{13} > 0$. The deformation field is then described by

$$x_1 = \lambda_1 X_1 + \kappa X_3, \quad x_2 = X_2, \quad x_3 = \lambda_1^{-1} X_3, \tag{7.54}$$

where κ is a shear strain in the plane Ox_1x_3, with $\kappa = 0$ at the start of shearing. Note that λ_1 is here a parameter that describes the fabric prior to shearing. As the shear deformation develops, the principal stretch directions rotate in the plane Ox_1x_3, and hence their magnitudes also change. The deformation and strain-rate fields are now expressed by the tensors

$$\boldsymbol{B} = \begin{pmatrix} \lambda_1^2 + \kappa^2 & 0 & \lambda_1^{-1}\kappa \\ 0 & 1 & 0 \\ \lambda_1^{-1}\kappa & 0 & \lambda_1^{-2} \end{pmatrix}, \quad \boldsymbol{D} = \begin{pmatrix} 0 & 0 & \frac{1}{2}\dot{\gamma} \\ 0 & 0 & 0 \\ \frac{1}{2}\dot{\gamma} & 0 & 0 \end{pmatrix}, \quad \dot{\gamma} = \lambda_1 \dot{\kappa}. \tag{7.55}$$

The principal stretch squares b_i $(i = 1, 2, 3)$, the eigenvalues of $\boldsymbol{B}$, see relations (7.10), are given by

$$b_2 = 1, \quad b_3 = b_1^{-1}, \quad 2b_1 = \lambda_1^2 + \lambda_1^{-2} + \kappa^2 + \sqrt{(\lambda_1^2 + \lambda_1^{-2} + \kappa^2)^2 - 4}\,, \tag{7.56}$$

and the associated principal unit vectors $\boldsymbol{e}^{(r)}$ are defined by

$$\boldsymbol{e}^{(2)} = (0, 1, 0)^T, \quad e_2^{(s)} = 0, \quad \lambda_1^{-1}\kappa\, e_1^{(s)} + \left(\lambda_1^{-2} - b_s\right) e_3^{(s)} = 0, \\ \left[e_1^{(s)}\right]^2 + \left[e_3^{(s)}\right]^2 = 1 \quad (s = 1, 3). \tag{7.57}$$

With the vector components given by (7.57), the structure tensors are expressed by

$$\boldsymbol{M}^{(2)} = \begin{pmatrix} 0 & 0 & 0 \\ 0 & 1 & 0 \\ 0 & 0 & 0 \end{pmatrix}, \quad \boldsymbol{M}^{(s)} = \begin{pmatrix} e_1^{(s)}e_1^{(s)} & 0 & e_1^{(s)}e_3^{(s)} \\ 0 & 0 & 0 \\ e_1^{(s)}e_3^{(s)} & 0 & e_3^{(s)}e_3^{(s)} \end{pmatrix} \quad (s = 1, 3), \tag{7.58}$$

and the invariants entering the constitutive law (7.43) are given by

$$\text{tr}\,(\boldsymbol{M}^{(2)}\boldsymbol{D}) = 0, \qquad \text{tr}\,(\boldsymbol{M}^{(s)}\boldsymbol{D}) = \dot{\gamma}\, e_1^{(s)}e_3^{(s)} \quad (s = 1, 3), \\ \text{tr}\,(\boldsymbol{B}\boldsymbol{D}) = \dot{\gamma}\,\lambda_1^{-1}\kappa, \quad K = \text{tr}\,\boldsymbol{B} = b_1 + 1 + b_1^{-1}. \tag{7.59}$$

In view of (7.55), (7.58) and (7.59), the orthotropic flow law (7.43) yields the following expression for the shear stress component S_{13}:

$$\frac{S_{13}}{2\mu_0 D_{13}} = \frac{1}{2}\left[f(b_1) + f(b_1^{-1}) + \frac{G(K)}{K}(\lambda_1^2 + \lambda_1^{-2} + \kappa^2)\right] = \frac{\mu_{13}}{\mu_0}, \tag{7.60}$$

which describes the variation of the shear viscosity with shear strain κ.

7.3.2 Material Response Functions

The viscosity ratios given by (7.53) and (7.60) describe the evolution of the axial and shear responses of anisotropic fabric with increasing strains, being the arguments of the material response functions f and G. At this stage, the specific properties of the latter functions are unknown. Ideally, the properties of f and G should be inferred from experimental results covering the whole range of axial and shear deformations that an ice sample undergoes as it develops anisotropy from the initially isotropic state. Unfortunately, it seems that the task of achieving the full correlation between the observed polar ice behaviour and the constitutive model functions and parameters is hardly viable yet, in spite of the considerable amount of experimental work already done (Jacka 1984; Jacka and Maccagnan 1984; Li et al. 1996; Treverrow et al. 2012). For this reason, it is attempted here to construct the forms of fabric response functions which fulfil a limited number of conditions derived analytically, with the aim to capture in the proposed model the most important properties of polar ice creep. Such an approach, unavoidably, requires further simplifications to be introduced, thus restricting the theory, but it is believed that the model will still retain sufficient flexibility to enable future correlations with detailed experimental data.

The minimum requirement that a constitutive model for anisotropic polar ice should satisfy is its capability of properly predicting the viscous material behaviour of ice at very large deformations. Such limit behaviour of ice is described quantitatively in terms of the axial and shear enhancement factors E_a and E_s, respectively (Budd and Jacka 1989), which were already used in the micro-mechanical constitutive models discussed in Sect. 6.3 to determine the ice crystal microscopic rheological parameters. Accordingly, for the indefinite axial stretch $\lambda_1 \to \infty$ (and hence $\lambda_3 \to 0$) in uniaxial compression, implying $b_1 \to \infty$ and $K \sim 2b_1$, the viscosity relation (7.53) yields

$$\frac{\mu_{33}}{\mu_0} \to \frac{2}{3} f(0) + \frac{1}{3} f(\infty) + \frac{1}{6} G(\infty) = E_a^{-1}. \tag{7.61}$$

Similarly, in the case of the simple shear viscosity ratio (7.60), as $\kappa \to \infty$ with λ_1 being finite, then $b_1 \sim \kappa^2$ and $K \sim \kappa^2$, and further, $\boldsymbol{e}^{(1)} \to (1, 0, 0)^T$ and $\boldsymbol{e}^{(3)} \to (0, 0, 1)^T$. This implies that

$$\frac{\mu_{13}}{\mu_0} \to \frac{1}{2} f(0) + \frac{1}{2} f(\infty) + \frac{1}{2} G(\infty) = E_s^{-1}. \tag{7.62}$$

Equations (7.61) and (7.62) relate the three limit values $f(0)$, $f(\infty)$ and $G(\infty)$ of the response functions entering the orthotropic flow law (7.43). To determine these limit values uniquely, a third equation is needed. This additional relation is derived by following the concept developed by Staroszczyk and Morland (2000a). Based on this concept, a set of equalities and inequalities connecting the directional shear viscosities μ_{ij} $(i, j = 1, 2, 3, i \neq j)$ given by (7.41) has been formulated. To present this concept, let us consider the principal stretch plane Ox_ix_j, with the principal stretches λ_i and λ_j describing the deformations from the initially isotropic state when $\lambda_i = \lambda_j = 1$, see Fig. 7.3. The figure illustrates the rotations of the diagonals of an initially unit square when $\lambda_j < \lambda_i$ and $\alpha_{ij} < \pi/2$, where $\tan(\alpha_{ij}/2) = \lambda_j/\lambda_i$. It is evident that each intersection line of any set of symmetric glide planes with the plane Ox_ix_j undergoes rotation towards the Ox_i axis which increases as α_{ij} decreases;

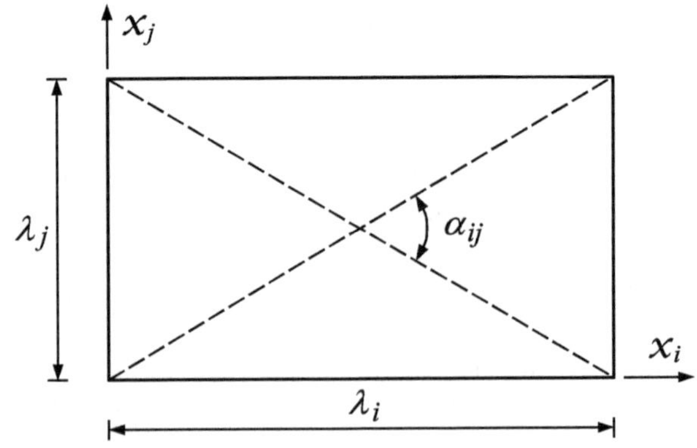

Fig. 7.3 Principal stretches λ_i and λ_j in the principal stretch plane Ox_ix_j

that is, there is increasing alignment of the crystal c-axes towards the direction of a smaller principal stretch as it decreases relative to the other stretches. Thus, the fluidity μ_{ij}^{-1} increases as α_{ij} decreases, or equivalently, the viscosity μ_{ij} increases as the angle α_{ij} increases; that is, as the ratio λ_j/λ_i increases. In other words, the smaller a given principal stretch is compared to the other two stretches, the stronger is the alignment of c-axes towards the direction of this stretch and, therefore, the easier is the crystal basal gliding on the plane normal to this principal stretch axis (that is, the smaller is the corresponding shear viscosity). For any ordering of stretches λ_i, say $\lambda_1 \geq \lambda_2 \geq \lambda_3$, there are six distinct sets of relative values of λ_i, and for each of them corresponding relations order the instantaneous directional shear viscosities μ_{12}, μ_{13} and μ_{23} in the coordinate frame of the principal stretch axes λ_i :

$$\lambda_1 = \lambda_2 = \lambda_3 = 1: \quad \mu_{ij} = \mu_0 \quad (i, j = 1, 2, 3)\,, \tag{7.63}$$
$$\lambda_1 = \lambda_2 > 1 > \lambda_3: \quad 0 < \mu_{13} = \mu_{23} < \mu_{12}\,, \tag{7.64}$$
$$\lambda_1 > \lambda_2 > 1 > \lambda_3: \quad 0 < \mu_{13} < \mu_{23} < \mu_{12}\,, \tag{7.65}$$
$$\lambda_1 > \lambda_2 = 1 > \lambda_3: \quad 0 < \mu_{13} < \mu_{23} = \mu_{12}\,, \tag{7.66}$$
$$\lambda_1 > 1 > \lambda_2 > \lambda_3: \quad 0 < \mu_{13} < \mu_{12} < \mu_{23}\,, \tag{7.67}$$
$$\lambda_1 > 1 > \lambda_2 = \lambda_3: \quad 0 < \mu_{13} = \mu_{12} < \mu_{23}\,. \tag{7.68}$$

Obviously, due to the identities $b_i = \lambda_i^2$ $(i = 1, 2, 3)$, the above viscosity equalities and inequalities also apply for the orderings $b_1 \geq b_2 \geq b_3$. Relations (7.63)–(7.68) can be used to assess the validity of the material responses predicted by the constitutive model for particular choices of the fabric response functions $f(b_r)$ and $G(K)$ (Staroszczyk and Morland 2000a).

Now, by employing relation (7.66) corresponding to the plane flow, in which case $\lambda_2 = b_2 = 1$, and hence $b_3 = b_1^{-1}$ and $K = b_1 + 1 + b_1^{-1}$, it is possible to relate $G(K)$ to $f(b_r)$ $(r = 1, 2, 3)$ explicitly by

$$G(K) = -\frac{K}{b_1 - b_1^{-1}} \left[f(b_1) - f(b_1^{-1}) \right], \quad K \geq 3, \tag{7.69}$$

where

$$2b_1 = K - 1 + \sqrt{(K-1)^2 - 4}\,. \tag{7.70}$$

The limit of (7.69) as $b_1 \to 1$, and hence $K \to 3$, combined with the normalization relation $(7.39)_2$, yields

$$f(1) - f'(1) = 1, \tag{7.71}$$

which is an additional restriction on the function $f(b_r)$ at $b_r = 1$. Furthermore, the limit of (7.69) as $b_1 \to \infty$, when $K \sim b_1$, gives the relation

$$f(0) - f(\infty) - G(\infty) = 0. \tag{7.72}$$

The system of three linear equations (7.61), (7.62) and (7.72) for $f(0)$, $f(\infty)$ and $G(\infty)$ has the solutions

$$f(0) = E_s^{-1}, \quad f(\infty) = 6E_a^{-1} - 5E_s^{-1}, \quad G(\infty) = 6\,(E_s^{-1} - E_a^{-1}), \tag{7.73}$$

which, together with relations (7.71) and $(7.39)_2$, with the latter determining the limit value $G(3)$, define the general properties of the fabric response functions $f(b_r)$ and $G(K)$ and, in particular, ensure that the constitutive model results agree with the observed limit viscosities defined by the axial and shear enhancement factors E_a and E_s . For the values of E_a and E_s pertinent to cold and warm ice (see p. 184), the limit values of the response functions f and G defined by (7.73) are given, respectively, by

$$E_a = \tfrac{1}{3}, \qquad E_s = 5: \quad f(0) = \tfrac{1}{5}, \qquad f(\infty) = 17, \qquad G(\infty) = -\tfrac{84}{5}, \tag{7.74}$$

$$E_a = 3, \qquad E_s = 8: \quad f(0) = \tfrac{1}{8}, \qquad f(\infty) = \tfrac{11}{8}, \qquad G(\infty) = -\tfrac{5}{4}. \tag{7.75}$$

The following forms of monotonic increasing response functions $f(b_r)$ were adopted (Staroszczyk and Morland 1999; Staroszczyk 2004) to investigate the creep behaviour of ice in uniaxial compression and simple shear configurations:

$$f(b_r) = f(\infty) - [f(\infty) - f(0)]\exp(-\alpha b_r^m), \qquad \alpha > 0, \quad m > 0, \tag{7.76}$$

$$f(b_r) = f(0) + [f(\infty) - f(0)]\tanh(\alpha b_r^m), \qquad \alpha > 0, \quad m > 0, \tag{7.77}$$

$$f(b_r) = f(\infty) - [f(\infty) - f(0)]\,\frac{\alpha}{\alpha + b^m}, \qquad \alpha > 0, \quad m > 0, \tag{7.78}$$

where m is a free parameter, and α is determined by the restriction (7.71). The forms of the other response function, $G(K)$, associated with $f(b_r)$, are prescribed by relations (7.69) and (7.70). Plots of the chosen fabric response functions $f(b_r)$ for cold ice (for warm ice they are of similar shapes) are presented in Fig. 7.4. The curves labelled (1) and (2) correspond to the function (7.76) with $m = 1.5$ and $m = 2$, respectively, the curves labelled (3) and (4) correspond to the function (7.77) with $m = 1$ and $m = 1.5$, respectively, and the label (5) indicates the curve corresponding to the function (7.78) with $m = 2$. The same labelling is applied in the plots illustrating the viscous response of ice to compression and simple shear in the next Sect. 7.3.3. The above particular values of the free parameter m in (7.76) and (7.77) were used in numerical simulations, since they provide good correlation with the micro-macroscopic orthotropic model developed by Gagliardini and Meyssonnier (1999), as demonstrated by Staroszczyk and Gagliardini (1999). The shapes of the response functions plotted in Fig. 7.4 are also consistent with those predicted by Morland and Staroszczyk (2009) by analysing the mechanism of the crystal easy glide planes rotation in polycrystalline ice without assuming any particular form of a constitutive law.

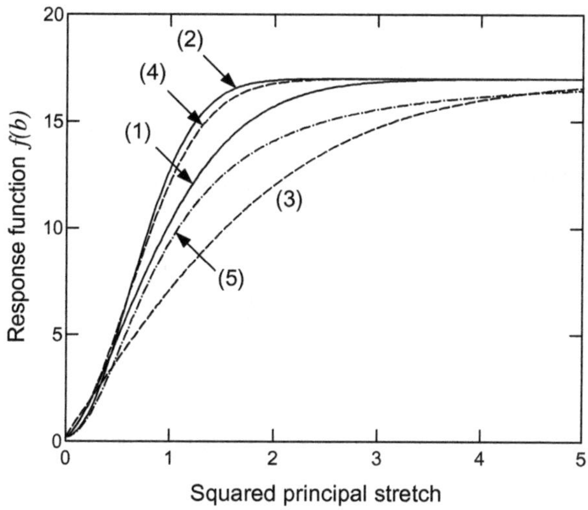

Fig. 7.4 Adopted forms of the response function $f(b_r)$ for cold ice

7.3.3 Model Illustrations

The orthotropic constitutive model defined by (7.43), with the response functions given by (7.76)–(7.78) and depicted in Fig. 7.4, was used to simulate the viscous behaviour of ice in unconfined uniaxial compression and simple shear flows. The results of simulations carried out for cold ice, described by the enhancement factors and limit values of the response functions given by (7.74), are presented in Figs. 7.5, 7.6 and 7.7. The viscous response of ice to uniaxial compression is illustrated in Fig. 7.5, showing the evolution of the normalized axial viscosity μ_{33}/μ_0, described by Eq. (7.53), with increasing lateral stretch λ_1. One can see that the functions (7.76) and (7.77), the solid and dashed lines in the figure, predict a monotonic increase in the axial viscosity with increasing axial deformation, while the function (7.78), the dash-dotted line, yields an initial softening of ice. Such ice softening effects are also predicted by the micro-mechanical models discussed in Chap. 6, see, for example, Fig. 6.7 on p. 188.

The response of cold ice to simple shearing, described by (7.60), is illustrated in Figs. 7.6 and 7.7, showing the evolution of the dimensionless shear viscosity μ_{13}/μ_0 with increasing shear strain κ. Fig. 7.6 displays, for different response functions, the results obtained for shearing started from an initially isotropic state. It is seen that, again, the functions (7.76) and (7.77) yield creep responses which are qualitatively different from the response given by (7.78). While the latter function (dash-dotted line) predicts monotonic softening of the material from the isotropic ice viscosity to the limit viscosity determined by the reciprocal of the shear enhancement factor E_s, the former two functions give responses in which ice initially hardens, with the maximum shear viscosity occurring at strains $\kappa \sim 1$, followed by the stage of progressive softening of the material until the limit shear viscosity $1/E_s = 0.2$ is

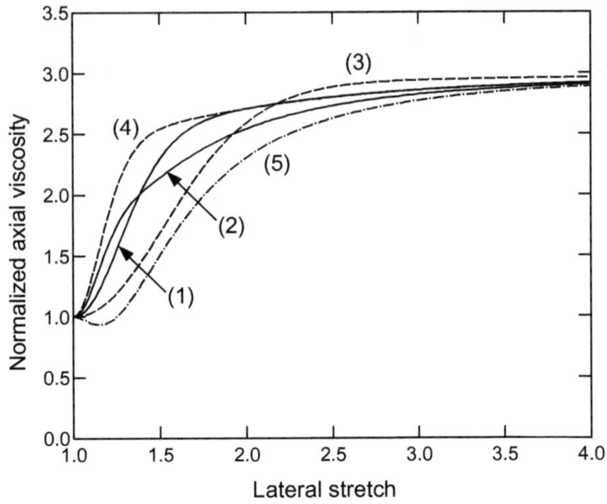

Fig. 7.5 Evolution of the normalized axial viscosity μ_{33}/μ_0 with the lateral stretch λ_1 in uniaxial compression for different response functions $f(b_r)$ (cold ice)

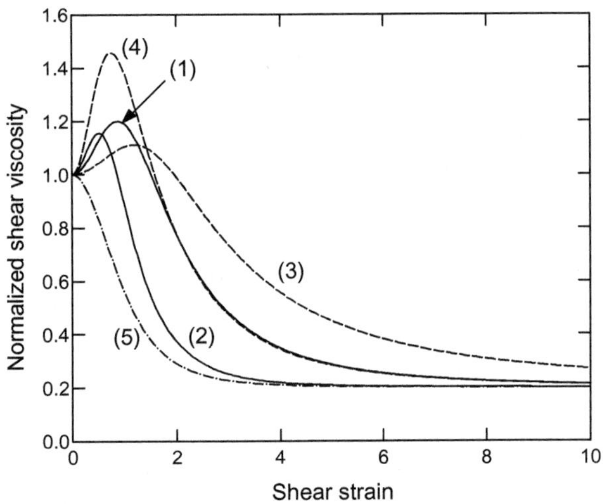

Fig. 7.6 Evolution of the normalized shear viscosity μ_{13}/μ_0 with the strain κ in simple shear started from an isotropic state ($\lambda_3 = 1$), for different response functions $f(b_r)$ (cold ice)

reached. Recall that the micro-mechanical models predict a very similar behaviour of ice, see the plot of μ_{13} in Fig. 6.10 on p. 193. The predictions of the phenomenological orthotropic flow law are also consistent with the results of micro-macroscopic model (Staroszczyk and Gagliardini 1999).

Corresponding to Fig. 7.6 are the plots in Fig. 7.7, displaying the behaviour of ice in simple shearing started from different anisotropic states induced in the plane flow $\lambda_2 = 1$ by performing the initial pre-compression along the x_3-axis defined by the stretch $\lambda_3 \leq 1$. Presented in the figure are the results obtained for the response function (7.76) with the parameter $m = 1.5$, demonstrating the effect of the initial anisotropy on the shear response of ice.

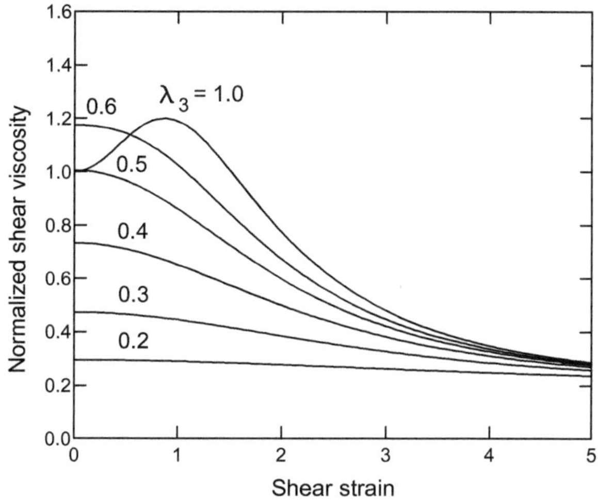

Fig. 7.7 Evolution of the normalized shear viscosity μ_{13}/μ_0 with the strain κ in simple shear started from different anisotropic states defined by the initial stretch λ_3 (cold ice)

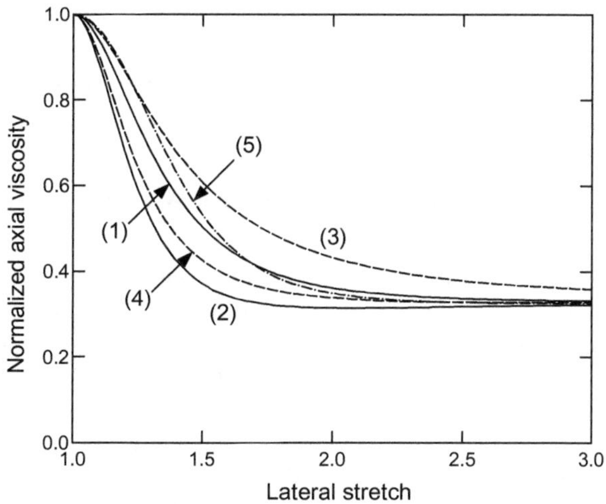

Fig. 7.8 Evolution of the normalized axial viscosity μ_{33}/μ_0 with the lateral stretch λ_1 in uniaxial compression for different response functions $f(b_r)$ (warm ice)

The results of simulations for warm ice, described by the enhancement factors and limit values of the response functions given by (7.75), are presented in Figs. 7.8 and 7.9. The plots illustrate the evolution of the dimensionless axial and shear viscosities with strains for the response functions plotted in Fig. 7.4. We observe that for warm ice, contrary to cold ice, the orthotropic constitutive model (7.43) predicts, for both uniaxial compression and simple shear deformations, progressive weakening of the ice creep response over the whole range of strains, with the normalized viscosities decreasing monotonically from unity in the initial isotropic state to the limit values, defined by enhancement factors E_a and E_s, for fully developed anisotropic fabrics. Such monotonic softening of warm ice, predicted by the model, is in good qualitative

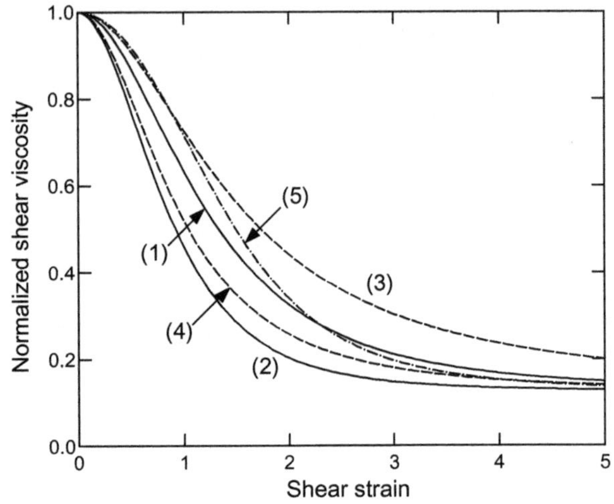

Fig. 7.9 Evolution of the normalized shear viscosity μ_{13}/μ_0 with the strain κ in simple shear started from an isotropic state ($\lambda_3 = 1$) for different response functions $f(b_r)$ (warm ice)

agreement with the experimental results reported by Jacka and Maccagnan (1984), Budd and Jacka (1989) and Treverrow et al. (2012).

7.4 Inverse Orthotropic Flow Law

In Sects. 7.2 and 7.3 the constitutive equations for polar ice treated as an orthotropic material are formulated, in which deviatoric stress is expressed in terms of strain-rate, strain and three structure tensors based on the current principal deformation axes. Such a stress–strain-rate form is the most convenient one for use in the momentum balance equations. However, a more common approach in theoretical glaciology is to express strain-rate in terms of deviatoric stress. For this reason, an inverse (that is, strain-rate–stress) form of the orthotropic flow law is discussed in this section, as it is possible that such a form will reveal different ice creep response features than the direct stress–strain-rate prescription, which can help improve the constitutive model correlations with experimental data.

The method of derivation of the strain-rate–stress form of the orthotropic law (Staroszczyk 2001) is analogous to that described in Sects. 7.2 and 7.3. Again, the constitutive model is obtained from the general, frame-indifferent tensor representation for orthotropic materials, which is subsequently reduced to a form which retains only those tensor generators that can be detected by measuring directional viscosities in simple shear tests conducted in different shear planes. The reduced constitutive equation involves again two material response functions with dependence on the principal stretches and an invariant measure of current deformation. These response functions are constructed by matching the model predictions with the observed behaviour of ice at large strains. The model is then used to illustrate the evolution of the creep response of initially isotropic ice during sustained uniaxial compression and simple shearing.

7.4.1 General Strain-Rate–Stress Formulation

As earlier in Sect. 7.2, the orthotropic constitutive law relating the strain-rate tensor $\boldsymbol{D}$ to the deviatoric Cauchy stress tensor $\boldsymbol{S}$ and the left Cauchy-Green deformation tensor $\boldsymbol{B}$ is derived from the general frame-indifferent representation given by Eqs. (B.10) and (B.11) (on p. 325) in Appendix B. Hence, we have

$$
\begin{aligned}
\boldsymbol{D} = \sum_{r=1}^{3} \left[\phi_r \boldsymbol{M}^{(r)} + \phi_{r+3}\left(\boldsymbol{M}^{(r)}\boldsymbol{S} + \boldsymbol{S}\boldsymbol{M}^{(r)}\right) + \phi_{r+6}\left(\boldsymbol{M}^{(r)}\boldsymbol{B} + \boldsymbol{B}\boldsymbol{M}^{(r)}\right)\right] + \\
+ \ \phi_{10}\boldsymbol{S}^2 + \phi_{11}\boldsymbol{B}^2 + \phi_{12}\left(\boldsymbol{B}\boldsymbol{S} + \boldsymbol{S}\boldsymbol{B}\right),
\end{aligned}
\tag{7.79}
$$

where the structure tensors $\boldsymbol{M}^{(r)}$ $(r = 1, 2, 3)$ are defined by relation $(7.8)_2$, and the 12 response coefficients ϕ_i $(i = 1, \ldots, 12)$ are functions of the 19 invariants J_k $(k = 1, \ldots, 19)$ of the tensors $\boldsymbol{M}^{(r)}$, $\boldsymbol{S}$ and $\boldsymbol{B}$ and their products:

$$
\begin{aligned}
&J_r = \operatorname{tr} \boldsymbol{M}^{(r)}\boldsymbol{S}, \quad J_{r+3} = \operatorname{tr} \boldsymbol{M}^{(r)}\boldsymbol{B}, \quad J_{r+6} = \operatorname{tr} \boldsymbol{M}^{(r)}\boldsymbol{S}^2, \\
&J_{r+9} = \operatorname{tr} \boldsymbol{M}^{(r)}\boldsymbol{B}^2, \quad J_{r+12} = \operatorname{tr} \boldsymbol{M}^{(r)}\boldsymbol{B}\boldsymbol{S} \quad (r = 1, 2, 3), \\
&J_{16} = \operatorname{tr} \boldsymbol{B}\boldsymbol{S}^2, \quad J_{17} = \operatorname{tr} \boldsymbol{B}^2\boldsymbol{S}, \quad J_{18} = \det \boldsymbol{S}, \quad J_{19} = \det \boldsymbol{B}.
\end{aligned}
\tag{7.80}
$$

Due to the ice incompressibility assumption, expressed by the conditions $\operatorname{tr} \boldsymbol{D} = 0$ and $\det \boldsymbol{B} = \det \boldsymbol{F} = 1$, only 11 response coefficients ϕ_i are independent, and only 18 invariants are non-trivial. It is assumed that the strain-rate $\boldsymbol{D}$ vanishes when the deviatoric stress $\boldsymbol{S}$ vanishes. Thus, we require that the coefficients $\phi_1, \phi_2, \phi_3, \phi_7, \phi_8, \phi_9, \phi_{11}$ vanish when $\boldsymbol{S} = \boldsymbol{O}$; that is, when the invariants $J_1, J_2, J_3, J_7, J_8, J_9, J_{13}, J_{14}, J_{15}, J_{16}, J_{17}, J_{18}$ are all equal to zero.

In order to simplify the general orthotropic flow law (7.79), including 11 independent response coefficients and 18 invariants as their arguments, the same approach is followed as for the direct, stress–strain-rate formulation considered in Sect. 7.2, which is based on the method of detection of instantaneous directional viscosities that can be measured in a series of simple shear tests. Accordingly, the flow law equation (7.79) is set in the reference frame, the axes of which coincide with the principal stretch axes, so that the current deformation field is described by the axial stretches λ_1, λ_2 and λ_3. In this frame, after removing the stress and strain-rate, so that the fabric defined by the current stretches is 'frozen', simple shearing by an infinitesimal strain κ in the direction of, say, x_i-axis on the plane normal to the x_j axis is performed. For this deformation field, defined by (7.20) on p. 225, the tensor generators in the general flow law (7.79) have, for $i \neq j$, the following instantaneous (ij) components, equal to the (ji) components:

$$\left(\boldsymbol{M}^{(r)}\boldsymbol{S}+\boldsymbol{S}\boldsymbol{M}^{(r)}\right)_{ij}=\begin{cases}S_{ij} & (r=i \text{ or } r=j)\\ 0 & (r\neq i \text{ and } r\neq j)\end{cases}, \tag{7.81}$$

$$\left(\boldsymbol{M}^{(r)}\boldsymbol{B}+\boldsymbol{B}\boldsymbol{M}^{(r)}\right)_{ij}=0, \quad \left(\boldsymbol{S}^2\right)_{ij}=0, \quad \left(\boldsymbol{B}^2\right)_{ij}=0, \tag{7.82}$$

$$(\boldsymbol{B}\boldsymbol{S}+\boldsymbol{S}\boldsymbol{B})_{ij}=(b_i+b_j)S_{ij}\,. \tag{7.83}$$

There are non-zero diagonal components of the instantaneous strain-rate $\boldsymbol{D}$, but these are of no interest at this point, since they cannot be detected in the simple shear tests. It is clear from the above relations that the only tensor generators which give non-zero off-diagonal components of the strain-rate tensor $\boldsymbol{D}$ are those appearing in (7.81) and (7.83). This implies that only four response coefficients: ϕ_4, ϕ_5, ϕ_6 and ϕ_{12} can be detected by instantaneous shear response of the orthotropic material. This is the same result as that deduced for the stress–strain-rate formulation discussed in Sect. 7.2. Accordingly, only the four above-indicated response coefficients are retained for further considerations.

7.4.2 Reduced Flow Law

Since both tensors $\boldsymbol{D}$ and $\boldsymbol{S}$ entering the direct and the inverse formulations of the orthotropic constitutive law are traceless, and both formulations contain the same four response functions ϕ_4, ϕ_5, ϕ_6 and ϕ_{12}, the process of the derivation of the reduced form of the inverse law essentially repeats, with small exceptions, the main steps described previously in Sect. 7.3 devoted to the direct law formulation. For this reason, most of the details involved in the derivation of the inverse (strain-rate–stress) form are be omitted here (these details can be found in the papers by Staroszczyk 2001, 2004).

Hence, the following form of the reduced orthotropic constitutive relation is considered:

$$\begin{aligned}\boldsymbol{D}=\sum_{r=1}^{3}\phi_{r+3}\left[\boldsymbol{M}^{(r)}\boldsymbol{S}+\boldsymbol{S}\boldsymbol{M}^{(r)}-\tfrac{2}{3}\mathrm{tr}\,(\boldsymbol{M}^{(r)}\boldsymbol{S})\boldsymbol{I}\right]+\\ +\;\phi_{12}\left[\boldsymbol{B}\boldsymbol{S}+\boldsymbol{S}\boldsymbol{B}-\tfrac{2}{3}\mathrm{tr}\,(\boldsymbol{B}\boldsymbol{S})\boldsymbol{I}\right],\end{aligned} \tag{7.84}$$

where the included scalars are expressed in terms of the invariants (7.80) by $\mathrm{tr}\,(\boldsymbol{M}^{(r)}\boldsymbol{S})=J_r$ $(r=1,2,3)$ and $\mathrm{tr}\,(\boldsymbol{B}\boldsymbol{S})=J_{13}+J_{14}+J_{15}$. By assuming again a separable dependence which factors out the invariants depending only on the deformation $\boldsymbol{B}$ and retains a common dependence on the invariants involving the stress $\boldsymbol{S}$, it can be concluded that the four fabric response functions have the forms

$$\begin{aligned}\phi_{r+3}\,(b_r,\mathrm{tr}\,\boldsymbol{B},\mathrm{tr}\,\boldsymbol{S}^2)&=\tfrac{1}{4}\,\eta_0\,(\mathrm{tr}\,\boldsymbol{S}^2)\,\hat{f}(b_r)\\ \phi_{12}\,(b_r,\mathrm{tr}\,\boldsymbol{B},\mathrm{tr}\,\boldsymbol{S}^2)&=\tfrac{1}{4}\,\eta_0\,(\mathrm{tr}\,\boldsymbol{S}^2)\,\hat{g}(\mathrm{tr}\,\boldsymbol{B})\end{aligned}\qquad (r=1,2,3), \tag{7.85}$$

where $\hat{f}(b_r)$ and $\hat{g}(\text{tr}\,\boldsymbol{B})$ are single-valued response functions, and $\eta_0\,(\text{tr}\,\boldsymbol{S}^2)$ is the fluidity of isotropic ice at a given magnitude of the deviatoric stress invariant $\boldsymbol{S}^2$. The factor $\frac{1}{4}$ in (7.85) is introduced to yield the standard form of the linearly viscous fluid flow law $\boldsymbol{S} = 2\mu_0\boldsymbol{D}$ ($\mu_0 = 1/\eta_0$) for isotropic ice (for which $\boldsymbol{B} = \boldsymbol{I}$ and $\text{tr}\,\boldsymbol{B} = 3$). The functions $\hat{f}$ and $\hat{g}$ are normalized by the relation

$$\hat{f}(1) + \hat{g}(3) = 1, \tag{7.86}$$

analogous to $(7.37)_3$. In view of (7.85), the instantaneous shear fluidity relation is defined by

$$\eta_{ij} = \tfrac{1}{2}\,\eta_0\left[\hat{f}(b_i) + \hat{f}(b_j) + (b_i + b_j)\,\hat{g}(K)\right] \quad (i \neq j), \tag{7.87}$$

where η_0 is the isotropic ice fluidity (reciprocal of the viscosity μ_0). Since η_{ij} must be bounded for the axial stretches $b_r \to \infty$, we rewrite the function $\hat{g}$ and the normalization condition (7.86) as

$$\hat{g}(K) = K^{-1}\hat{G}(K), \quad \hat{f}(1) + \tfrac{1}{3}\,\hat{G}(3) = 1, \tag{7.88}$$

where $\hat{G}$ is bounded and $K = \text{tr}\,\boldsymbol{B} = b_1 + b_2 + b_3$. Thus, the instantaneous fluidity relation (7.87) becomes

$$\eta_{ij} = \tfrac{1}{2}\,\eta_0\left[\hat{f}(b_i) + \hat{f}(b_j) + (b_i + b_j)\,K^{-1}\hat{G}(K)\right], \tag{7.89}$$

and, on account of (7.85) and $(7.88)_1$, the reduced inverse orthotropic flow law (7.84) takes the form

$$\begin{aligned} \boldsymbol{D} = \frac{\eta_0}{4}\Big\{ &\sum_{r=1}^{3} \hat{f}(b_r)\left[\boldsymbol{M}^{(r)}\boldsymbol{S} + \boldsymbol{S}\boldsymbol{M}^{(r)} - \tfrac{2}{3}\text{tr}\,(\boldsymbol{M}^{(r)}\boldsymbol{S})\boldsymbol{I}\right] + \\ &+ K^{-1}\hat{G}(K)\left[\boldsymbol{B}\boldsymbol{S} + \boldsymbol{S}\boldsymbol{B} - \tfrac{2}{3}\text{tr}\,(\boldsymbol{B}\boldsymbol{S})\boldsymbol{I}\right]\Big\}. \end{aligned} \tag{7.90}$$

Application of the reduced constitutive law (7.90) to uniaxial and simple shear flows yields the instantaneous axial and shear viscosities as functions of current deformations. For indefinitely large strains, these viscosity relations enable correlations of the model predictions with the observed creep behaviour of ice, in a way analogous to that already followed in Sect. 7.3. In the case of unconfined uniaxial compression flow started from an initially isotropic state, with the deformation field defined by relations (7.17) and the components of left Cauchy-Green deformation, strain-rate and deviatoric stress tensors given by (7.50) and (7.51) on p. 231, the flow law (7.90) determines the viscous response of ice under compression as

$$\frac{2\mu_0 D_{33}}{S_{33}} = \frac{1}{3}\left[\hat{f}(b_1) + 2\hat{f}(b_1^{-2}) + \frac{\hat{G}(K)}{K}(b_1 + 2b_1^{-2})\right] = \frac{\eta_{33}}{\eta_0}, \tag{7.91}$$

where η_{33}/η_0 defines the ratio of the fabric induced axial fluidity to isotropic ice fluidity. The limit value of this ratio for indefinitely large axial compression, when $b_1 = \lambda_1^2 \to \infty$ and then $K \sim 2b_1$, is defined by

$$\frac{\eta_{33}}{\eta_0} \to \frac{2}{3}\hat{f}(0) + \frac{1}{3}\hat{f}(\infty) + \frac{1}{6}\hat{G}(\infty) = E_a\,, \tag{7.92}$$

where E_a in the axial enhancement factor.

The analysis of the simple shear flow configuration, with the deformation field defined by relations (7.54) and the Cauchy-Green deformation and the strain-rate tensor components given by (7.55) on p. 232, is more complex than that in the case of the direct flow law formulation investigated in Sect. 7.3. The reason for this is the structure of the stress tensor $\boldsymbol{S}$, which in the simple shear flow has not only off-diagonal non-zero components S_{ij} $(i \neq j)$, but, due to the Poynting effect, has also the diagonal non-vanishing components S_{ii}. Hence, for simple shearing assumed to occur in the Ox_1x_3 plane, the stress tensor has the components

$$\boldsymbol{S} = \begin{pmatrix} S_{11} & 0 & S_{13} \\ 0 & S_{22} & 0 \\ S_{13} & 0 & S_{33} \end{pmatrix}, \quad \operatorname{tr}\boldsymbol{S} = S_{11} + S_{22} + S_{33} = 0. \tag{7.93}$$

With the tensor $\boldsymbol{S}$ given by (7.93), the flow law (7.90) expresses the strain-rate component D_{13} in terms of the three stress components: S_{13}, S_{11} and S_{33} (recall, that in the direct stress–strain-rate formulation of the flow law, a single stress component S_{13} is expressed in terms of only one strain-rate component D_{13}, which makes the calculation of the instantaneous shear viscosity μ_{13} straightforward, see relation (7.60)). Therefore, in order to express the shear strain-rate in terms of the corresponding shear stress alone in the case of the strain-rate–stress formulation of the constitutive law, all three stress components S_{13}, S_{11} and S_{33}, required to enforce simple shear flow, must be calculated. This can be accomplished by solving the following three equations for the three strain-rates D_{13}, D_{11} and D_{33} in terms of the three deviatoric stress components S_{13}, S_{11} and S_{33} (Staroszczyk 2001):

$$\begin{aligned} 2\mu_0 D_{13} =& \frac{1}{2}(S_{11} + S_{33})\left[\hat{f}(b_1)e_1^{(1)}e_3^{(1)} + \hat{f}(b_1^{-1})e_1^{(3)}e_3^{(3)} + \frac{\hat{G}(K)}{K}\lambda^{-1}\kappa\right] \\ &+ \frac{1}{2}S_{13}\left[\hat{f}(b_1) + \hat{f}(b_1^{-1}) + \frac{\hat{G}(K)}{K}(\lambda_1^2 + \lambda_1^{-2} + \kappa^2)\right], \end{aligned} \tag{7.94}$$

$$
\begin{aligned}
2\mu_0 D_{11} = &\frac{1}{3}S_{11}\left[2\hat{f}(b_1)e_1^{(1)}e_1^{(1)} + \hat{f}(1) + 2\hat{f}(b_1^{-1})e_1^{(3)}e_1^{(3)} + \frac{\hat{G}(K)}{K}(2\lambda_1^2 + 2\kappa^2 + 1)\right] \\
&+ \frac{1}{3}S_{33}\left[-\hat{f}(b_1)e_3^{(1)}e_3^{(1)} + \hat{f}(1) - \hat{f}(b_1^{-1})e_3^{(3)}e_3^{(3)} - \frac{\hat{G}(K)}{K}(\lambda_1^{-2} - 1)\right] \\
&+ \frac{1}{3}S_{13}\left[\hat{f}(b_1)e_1^{(1)}e_3^{(1)} + \hat{f}(b_1^{-1})e_1^{(3)}e_3^{(3)} + \frac{\hat{G}(K)}{K}\lambda^{-1}\kappa\right],
\end{aligned}
\tag{7.95}
$$

$$
\begin{aligned}
2\mu_0 D_{33} = &\frac{1}{3}S_{11}\left[-\hat{f}(b_1)e_1^{(1)}e_1^{(1)} + \hat{f}(1) - \hat{f}(b_1^{-1})e_1^{(3)}e_1^{(3)} - \frac{\hat{G}(K)}{K}(\lambda_1^2 + \kappa^2 - 1)\right] \\
&+ \frac{1}{3}S_{33}\left[2\hat{f}(b_1)e_3^{(1)}e_3^{(1)} + \hat{f}(1) + 2\hat{f}(b_1^{-1})e_3^{(3)}e_3^{(3)} + \frac{\hat{G}(K)}{K}(2\lambda_1^{-2} + 1)\right] \\
&+ \frac{1}{3}S_{13}\left[\hat{f}(b_1)e_1^{(1)}e_3^{(1)} + \hat{f}(b_1^{-1})e_1^{(3)}e_3^{(3)} + \frac{\hat{G}(K)}{K}\lambda^{-1}\kappa\right].
\end{aligned}
\tag{7.96}
$$

By setting to zero the axial strain-rates D_{11} and D_{33} in (7.95) and (7.96) (since in simple shear flow all axial strain-rates are zero), the stress components S_{11} and S_{33} can be eliminated to relate D_{13} in terms of only S_{13}, and thus to provide an expression for the instantaneous shear viscosity μ_{13}. The resulting relation, expressed in the form $2\mu_0 D_{13}/S_{13} = [\dots] = \eta_{13}/\eta_0$, analogous to (7.92), describes the evolution of the normalized shear fluidity in terms of the shear strain κ. In the limit, as $\kappa \to \infty$ with λ_1 finite, then $b_1 \sim \kappa^2$ and $K \sim \kappa^2$, and, further, $\boldsymbol{e}^{(1)} \to (1, 0, 0)^T$ and $\boldsymbol{e}^{(3)} \to (0, 0, 1)^T$, Eq. (7.94) implies that

$$
\frac{\eta_{13}}{\eta_0} \to \frac{1}{2}\hat{f}(0) + \frac{1}{2}\hat{f}(\infty) + \frac{1}{2}\hat{G}(\infty) = E_s\,, \tag{7.97}
$$

where E_s is the shear enhancement factor.

The two viscosity relations (7.92) and (7.97) express the three limit values of the response functions $\hat{f}(0)$, $\hat{f}(\infty)$ and $\hat{G}(\infty)$ in terms of the two enhancement factors for compression and shear. In order to derive a third relation necessary for the unique determination of the above three limit values, the approach described in Sect. 7.3.2 is followed, and, in particular, use is made of Eq. (7.66) on p. 235 for the shear viscosities in the plane flow defined by $\lambda_2 = b_2 = 1$ and $b_3 = b_1^{-1}$. Hence, bearing in mind that $\mu_{ij} = \eta_{ij}^{-1}$, Eqs. (7.66) (implying $\mu_{12} = \mu_{23}$) and (7.89) yield the relation

$$
\hat{G}(K) = -\frac{K}{b_1 - b_1^{-1}}\left[\hat{f}(b_1) - \hat{f}(b_1^{-1})\right], \quad K \geq 3, \tag{7.98}
$$

where b_1 is given in terms of K by (7.70). The limit of (7.98) as $b_1 \to 1$ and $K \to 3$, combined with the normalization $(7.88)_2$, gives

$$\hat{f}(1) - \hat{f}'(1) = 1, \tag{7.99}$$

which is a restriction on the function $\hat{f}(b_r)$ at $b_r = 1$. In turn, the limit of (7.98) as $b_1 \to \infty$, and then $K \sim b_1$, provides the relation

$$\hat{f}(0) - \hat{f}(\infty) - \hat{G}(\infty) = 0. \tag{7.100}$$

The system of three linear equations (7.92), (7.97) and (7.100) has the solution

$$\hat{f}(0) = E_s\,, \quad \hat{f}(\infty) = 6E_a - 5E_s\,, \quad \hat{G}(\infty) = 6\,(E_s - E_a), \tag{7.101}$$

which defines the limit properties of the response coefficients $\hat{f}$ and $\hat{G}$. For the adopted enhancement factors, $E_a = \frac{1}{3}$ and $E_s = 5$ for cold ice and $E_a = 3$ and $E_s = 8$ for warm ice, the limit values of the fabric response coefficients $\hat{f}$ and $\hat{G}$, as prescribed by (7.101), are given, respectively, by

$$E_a = \tfrac{1}{3}, \qquad E_s = 5\,: \quad \hat{f}(0) = 5, \qquad \hat{f}(\infty) = -23, \qquad \hat{G}(\infty) = 28, \tag{7.102}$$

$$E_a = 3, \qquad E_s = 8\,: \quad \hat{f}(0) = 8, \qquad \hat{f}(\infty) = -22, \qquad \hat{G}(\infty) = 30. \tag{7.103}$$

It can be noted that $\hat{f}(\infty) < \hat{f}(0)$ for both types of ice, in contrast to the direct model for which $f(\infty) > f(0)$, see definitions (7.74) and (7.75) on p. 236. Therefore, for illustration purposes, the following monotonic decreasing response functions $\hat{f}$, of the forms analogous to those previously used for the direct model, are adopted:

$$\hat{f}(b_r) = \hat{f}(\infty) - [\hat{f}(\infty) - \hat{f}(0)]\exp(-\alpha b_r^m), \qquad \alpha > 0, \quad m > 0, \tag{7.104}$$

$$\hat{f}(b_r) = \hat{f}(0) + [\hat{f}(\infty) - \hat{f}(0)]\tanh(\alpha b_r^m), \qquad \alpha > 0, \quad m > 0, \tag{7.105}$$

$$\hat{f}(b_r) = \hat{f}(\infty) - [\hat{f}(\infty) - \hat{f}(0)]\frac{\alpha}{\alpha + b^m}, \qquad \alpha > 0, \quad m > 0, \tag{7.106}$$

where m is a free parameter, and α is determined by (7.99). The other response function, $\hat{G}(K)$, is related to $\hat{f}(b_r)$ by (7.98).

7.4.3 *Simple Flow Simulations*

The creep behaviour of cold ice predicted by the inverse orthotropic constitutive model (7.90), with the response coefficients defined by Eqs. (7.104)–(7.106), is illustrated in Figs. 7.10 and 7.11. The response of ice to uniaxial compression, described by (7.91), is presented in Fig. 7.10, showing the evolution of the normalized axial viscosity $\mu_{33}/\mu_0 = \eta_0/\eta_{33}$ with increasing lateral stretch λ_1 for different forms of the response functions. The curves plotted in the figure have been obtained for the response function (7.104) with $m = 1$ and $m = 2$, the labels (1) and (2), respectively,

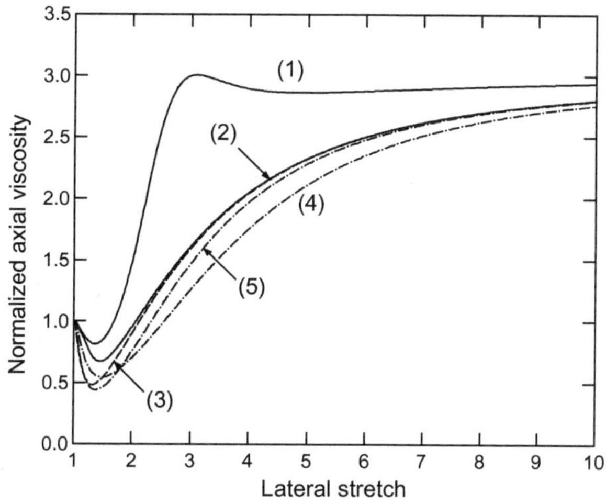

Fig. 7.10 Evolution of the normalized axial viscosity μ_{33}/μ_0 with the lateral stretch λ_1 in uniaxial compression for different response functions $\hat{f}(b_r)$ (cold ice)

the function (7.105) with $m = 2$, the label (3), and the function (7.106) with $m = 1.5$ and $m = 2$, the labels (4) and (5), respectively. It is seen that, with exception of the response function represented by the curve (1), all the functions predict qualitatively similar behaviour of ice. By comparing the results shown in Fig. 7.10 with those in Fig. 7.5 one can note two features. First, the inverse flow law predicts initial softening of cold ice under compression for all adopted forms of the response functions, whereas the direct formulation yields such a behaviour only for the function of the type (7.106). And second, the limit axial viscosities are now reached at significantly lower rates than in the case of the stress–strain-rate prescription.

The viscous response of cold ice to simple shearing started from an initially isotropic state ($\lambda_1 = \lambda_3 = 1$), described by Eqs. (7.94)–(7.96), is illustrated in Fig. 7.11. The plots show the variation of the dimensionless shear viscosity μ_{13}/μ_0 with the shear strain κ for the response functions (7.104)–(7.106); the same labelling as in the previous figure applies. Contrary to the corresponding results given by the direct constitutive model, see Fig. 7.6, which predicts the initial hardening of cold ice during its shearing for the fabric response functions of the types (7.104) and (7.105), the inverse formulation predicts monotonic decrease in the shear viscosity for all the functions used in the simulations. It is also observed that the limit viscosities given by the inverse model are approached at much smaller strains κ than those predicted by the direct model simulations.

The results of the simulations carried out for warm ice are presented in Figs. 7.12 and 7.13. In these figures, the curves labelled (1) and (2) have been obtained for the function (7.104) with $m = 1$ and $m = 1.5$, respectively, the curves (3) and (4) correspond to (7.105) with $m = 1$ and $m = 1.5$, respectively, and the curve (5) corresponds to the function (7.106) with $m = 1$. Comparing the evolution of the normalized viscosities with increasing strain in uniaxial compression (Fig. 7.12) and simple shear (Fig. 7.13) with the corresponding results given by the stress–strain-rate formula-

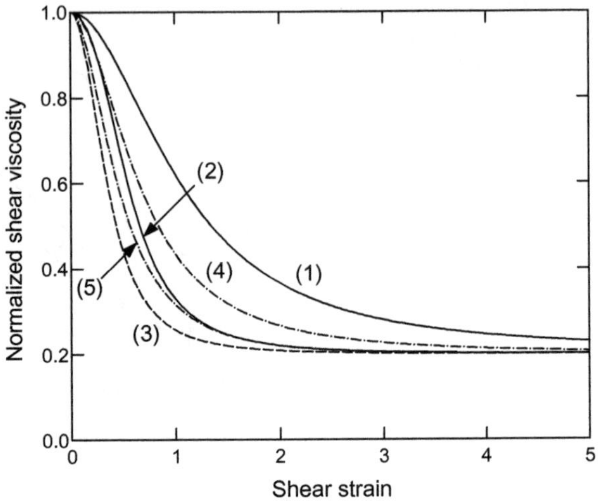

Fig. 7.11 Evolution of the normalized shear viscosity μ_{13}/μ_0 with the strain κ in simple shear started from an isotropic state ($\lambda_1 = \lambda_3 = 1$) for different response functions $\hat{f}(b_r)$ (cold ice)

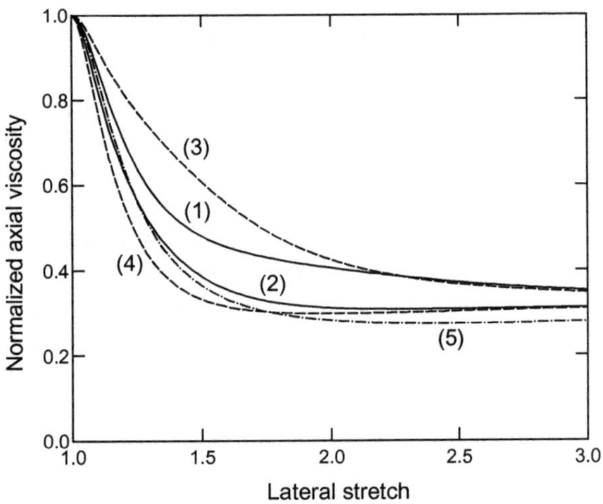

Fig. 7.12 Evolution of the normalized axial viscosity μ_{33}/μ_0 with the lateral stretch λ_1 in uniaxial compression for different response functions $\hat{f}(b_r)$ (warm ice)

tion, see Figs. 7.8 and 7.9, we note that both, the inverse and the direct, constitutive models yield qualitatively similar behaviour of warm ice, with monotonic softening of the material with increasing deformations, irrespective of the adopted forms of the response functions. It is also seen that for both warm and cold ice the inverse model predicts higher rates of ice softening with increasing deformation than the direct model considered in Sects. 7.2 and 7.3.

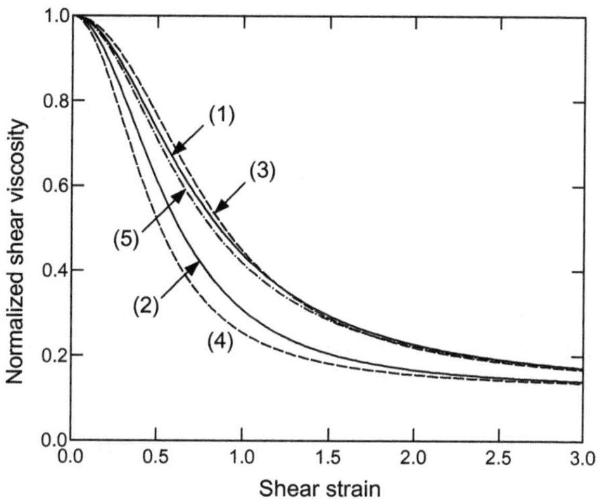

Fig. 7.13 Evolution of the normalized shear viscosity μ_{13}/μ_0 with the strain κ in simple shear started from an isotropic state ($\lambda_3 = 1$) for different response functions $\hat{f}(b_r)$ (warm ice)

7.5 Additive Forms of Constitutive Laws

In the previous sections of this chapter two versions of the orthotropic constitutive laws are formulated to describe the mechanism of induced anisotropy of polar ice arising from the rotation of crystal c-axes during creep deformation of the material. In the first version of the law, called the direct formulation, the deviatoric stress is expressed in terms of the strain-rate, strain and three structure tensors based on the principal stretch axes (see Sects. 7.2 and 7.3). In the analogous second version of the flow law, called the inverse formulation, the strain-rate is expressed in terms of the stress, strain and three structure tensors (see Sect. 7.4). In both formulations, the constitutive equations can be expressed by similar general representations

$$\boldsymbol{S} = \boldsymbol{H}(\boldsymbol{D}, \boldsymbol{B}, \boldsymbol{M}^{(r)}), \quad \boldsymbol{D} = \hat{\boldsymbol{H}}(\boldsymbol{S}, \boldsymbol{B}, \boldsymbol{M}^{(r)}) \quad (r = 1, 2, 3), \tag{7.107}$$

where $\boldsymbol{H}$ and $\hat{\boldsymbol{H}}$ are tensor-valued functions defining the viscous response of the material. The above representations are of a common form which henceforth will be called the *multiplicative form* of a constitutive law. This form was successfully used in numerical simulations to model polar ice sheet flows (Staroszczyk and Morland 1999, 2000b; Staroszczyk 2003). However, when the first representation in (7.107), with a relevant extension, was applied to model the process of dynamic recrystallization (considered further in this chapter in Sect. 7.6), a somewhat unexpected response to continuous shearing was predicted (Staroszczyk and Morland 2001). This prompted reformulation of the original orthotropic viscous law into an alternative *additive form*, in which the material response is decomposed into isotropic and anisotropic parts, with the latter describing the strength of macroscopic anisotropy developing in the material due to the formation and subsequent evolution of the ice fabric from its initially isotropic state. Thus, two alternative formulations of the flow laws, which seem to be conceptually simpler than the conventionally used multiplicative forms (7.107), are now considered:

$$\boldsymbol{S} = \boldsymbol{H}_I(\boldsymbol{D}) + \boldsymbol{H}_A(\boldsymbol{D}, \boldsymbol{B}, \boldsymbol{M}^{(r)}), \tag{7.108}$$

$$\boldsymbol{D} = \hat{\boldsymbol{H}}_I(\boldsymbol{S}) + \hat{\boldsymbol{H}}_A(\boldsymbol{S}, \boldsymbol{B}, \boldsymbol{M}^{(r)}), \tag{7.109}$$

where $\boldsymbol{H}_I$ and $\boldsymbol{H}_A$ define, respectively, the isotropic and the additional anisotropic response in the direct (stress–strain-rate) formulation, and $\hat{\boldsymbol{H}}_I$ and $\hat{\boldsymbol{H}}_A$ define, respectively, the isotropic and anisotropic responses in the inverse (strain-rate–stress formulation). The vanishing of the anisotropic parts H_A and $\hat{H}_A$ in an isotropic state $\boldsymbol{B} = \boldsymbol{I}$ requires that

$$\boldsymbol{H}_A(\boldsymbol{D}, \boldsymbol{I}, \boldsymbol{M}^{(r)}) = \boldsymbol{O}, \quad \hat{\boldsymbol{H}}_A(\boldsymbol{S}, \boldsymbol{I}, \boldsymbol{M}^{(r)}) = \boldsymbol{O}. \tag{7.110}$$

The additive representations (7.108) and (7.109) are used in this section to construct the orthotropic viscous flow laws by following the same method, based on the concept of instantaneous directional shear viscosities, as that applied in the derivation of the multiplicative laws in Sects. 7.2–7.4. Compared to the constitutive equations developed in the preceding sections, different deformation invariants as arguments of the response coefficients are used in the additive formulations. By making use of one of the viscosity relations discussed in Sect. 7.3.2, an explicit algebraic relation which connects two material response coefficients is derived, thus enabling the simplification of the constitutive description to a single fabric response function. The new constitutive laws are applied to investigate uniaxial and simple shear responses, and it turns out that, for both stress (direct) and strain-rate (inverse) formulations, a given shear response can be satisfied identically by relating the odd part of the single fabric response function to the even part, or to the full, of the fabric response function (Morland and Staroszczyk 2003a, b). Specific forms of the new response functions are determined by correlations with complete idealized uniaxial compression and simple shear responses. These new functions are then used to illustrate the evolution of the axial and shear viscosities with increasing deformations in simple flow configurations.

7.5.1 Stress and Strain-Rate Formulations

A new feature, compared to the constitutive models discussed in Sects. 7.2–7.4, is the choice of new invariant arguments for the fabric response coefficients. The new arguments, ξ_r and ζ, are related to the original invariant arguments, b_r and K, as follows:

$$\xi_r = b_r - 1/b_r \ \ (r = 1, 2, 3), \quad \zeta = \sqrt{(K-1)^2 - 4}, \quad K = b_1 + b_2 + b_3, \tag{7.111}$$

where ξ_r replace b_r and ζ replaces K. Henceforth, omitting the subscript r for common relations, the new invariants have the properties

$$\xi \to -\infty \ \text{as} \ b \to 0, \quad \xi = 0 \ \text{at} \ b = 1, \quad \xi \to \infty \ \text{as} \ b \to \infty, \tag{7.112}$$

and

$$\zeta \geq 0, \quad \zeta = 0 \ \text{at} \ K = 3, \quad \zeta \sim K \to \infty \ \text{as} \ K \to \infty. \tag{7.113}$$

Without loss of generality, we adopt the ordering

$$\xi_1 \geq \xi_2 \geq \xi_3 \,. \tag{7.114}$$

The additive representation (7.108) for the stress–strain-rate formulation is adopted in the form

$$\begin{aligned} \boldsymbol{S} = 2\mu_0 \Big\{ & \boldsymbol{D} + \sum_{r=1}^{3} \bar{f}(\xi_r) \left[\boldsymbol{M}^{(r)}\boldsymbol{D} + \boldsymbol{D}\boldsymbol{M}^{(r)} - \tfrac{2}{3}\text{tr}\,(\boldsymbol{M}^{(r)}\boldsymbol{D})\boldsymbol{I} \right] + \\ & + \bar{g}(\zeta) \left[\boldsymbol{B}\boldsymbol{D} + \boldsymbol{D}\boldsymbol{B} - \tfrac{2}{3}\text{tr}\,(\boldsymbol{B}\boldsymbol{D})\boldsymbol{I} \right] \Big\}, \end{aligned} \tag{7.115}$$

where the isotropic part of the viscous response is determined by the first term in the braces, while the anisotropic part is defined by the second and third terms in the braces. As before, μ_0 in (7.115) is the isotropic ice viscosity, which is a function of ice temperature and either the deviatoric stress or strain-rate invariant (see definitions (7.44)–(7.46) in Sect. 7.3). The functions $\bar{f}(\xi_r)$ and $\bar{g}(\zeta)$ are the new fabric response coefficients for the stress formulation, analogous to the functions $f(b_r)$ and $g(K)$ used in Sect. 7.3. The vanishing of the anisotropic part in the initial undeformed state $\boldsymbol{B} = \boldsymbol{I}$ yields the normalization condition

$$\bar{f}(0) + \bar{g}(0) = 0. \tag{7.116}$$

The requirement of the bounded response as $b_1 \to \infty$, and hence both K and $\zeta \to \infty$, implies $\bar{g}(\zeta) \sim \zeta^{-1}$ as $\zeta \to \infty$. Therefore, it is convenient to introduce an alternative response function $\bar{G}(\zeta)$ defined by

$$\bar{G}(\zeta) = \zeta \bar{g}(\zeta), \quad \bar{G}(\zeta) \sim \bar{g}(0)\zeta = -\bar{f}(0)\zeta \ \text{as} \ \zeta \to 0, \tag{7.117}$$

where $\bar{G}(\zeta)$ is finite and non-zero as $\zeta \to \infty$.

The additive form of the constitutive law for the strain-rate–stress formulation (7.109) is, in turn, expressed by the equation

$$D = \frac{1}{2\mu_0} \left\{ S + \sum_{r=1}^{3} \check{f}(\xi_r) \left[M^{(r)} S + S M^{(r)} - \tfrac{2}{3} \mathrm{tr}\,(M^{(r)} S) I \right] + \right.$$
$$\left. + \check{g}(\zeta) \left[BS + SB - \tfrac{2}{3} \mathrm{tr}\,(BS) I \right] \right\}. \tag{7.118}$$

The functions $\check{f}(\xi_r)$ and $\check{g}(\zeta)$ are the new fabric response coefficients for the strain-rate formulation. They depend on the same invariant arguments as $\bar{f}$ and $\bar{g}$, respectively, and are the analogues of the functions $\hat{f}(b_r)$ and $\hat{g}(K)$ used in Sect. 7.4. The vanishing of the anisotropic part in the relation (7.118) in the initial isotropic state $B = I$ gives the condition

$$\check{f}(0) + \check{g}(0) = 0. \tag{7.119}$$

Again, an alternative fabric response function $\check{G}(\zeta)$ is introduced

$$\check{G}(\zeta) = \zeta \check{g}(\zeta), \quad \check{G}(\zeta) \sim \check{g}(0)\zeta = -\check{f}(0)\zeta \ \text{ as } \ \zeta \to 0, \tag{7.120}$$

where $\check{G}(\zeta)$ is bounded and non-zero as $\zeta \to \infty$.

In the coordinate axes x_1, x_2 and x_3 coinciding with the current directions of the principal stretches λ_1, λ_2 and λ_3, the two constitutive laws (7.115) and (7.118) prescribe the instantaneous viscous responses as

$$S_{ij} = 2\mu_0 \left[1 + \bar{f}(\xi_i) + \bar{f}(\xi_j) + (b_i + b_j)\bar{g}(\zeta) \right] D_{ij}\,, \tag{7.121}$$

$$2\mu_0 D_{ij} = \left[1 + \check{f}(\xi_i) + \check{f}(\xi_j) + (b_i + b_j)\check{g}(\zeta) \right] S_{ij}\,. \tag{7.122}$$

Now, we apply again the concept of instantaneous directional shear viscosities described in Sect. 7.2. That is, let us remove the stress and strain-rate, so that the fabric is 'frozen' at given stretches λ_1, λ_2 and λ_3, and the deformation is given by relations (7.17) on p. 225. By performing a series of simple shear tests in the three planes Ox_ix_j $(i \neq j)$, with the new deformation defined by (7.20), the above Eqs. (7.121) and (7.122) determine the following instantaneous directional viscosities $\mu_{ij} = S_{ij}/(2D_{ij})$ $(i \neq j)$ predicted by the stress and strain-rate formulations, respectively:

$$\mu_{ij} = \mu_0 \left[1 + \bar{f}(\xi_i) + \bar{f}(\xi_j) + (b_i + b_j)\bar{g}(\zeta) \right], \tag{7.123}$$

$$\mu_{ij} = \mu_0 \left[1 + \check{f}(\xi_i) + \check{f}(\xi_j) + (b_i + b_j)\check{g}(\zeta) \right]^{-1}, \tag{7.124}$$

which are analogous to relations (7.41) and (7.89). Recall that the latter relation is expressed in terms of the fluidities η_{ij} and η_0 which are equal to the reciprocals of the corresponding viscosities μ_{ij} and μ_0.

Examination of the viscosity conditions (7.63)–(7.68) formulated in Sect. 7.3.2 shows that equality (7.63) for the isotropic state defined by $b_1 = b_2 = b_3 = 1$, implying $\xi_1 = \xi_2 = \xi_3 = 0$, follows immediately from both relations (7.123) and

(7.124), provided that the respective normalization conditions (7.116) and (7.119) are satisfied. The viscosity equalities (7.64) for $b_1 = b_2$, hence $\xi_1 = \xi_2$, and (7.68) for $b_2 = b_3$, hence $\xi_2 = \xi_3$, also follow from both expressions (7.123) and (7.124). Finally, the equality $\mu_{23} = \mu_{12}$ in (7.66) for the plane flow, when $b_2 = 1$, hence $b_3 = 1/b_1$, so that $\xi_2 = 0$, $\xi_3 = -\xi_1$ and $\zeta = \xi_1$, yields identical explicit expressions for $\bar{G}(\zeta)$ in terms of $\bar{f}(\zeta)$, and $\check{G}(\zeta)$ in terms of $\check{f}(\zeta)$; that is,

$$G(\zeta) = \zeta g(\zeta) = -f(\zeta) + f(-\zeta) = -2f^o(\zeta), \quad \zeta \geq 0, \tag{7.125}$$

where f, g and G denote either the over-bar or over-check functions, and $f(\xi)$ has the decomposition into even, f^e, and odd, f^o, parts defined by

$$f(\xi) = f^e(\xi) + f^o(\xi), \tag{7.126}$$

with the relations

$$2f^e(\xi) = f(\xi) + f(-\xi), \quad 2f^o(\xi) = f(\xi) - f(-\xi). \tag{7.127}$$

Accordingly, in terms of the new invariant arguments, in either formulation $g(\zeta)$ is determined explicitly in terms of $f^o(\xi)$, and each of the two additive constitutive laws can be expressed directly in terms of a single fabric response function $f(\xi)$. The limit of (7.125) as $b_1 \to 1$, $\zeta \to 0$, along with the normalization conditions (7.116) and (7.119), gives

$$G(0) = 0, \quad g(0) = -f(0) = -2f'(0), \tag{7.128}$$

which is a restriction on $f(\xi)$ at $\xi = 0$. The limit of (7.125) as $b_1 \to \infty$, $\zeta \sim b_1 \to \infty$, yields in turn

$$G(\infty) = -2f^o(\infty), \quad g(\infty) = 0. \tag{7.129}$$

The remaining two directional viscosity relations, (7.65) and (7.67), must be examined numerically for any specific forms of the fabric response coefficients in order to confirm the validity of viscous responses predicted by the constitutive theory.

7.5.2 *Material Response Functions*

A realistic constitutive model should predict observed responses of polar ice in any, physically feasible, flow configuration. This, however, is hardly viable at this stage, and it seems that the matching of uniaxial and shear responses for a wide range of strains is the most that can be achieved at present. Hence, we focus here on the determination of the fabric response coefficients which ensure the correct reproduction of the unconfined uniaxial and simple shear responses measured in experiments. As earlier in Sects. 7.3 and 7.4, the starting point is the derivation of the limit properties of the response functions at the initially isotropic state of the material, and at very large axial/shear strains. At the next stage, assuming certain forms of the response functions which satisfy the above limit conditions, correlations of the model results with complete (idealized) viscous responses to compression and shearing will be made for the stress–strain-rate formulation, in order to demonstrate the validity of the constitutive theory predictions.

The unconfined uniaxial compression flow field, assuming compression taking place along the x_3-axis, is defined by Eqs. (7.17)–(7.19) on p. 225. The additive flow law (7.115) then predicts the axial viscosity relation given by the stress formulation as

$$\tilde{\mu}_{33}(\xi_1) = \frac{S_{33}}{2\mu_0 D_{33}} = 1 + \frac{2}{3}\bar{f}(\xi_1) + \frac{4}{3}\bar{f}(\xi_3) + \frac{2}{3}(b_1 + 2b_1^{-2})\,\bar{g}(\zeta), \tag{7.130}$$

while the law (7.118) yields a similar relation for the strain-rate formulation

$$\tilde{\mu}_{33}(\xi_1) = \left[1 + \frac{2}{3}\check{f}(\xi_1) + \frac{4}{3}\check{f}(\xi_3) + \frac{2}{3}(b_1 + 2b_1^{-2})\,\check{g}(\zeta)\right]^{-1}, \tag{7.131}$$

where $\tilde{\mu}_{33} = \mu_{33}/\mu_0$ denotes the dimensionless axial viscosity. The arguments of the response functions in the above two expressions are

$$\begin{aligned} \xi_1 = b_1 - b_1^{-1} \geq 0, \quad \xi_3 = b_1^{-2} - b_1^2 \leq 0, \\ \zeta = \sqrt{(2b_1 + b_1^{-2} - 1)^2 - 4} \geq 0. \end{aligned} \tag{7.132}$$

In the limit of indefinitely large axial strains, when $b_1 \to \infty$, we have $\xi_1 \to \infty$, $\xi_3 \to -\infty$ and $\zeta \sim 2b_1 \to \infty$, so $\tilde{\mu}_{33}(\xi_1) \to E_a^{-1}$, where E_a is the enhancement factor for compression. Thus, eliminating $g(\zeta)$ by (7.125), Eqs. (7.130) and (7.131) supply

$$\tfrac{2}{3}\bar{f}(\infty) + \tfrac{4}{3}\bar{f}(-\infty) - \tfrac{2}{3}\bar{f}^o(\infty) = 2\bar{f}^e(\infty) - \tfrac{4}{3}\bar{f}^o(\infty) \quad = E_a^{-1} - 1, \tag{7.133}$$

$$\tfrac{2}{3}\check{f}(\infty) + \tfrac{4}{3}\check{f}(-\infty) - \tfrac{2}{3}\check{f}^o(\infty) = 2\check{f}^e(\infty) - \tfrac{4}{3}\check{f}^o(\infty) \quad = E_a - 1, \tag{7.134}$$

which are analogous relations for the limit values of the response coefficients in the stress and strain-rate formulations, respectively, with E_a^{-1} in the first expression replaced by E_a in the second expression.

In the simple shear flow, assumed to occur in the plane Ox_1x_3, the deformation field is described by Eq. (7.54) on p. 232, with the deformation and strain-rate tensors $\boldsymbol{B}$ and $\boldsymbol{D}$ defined by (7.55). For the sake of simplicity, it is assumed that the flow starts from the isotropic state $\lambda_1 = \lambda_2 = \lambda_3 = 1$ (in Sects. 7.3 and 7.4, for more generality, we had $\lambda_1 = \lambda_3^{-1} \neq 1$, $\lambda_2 = 1$). In terms of the strain κ measuring the shear deformation in the plane Ox_1x_3, the new response function arguments are defined by

$$\xi_1 = -\xi_3 = \zeta_s = \sqrt{(2+\kappa^2)^2 - 4}\,, \quad \xi_2 = 0. \tag{7.135}$$

With the above-defined variables, the stress formulation of the flow law (7.115) gives the following relation for the normalized shear viscosity:

$$\tilde{\mu}_{13}(\kappa) = \frac{S_{13}}{2\mu_0 D_{13}} = 1 + 2\left[\bar{f}^e(\zeta_s) - (2+\kappa^2)\,\bar{f}^o(\zeta_s)/\zeta_s\right], \tag{7.136}$$

with $\tilde{\mu}_{13}(0) = 1$ in the initial undeformed, and isotropic, state. In the limit of $\kappa \to \infty$, we have $\zeta_s \sim \kappa^2 \to \infty$, so $\tilde{\mu}_{13}(\kappa) \to E_s^{-1}$, where E_s is the shear enhancement factor. Thus, relation (7.136) yields

$$2\left[\bar{f}^e(\infty) - \bar{f}^o(\infty)\right] = E_s^{-1} - 1. \tag{7.137}$$

In the case of the strain-rate formulation (7.118) of the additive constitutive law, the derivation of the viscosity relation analogous to (7.136) is more difficult, as was already pointed out in Sect. 7.4.2. This is due to the fact that the enforcement of the simple shear deformation (with all the strain-rate tensor components equal to zero except $D_{13} = D_{31} \neq 0$) requires the determination of all the (non-vanishing) diagonal components of the deviatoric stress tensor $\boldsymbol{S}$. We omit here all the details (they can be found in the paper by Morland and Staroszczyk 2003b) and only present the final result for the limit normalized shear viscosity at indefinitely large strain κ:

$$2\left[\check{f}^e(\infty) - \check{f}^o(\infty)\right] = E_s - 1, \tag{7.138}$$

which is the analogue of the limit relation (7.137), with E_s^{-1} now replaced by E_s.

The two pairs of relations, (7.133) and (7.137) for the stress formulation, and (7.134) and (7.138) for the strain-rate formulation, allow us to express the limit values of the even and odd parts of the response functions in terms of the axial and shear enhancement factors E_a and E_s. Accordingly, the combination of the uniaxial and simple shear relations, (7.133) and (7.137), for the stress formulation, gives

$$\bar{f}^e(\infty) = \tfrac{1}{2}\,(3E_a^{-1} - 2E_s^{-1} - 1), \quad \bar{f}^o(\infty) = \tfrac{3}{2}(E_a^{-1} - E_s^{-1}). \tag{7.139}$$

In the strain-rate formulation, in turn, the corresponding relations derived from (7.134) and (7.138) are defined by

$$\check{f}^e(\infty) = \tfrac{1}{2}(3E_a - 2E_s - 1), \quad \check{f}^o(\infty) = \tfrac{3}{2}(E_a - E_s), \tag{7.140}$$

and are analogous to (7.139), with E_a^{-1} replaced by E_a, and E_s^{-1} by E_s. The limit values of the other response functions, $\bar{G}(\infty)$ and $\check{G}(\infty)$, are then obtained from (7.129).

For the previously used values of the axial and shear enhancement factors, $E_a = 1/3$ and $E_s = 5$ for cold ice, and $E_a = 3$ and $E_s = 8$ for warm ice, the limit expressions (7.139) for the stress formulation are

$$\text{cold ice}: \quad \bar{f}^e(\infty) = 3.8, \qquad \bar{f}^o(\infty) = 4.2, \qquad \bar{G}(\infty) = -8.4, \tag{7.141}$$

$$\text{warm ice}: \quad \bar{f}^e(\infty) = -0.125, \qquad \bar{f}^o(\infty) = 0.3125, \qquad \bar{G}(\infty) = -0.625, \tag{7.142}$$

while the corresponding values for the strain-rate formulation prescribed by (7.140) are

$$\text{cold ice}: \quad \check{f}^e(\infty) = -5, \qquad \check{f}^o(\infty) = -7, \qquad \check{G}(\infty) = 14, \tag{7.143}$$

$$\text{warm ice}: \quad \check{f}^e(\infty) = -4, \qquad \check{f}^o(\infty) = -7.5, \qquad \check{G}(\infty) = 15, \tag{7.144}$$

with the limits $f(\infty) = f^e(\infty) + f^o(\infty)$ and $f(-\infty) = f^e(\infty) - f^o(\infty)$.

The limit relations (7.141)–(7.144), together with the derivative restriction $(7.128)_2$ on $f(\xi)$ at $\xi = 0\,(b = 1)$, constitute a minimum set of conditions which fabric response functions in a valid constitutive model must satisfy. In order to develop a realistic theory, the response functions should predict observed ice creep responses over the whole range of relevant strains, not only at their selected values. Morland and Staroszczyk (2003b) adopted idealized monotonic response functions to approximate the real behaviour of ice during compression and shearing. These functions prescribe the variations of the normalized axial and shear viscosities with relevant strains as follows:

$$\bar{\mu}_{33}(b_1) = 1 + (E_a^{-1} - 1)\left\{1 - \exp[1 - (b_1 - 1)/b^*]^2\right\}, \tag{7.145}$$

$$\bar{\mu}_{13}(\kappa) = 1 + (E_s^{-1} - 1)\left\{1 - \exp(-\kappa^2/k^*)\right\}, \tag{7.146}$$

where b^* and k^* are the scales of b_1 and κ^2, respectively, over which $\bar{\mu}_{33}(b_1)$ and $\bar{\mu}_{13}(\kappa^2)$ make a significant change. Then, a variety of plausible response function representations with free parameters were explored, with the objective to correlate, to a good approximation, the model predicted material responses $\tilde{\mu}_{33}$ and $\tilde{\mu}_{13}$ with those prescribed by (7.145) and (7.146). The correlations were carried out by a weighted least squares matching. The details (the forms of the response functions and the specific values of the free parameters entering those functions) can be found in the afore-cited work. Here, we restrict ourselves to the presentation of the final results.

7.5.3 Illustrations

Following the discussion on the construction and the properties of the response functions in the additive stress and strain-rate constitutive formulations, some results of simulations are presented below.

Figures 7.14 and 7.15 show the variation of the fabric response functions for cold ice. These functions have been obtained by correlating the results given by the flow laws (7.115) and (7.118) with the assumed viscous responses (7.145) and (7.146). Fig. 7.14 displays the functions $\bar{f}(\xi)$, $\bar{g}(\xi)$ and $\bar{G}(\xi)$ in the stress formulation of the flow law, with $\bar{f}^o(\infty)$ as unit. It is seen that all the response functions are nearly monotonic. The analogous fabric response functions $\check{f}(\xi)$, $\check{g}(\xi)$ and $\check{G}(\xi)$ in the strain-rate formulation, all nearly monotonic again, are shown in Fig. 7.15, with $\check{f}^o(\infty)$ adopted as unit. It is seen that the corresponding functions, both qualitatively and quantitatively, are very similar in both formulations.

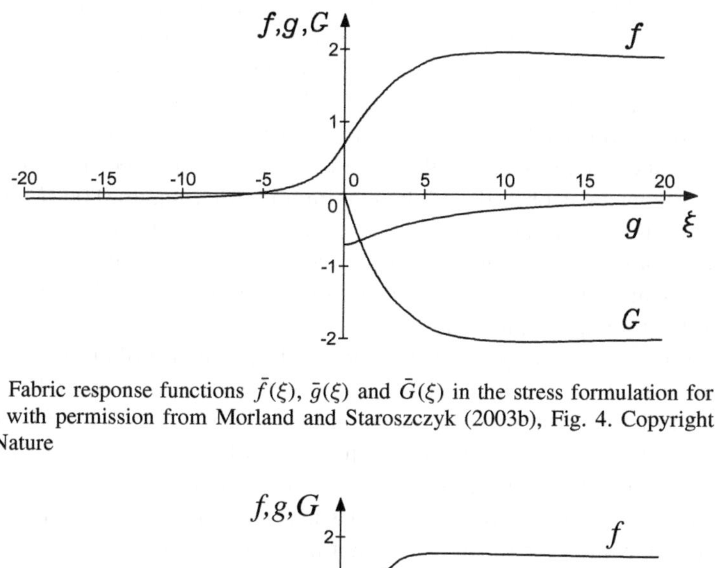

Fig. 7.14 Fabric response functions $\bar{f}(\xi)$, $\bar{g}(\xi)$ and $\bar{G}(\xi)$ in the stress formulation for cold ice. Reprinted with permission from Morland and Staroszczyk (2003b), Fig. 4. Copyright 2003 by Springer Nature

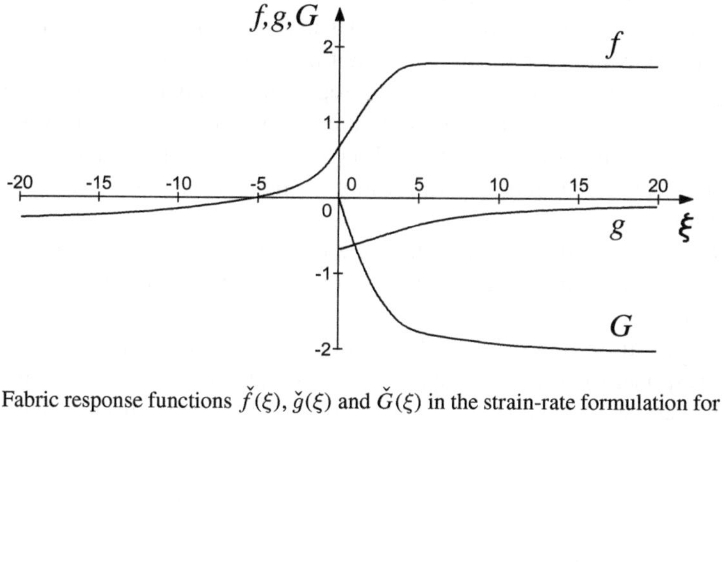

Fig. 7.15 Fabric response functions $\check{f}(\xi)$, $\check{g}(\xi)$ and $\check{G}(\xi)$ in the strain-rate formulation for cold ice

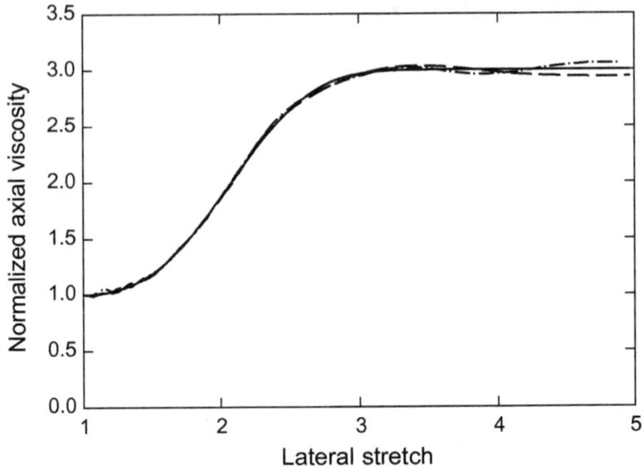

Fig. 7.16 Normalized uniaxial viscosities $\tilde{\mu}_{33}$ predicted by the stress (dashed line) and the strain-rate (dashed-dotted line) formulations compared with the prescribed response $\bar{\mu}_{33}$ (solid line) for cold ice

Figure 7.16 illustrates the variation of the normalized uniaxial viscosity with the lateral stretch λ_1 in unconfined compression performed along the x_3 axis. Compared with the prescribed response $\bar{\mu}_{33}$ (solid line), defined by (7.145), are the predicted responses $\tilde{\mu}_{33}$ given by Eqs. (7.130) and (7.131) for the stress (dashed line) and strain-rate (dashed-dotted line) forms of the constitutive law. We note a very close overall agreement achieved with the response functions plotted in Figs. 7.14 and 7.15, with the mean error 0.01 and the maximum error equal to 0.06 (Morland and Staroszczyk 2003b). A particularly good matching is seen for the lateral stretches λ_1 ranging from about 1 to about 3. The predicted simple shear responses $\tilde{\mu}_{13}$ are practically identical to the prescribed response $\bar{\mu}_{13}$ defined by (7.146), therefore their graphical illustration is omitted here.

The results for warm ice given by the additive forms of the orthotropic constitutive laws defined by Eqs. (7.115) and (7.118) are illustrated in Figs. 7.17 and 7.18. The plots in Fig. 7.17 show the variation of the axial viscosity $\tilde{\mu}_{33}$ with increasing deformation. Again, the prescribed response $\bar{\mu}_{33}$ (solid line), defined by (7.145), is compared with the predicted responses $\tilde{\mu}_{33}$ for the stress formulation (dashed line), (7.130), and the strain-rate formulation (dashed-dotted line), (7.131). It is seen that the correlation results for warm ice are not as good as those obtained for cold ice (see Fig. 7.16), though they can still be regarded as satisfactory for applications. The greatest differences between the predicted and the prescribed responses, equal to 0.07, occur at the moderate values of the lateral stretches λ_1, ranging from about 1.5 to about 2, with much better conformity taking place at either small or very large deformations.

Finally, Fig. 7.18 demonstrates how the normalized shear viscosity of warm ice evolves with increasing shear strain κ. The prescribed viscosity $\bar{\mu}_{13}$ (solid line), given by (7.146), is compared with the viscosities $\tilde{\mu}_{13}$ predicted by the stress (dashed line) and the strain-rate (dashed dotted line) formulations. As in the case of the uniaxial response illustrated in the previous figure, the best correlations are achieved for either very small or very large shear strains. The largest discrepancies between

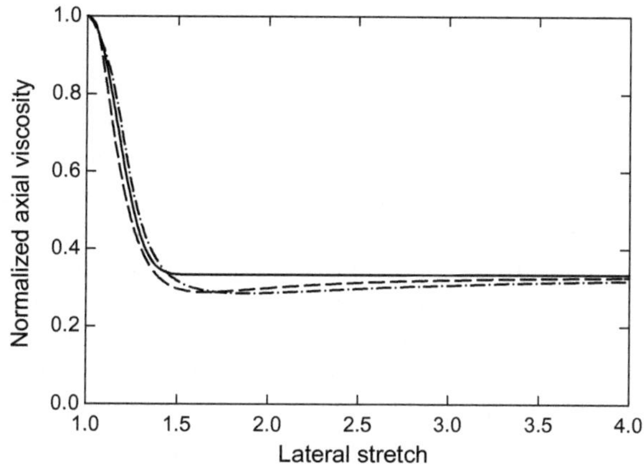

Fig. 7.17 Normalized uniaxial viscosities $\tilde{\mu}_{33}$ predicted by the stress (dashed line) and the strain-rate (dashed-dotted line) formulations compared with the prescribed response $\bar{\mu}_{33}$ (solid line) for warm ice

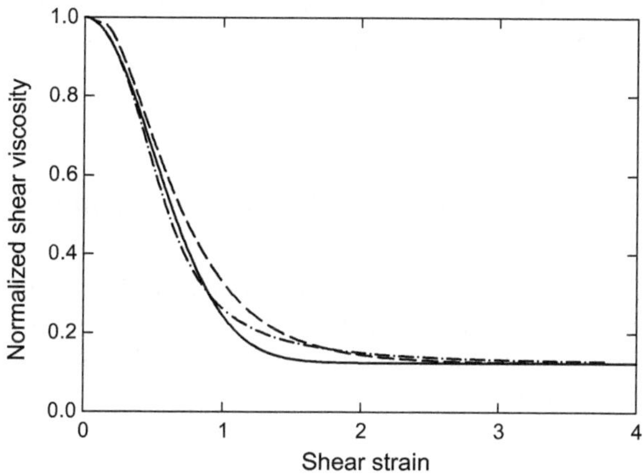

Fig. 7.18 Normalized shear viscosities $\tilde{\mu}_{13}$ predicted by the stress (dashed line) and the strain-rate (dashed-dotted line) formulations compared with the prescribed response $\bar{\mu}_{13}$ (solid line) for warm ice

the prescribed and the predicted responses, equal to 0.09 occur at strains $\kappa \sim 1.5$, though they can still be considered as acceptable for polar ice sheet flow modelling discussed in Chap. 8.

7.6 Dynamic Recrystallization Model

The micro-mechanism of dynamic (migration) recrystallization and its consequences for the macroscopic viscous creep behaviour of polycrystalline ice are described in Sect. 6.4, devoted to the micro-mechanical modelling of this important phenomenon. Here, the mechanism of dynamic recrystallization is treated by applying a phenomenological approach. Thus, the processes taking place on the crystal level are ignored, and only the macroscopic effects, consisting in the destruction of the macroscopic anisotropy of ice and return to its nearly isotropic response, are accounted for in the constitutive description developed further in this section. For this purpose, some of the orthotropic viscous flow laws formulated earlier in this chapter are extended (Staroszczyk and Morland 2001; Staroszczyk 2004) to incorporate the mechanism of continuous weakening of the anisotropic fabric over a finite range of shear strain-rate before the ice isotropy is restored. The fabric weakening at increasing strain-rates is captured by incorporating dependence on a smooth scaling function which, in turn, depends on a critical strain-rate invariant and temperature, and increases as temperature decreases. The effects of stress and strain magnitudes on ice recrystallization (analysed in Sect. 6.4) are not investigated, though they can be modelled in a similar fashion; that is, by scaling the ice fabric anisotropic strength by properly constructed smooth functions of stress/strain invariant arguments. Both multiplicative and additive forms of the phenomenological constitutive equations are used to derive their modified counterparts in order to incorporate the dynamic recrystallization process. The multiplicative and additive forms of the laws are then employed to simulate the viscous behaviour of ice in the uniaxial and simple shear flow configurations, and to compare the model predictions of ice viscosities for different magnitudes of critical strain-rates defining the onset of recrystallization.

7.6.1 Multiplicative Form of the Model

The starting point of our considerations is the multiplicative form of the orthotropic constitutive law formulated in Sect. 7.3, expressing the deviatoric stress $\boldsymbol{S}$ in terms of the strain-rate $\boldsymbol{D}$ and the left Cauchy-Green deformation tensor $\boldsymbol{B}$, and given by Eq. (7.43) on p. 229 as follows:

$$\boldsymbol{S} = \mu_0 \Big\{ \sum_{r=1}^{3} f(b_r) \big[\boldsymbol{M}^{(r)}\boldsymbol{D} + \boldsymbol{D}\boldsymbol{M}^{(r)} - \tfrac{2}{3}\mathrm{tr}\,(\boldsymbol{M}^{(r)}\boldsymbol{D})\boldsymbol{I} \big] + \\ + K^{-1}G(K) \big[\boldsymbol{B}\boldsymbol{D} + \boldsymbol{D}\boldsymbol{B} - \tfrac{2}{3}\mathrm{tr}\,(\boldsymbol{B}\boldsymbol{D})\boldsymbol{I} \big] \Big\}. \tag{7.147}$$

Recall that the arguments of the material response coefficients $f(b_r)$ and $G(K)$ are the three eigenvalues b_r of the tensor $\boldsymbol{B}$ defined by (7.10), and the trace of the latter tensor $K = b_1 + b_2 + b_3$.

The isotropic ice viscosity μ_0 in (7.147) is assumed again to be a function of the strain-rate invariant tr $\boldsymbol{D}^2$ and temperature T, see Sect. 7.3. Hence, we adopt

$$\mu_0 = \mu_0\,(\tilde{I}, T), \quad \tilde{I} = \tfrac{1}{2}\mathrm{tr}\,\boldsymbol{D}^2 = D_0^2\, I, \tag{7.148}$$

where I is defined by (7.46) on p. 230 and $D_0 = 1\ \mathrm{yr}^{-1} = 3.17 \times 10^{-8}\ \mathrm{s}^{-1}$ is the normalizing strain-rate unit describing a typical strain-rate magnitude in large polar ice sheets. The temperature-dependence of μ_0 is described by the relation

$$\mu_0(\tilde{I}, T) = \mu_0(\tilde{I}, T_m)/a(T), \tag{7.149}$$

where T_m denotes the ice melting point, and $a(T)$ is the temperature scaling factor given by (3.18) on p. 40, with the properties $a(T_m) = 1$ and $a(T) < 1$ for $T < T_m$.

Following Staroszczyk and Morland (2001), it is supposed that fabric weakening due to the ice recrystallization process depends on a single strain-rate invariant and temperature through an effective strain-rate invariant $\bar{I}$ defined by

$$\bar{I} = \frac{\tilde{I}}{[c(T)]^2}, \tag{7.150}$$

where the scaling factor $c(T)$ is a decreasing function of temperature, so that $\tilde{I}$ is a decreasing function of temperature at fixed $\bar{I}$. Further, we introduce a critical temperature-independent strain-rate invariant $\bar{I}_c$, with a corresponding temperature-dependent critical value $\tilde{I}_c$ of the strain-rate invariant, at the centre of the range over which fabric weakening occurs. This new invariant, by analogy to (7.150), is defined by

$$\bar{I}_c = \frac{\tilde{I}_c}{[c(T)]^2}. \tag{7.151}$$

Next we introduce a critical strain-rate invariant range, with its lower and upper critical values given by

$$\bar{I}_{cl} = \bar{I}_c(1 - \delta), \quad \bar{I}_{cu} = \bar{I}_c(1 + \delta), \tag{7.152}$$

where the full anisotropic ice fabric strength is retained below $\bar{I}_{cl}$, isotropy is restored above $\bar{I}_{cu}$, and there is a continuous monotonic decrease of fabric strength between $\bar{I}_{cl}$ and $\bar{I}_{cu}$. The critical centre level, $\bar{I}_c$, and relative half-span of the critical range, δ, are free parameters in the recrystallization model. The mechanism of fabric weakening is then expressed in terms of a continuous strength scaling factor $r(\bar{I})$ defined as follows:

$$\begin{aligned} r(\bar{I}) &= 1 \quad \text{if} \quad \bar{I} \le \bar{I}_{cl}, \\ r'(\bar{I}) &\le 0 \quad \text{if} \quad \bar{I}_{cl} \le \bar{I} \le \bar{I}_{cu}, \\ r(\bar{I}) &= 0 \quad \text{if} \quad \bar{I} \ge \bar{I}_{cu}. \end{aligned} \tag{7.153}$$

An example scaling function $r(\bar{I})$ is shown in Fig. 7.19. The plotted function is a third-order polynomial, with the property $\mathrm{d}r/\mathrm{d}\bar{I} = 0$ at both $\bar{I} = \bar{I}_{cl}$ and $\bar{I} = \bar{I}_{cu}$, and the half-span of the critical range of the effective strain-rate invariant adopted as $\delta = 0.2$.

In order to modify the viscous flow law (7.147) so that the full fabric response is reduced to the isotropic response continuously through the critical range of $\bar{I}$, the Cauchy-Green deformation tensor $\boldsymbol{B}$ is replaced by a modified tensor involving the scaling factor $r(\bar{I})$. The modified deformation measure changes continuously from $\boldsymbol{B}$ when $r(\bar{I})$ is unity to the unit tensor $\boldsymbol{I}$ when $r(\bar{I})$ is zero. Two modified tensors

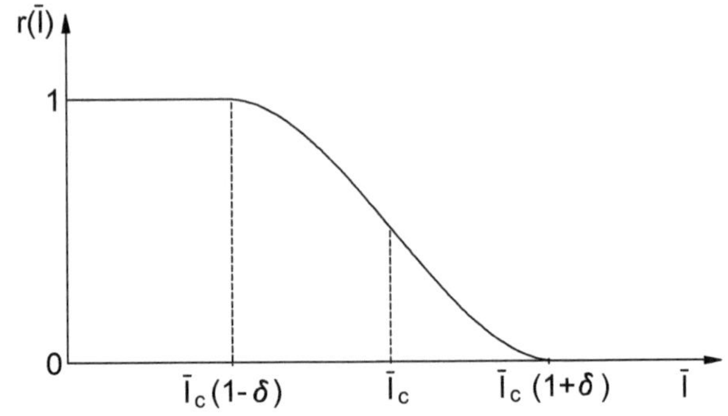

Fig. 7.19 Fabric strength scaling function $r(\bar{I})$

which are linear in both the tensor $\boldsymbol{B}$ and the scalar $r(\bar{I})$ are considered. The first is

$$\bar{\boldsymbol{B}} = \boldsymbol{I} + r(\bar{I})\,[\boldsymbol{B} - \boldsymbol{I}], \tag{7.154}$$

and the second is

$$\hat{\boldsymbol{B}} = J_0 \bar{\boldsymbol{B}} = J_0\,\{\boldsymbol{I} + r(\bar{I})\,[\boldsymbol{B} - \boldsymbol{I}]\}, \quad J_0 = (\det \bar{\boldsymbol{B}})^{-1/3}. \tag{7.155}$$

Both modified deformation measures have the properties

$$\begin{aligned} \bar{\boldsymbol{B}} = \hat{\boldsymbol{B}} = \boldsymbol{B}, \quad J_0 = 1 \quad \text{when} \quad r = 1, \\ \bar{\boldsymbol{B}} = \hat{\boldsymbol{B}} = \boldsymbol{I}, \quad J_0 = 1 \quad \text{when} \quad r = 0, \end{aligned} \tag{7.156}$$

and change continuously from $\boldsymbol{B}$ when $r(\bar{I}) = 1$ to $\boldsymbol{I}$ when $r(\bar{I}) = 0$. This change occurs, however, through a modification of the tensor $\boldsymbol{B}$, not strictly through a scaling of the strength of the anisotropy. Furthermore,

$$\boldsymbol{B} = \boldsymbol{I} \Rightarrow \bar{\boldsymbol{B}} = \boldsymbol{I}, \quad J_0 = 1, \quad \text{hence} \quad \hat{\boldsymbol{B}} = \boldsymbol{I}. \tag{7.157}$$

The eigenvectors (principal axes) of both $\bar{\boldsymbol{B}}$ and $\hat{\boldsymbol{B}}$ are those of $\boldsymbol{B}$, namely $\boldsymbol{e}^{(r)}$ ($r = 1, 2, 3$), and hence the corresponding structure tensors $\boldsymbol{M}^{(r)}$ are the same. The eigenvalues (squared principal stretches) are, respectively,

$$\bar{b}_r = 1 + r(\bar{I})\,[b_r - 1], \quad \hat{b}_r = J_0\,\bar{b}_r \quad (r = 1, 2, 3). \tag{7.158}$$

Note that

$$\det \boldsymbol{B} = 1, \quad \det \bar{\boldsymbol{B}} \neq 1, \quad \det \hat{\boldsymbol{B}} = 1, \tag{7.159}$$

$$K = \operatorname{tr} \boldsymbol{B}, \quad \bar{K} = \operatorname{tr} \bar{\boldsymbol{B}} \neq K, \quad \hat{K} = J_0\,\bar{K} \neq K, \tag{7.160}$$

Table 7.1 Six combinations of the modified deformation measures

$\bar{A}$	$\bar{B}$	$\bar{C}$	$\hat{A}$	$\hat{B}$	$\hat{C}$
$\bar{\boldsymbol{B}}$	$\bar{\boldsymbol{B}}$	$\boldsymbol{B}$	$\hat{\boldsymbol{B}}$	$\hat{\boldsymbol{B}}$	$\boldsymbol{B}$
$\bar{b}_r$	b_r	$\bar{b}_r$	$\hat{b}_r$	b_r	$\hat{b}_r$
$\bar{K}$	K	$\bar{K}$	$\hat{K}$	K	$\hat{K}$

and in a plane deformation, with $b_2 = 1$:

$$b_2 = 1 \Rightarrow \bar{b}_2 = 1, \quad \text{but} \quad \hat{b}_2 \neq 1. \tag{7.161}$$

Associated with each of the modified deformation measures $\bar{\boldsymbol{B}}$ and $\hat{\boldsymbol{B}}$ there are three combinations of a chosen modified tensor and two deformation invariants, labelled $\bar{A}$, $\bar{B}$, $\bar{C}$, and $\hat{A}$, $\hat{B}$, $\hat{C}$, respectively (see Table 7.1). These six combinations of the deformation measures lead to six modified viscous flow laws obtained from the constitutive equation (7.147) by replacing $\boldsymbol{B}$ and its invariants by each of the above sets in turn. These six new flow laws are then explored by applying the concept of instantaneous directional viscosities, previously applied in Sects. 7.2 and 7.5. Thus, we consider distinct axial stretches described by (7.17) and (7.18), with arbitrary strain-rate history, for each of the six laws in turn, ending with a deformation $\boldsymbol{B}$, associated tensors $\bar{\boldsymbol{B}}$ and $\hat{\boldsymbol{B}}$, and the respective invariants defined by (7.158) and (7.160). If the stress and strain-rate are now removed abruptly, subsequent deformation at any strain-rate has an instantaneous response governed by these 'frozen' fabric measures. Then, simple shearing on the principal stretch planes, described by (7.20), determines instantaneous viscosities for the six modified laws, analogous to (7.41). These viscosities were examined in detail by Staroszczyk and Morland (2001), and it turned out that only one of the six plausible constructions applied to the original orthotropic flow law consistently satisfies all the viscosity equalities and inequalities (7.63)–(7.68) on p. 235 derived by Staroszczyk and Morland (2000a), and yields physically acceptable material behaviour. This one particular construction, labelled here by $\hat{A}$, was therefore left for further applications.

Accordingly, the constitutive relation (7.147) with the replacements for the modification $\hat{A}$ shown in Table 7.1 gives the following expression for the normalized axial viscosity μ_{33}/μ_0:

$$\frac{S_{33}}{2\mu_0 D_{33}} = \frac{1}{3}\left[f(\hat{b}_1) + 2f(\hat{b}_3) + \frac{G(\hat{K})}{\hat{K}}\,(\hat{b}_1 + 2\,\hat{b}_3)\right], \tag{7.162}$$

which is analogous to relation (7.53) on p. 232 given by the unmodified flow law. In (7.162),

$$\begin{gathered}\hat{b}_1 = \hat{b}_2 = J_0\,\bar{b}_1, \quad \hat{b}_3 = J_0\,\bar{b}_3, \quad \hat{K} = J_0\,\bar{K}, \quad J_0 = (\bar{b}_1^2\,\bar{b}_3)^{-1/3}, \\ \bar{b}_1 = 1 + r(b_1 - 1), \quad \bar{b}_3 = 1 + r(b_1^{-2} - 1), \quad \bar{K} = 3 + r(2b_1 + b_1^{-2} - 3).\end{gathered} \tag{7.163}$$

In simple shear flow in the plane Ox_1x_3 started from an initially isotropic state ($\lambda_1 = \lambda_2 = \lambda_3 = 1$), with the deformation field described by (7.54) and (7.55), and the eigenvalues and eigenvectors of $\boldsymbol{B}$ determined by (7.56) and (7.57), the flow law (7.147), with the replacements corresponding to $\hat{A}$ in Table 7.1, defines the normalized shear viscosity μ_{13}/μ_0 in the form

$$\frac{S_{13}}{2\mu_0 D_{13}} = \frac{1}{2}\left[f(\hat{b}_1) + f(\hat{b}_3) + \frac{G(\hat{K})}{\hat{K}}\left(2+\kappa^2\right)\right], \tag{7.164}$$

analogous to (7.60) on p. 233. The invariants in (7.164) are

$$\begin{gathered}\hat{b}_1 = J_0\bar{b}_1, \quad \hat{b}_3 = J_0\bar{b}_3, \quad \hat{K} = J_0\bar{K}, \quad J_0 = \det(\bar{b}_1\bar{b}_3)^{-1/3}, \\ \bar{b}_1 = 1 + r(b_1 - 1), \quad \bar{b}_3 = 1 + r(b_1^{-1} - 1), \quad \bar{K} = 3 + r(b_1 + b_1^{-1} - 2).\end{gathered} \tag{7.165}$$

7.6.2 *Flow Simulations with the Multiplicative Law*

The response of ice to uniaxial compression and simple shearing, described by relations (7.162) and (7.164) respectively, was simulated numerically in order to examine the predictions of the multiplicative form of the modified constitutive model. It was assumed in the simulations that the flow of ice starts from an isotropic state, defined by $\boldsymbol{B} = \boldsymbol{I}$, and the deformation rates grow continuously with time from their respective values prescribed at the beginning of flow. Thus, as the deformation progresses and the ice fabric anisotropy develops, at some instant of time the effective strain-rate invariant $\bar{I}$ reaches its lower critical value, $\bar{I}_{cl}$ (see Fig. 7.19), marking the onset of the dynamic recrystallization process, and then, with the strain-rates still growing, the upper critical value of the invariant, $\bar{I}_{cu}$, is reached, beyond which the fabric response becomes isotropic again.

The simulations were carried out with the fabric response function $f(b)$ adopted in the form (7.76), with the free parameter value $m = 1.5$. The other response function, $G(K)$, is related to $f(b_r)$ by (7.69). The limit values of both response functions are determined in terms of the reciprocals of the enhancement factors E_a and E_s by Eq. (7.73) on p. 235, and their specific values for cold and warm ice are given, respectively, by (7.74) and (7.75). Note that the functions $f(b)$ and $G(K)$ are independent of temperature, so the functions determined by the two sets of the latter limits are modelling warm and cold ice treated as different materials, not the response of the same ice at warm and cold temperatures.

In the uniaxial unconfined compression simulations it is assumed that ice is compressed along the x_3-axis at a constant displacement-rate; that is, $\dot{\lambda}_3 = \text{constant} < 0$, with $\lambda_3\ (t = 0) = 1$. Accordingly, the evolution of the principal stretches is expressed by the relations

$$\lambda_3 = 1 + \dot{\lambda}_3 t \le 1, \quad \lambda_1 = \lambda_2 = \lambda_3^{-1/2} \ge 1. \tag{7.166}$$

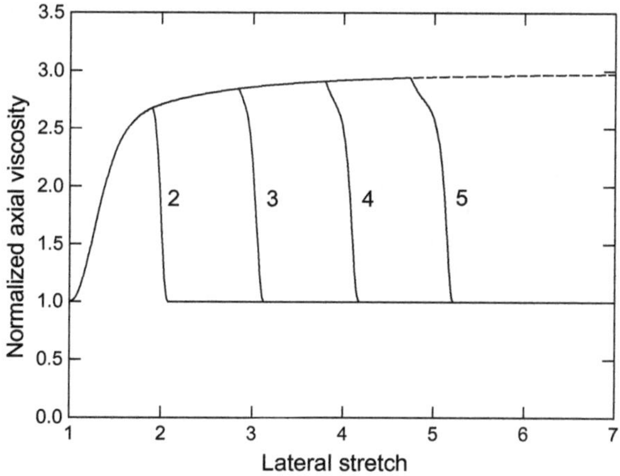

Fig. 7.20 Evolution of the normalized axial viscosity μ_{33}/μ_0 with lateral stretch λ_1 in uniaxial compression for different values of the critical stretch λ_1^c, predicted by the multiplicative form of the flow law, for cold ice. Reprinted from Staroszczyk and Morland (2001), Fig.2b, with permission of the Royal Society of London

Since the axial strain-rate in the assumed flow configuration is $D_{33} = \dot{\lambda}_3/\lambda_3$, and the strain-rate invariant is $\tilde{I} = \frac{3}{4}D_{33}^2$, it can be found that the ratio of the current value of the invariant $\tilde{I}$ to the critical value $\tilde{I}_c$ is given by

$$\frac{\tilde{I}}{\tilde{I}_c} = \frac{\bar{I}}{\bar{I}_c} = \left(\frac{D_{33}}{D_{33}^c}\right)^2 = \left(\frac{\lambda_3^c}{\lambda_3}\right)^2 = \left(\frac{\lambda_1}{\lambda_1^c}\right)^4, \tag{7.167}$$

where D_{33}^c is the axial strain-rate at which $\tilde{I} = \tilde{I}_c$, or equivalently $\bar{I} = \bar{I}_c$, and λ_1^c and λ_3^c are the principal stretches at which the critical value of the strain-rate invariant is reached. By selecting the value of the lateral stretch λ_1^c at which we wish the process of recrystallization to occur during a compression test, we can find, for a given value of $\bar{I}_c$, the required displacement-rate $\dot{\lambda}_3$. Using then Eq. (7.167) one can evaluate the value of the current strain-rate invariant relative to its critical level (which is needed for the calculation of the fabric strength scaling factor $r(\bar{I})$) in terms of the current stretch λ_1 relative to λ_1^c.

The variation of the normalized axial viscosity $\mu_{33}/\mu_0 = S_{33}/(2\,\mu_0 D_{33})$ with increasing lateral stretch λ_1, described by (7.162), is illustrated in Fig. 7.20, showing the results for cold ice. The curves in the figure (solid lines) are labelled by the corresponding values of the critical axial stretches λ_1^c, and for reference the result for the uniaxial compression with no dynamic recrystallization involved is also shown (dashed lines). It can be seen that, irrespective of the value of the critical stretch λ_1^c, the transition from the strongly anisotropic to isotropic response (for which the normalized axial viscosity is unity) takes place over a narrow range of strain, within which an abrupt change in the ice axial viscosity occurs.

In the simple shear simulations it is assumed that an ice sample is deformed in such a way that the rate of shear-strain increases linearly with time, and at the start of viscous flow both shear strain and its rate are zero. Thus,

$$\dot{\kappa} = ct, \quad \kappa = \tfrac{1}{2}ct^2, \tag{7.168}$$

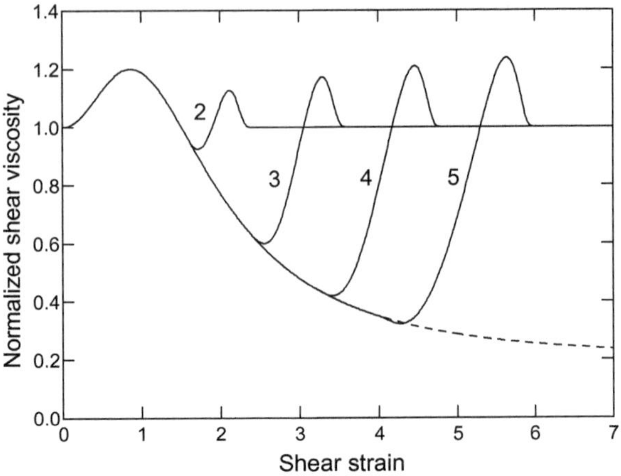

Fig. 7.21 Evolution of the normalized shear viscosity μ_{13}/μ_0 with strain κ in simple shear for different values of the critical strain κ_c, predicted by the multiplicative form of the flow law, for cold ice. Reprinted from Staroszczyk and Morland (2001), Fig.3b, with permission of the Royal Society of London

where $c > 0$ is a constant. The only non-zero strain-rate tensor components are $D_{13} = D_{31} = \frac{1}{2}\dot{\kappa}$, so the strain-rate invariant is $\tilde{I} = D_{13}^2 = \frac{1}{4}\dot{\kappa}^2 = \frac{1}{2}c\kappa$. Hence, the ratio of the current value of the invariant $\tilde{I}$ to the critical value value $\tilde{I}_c$ is expressed by

$$\frac{\tilde{I}}{\tilde{I}_c} = \frac{\bar{I}}{\bar{I}_c} = \left(\frac{D_{13}}{D_{13}^c}\right)^2 = \frac{\kappa}{\kappa_c}, \tag{7.169}$$

where D_{13}^c is the shear strain-rate at which $\tilde{I} = \tilde{I}_c$, hence $\bar{I} = \bar{I}_c$, and κ_c is the corresponding shear strain at which the critical value of the strain-rate invariant is reached. Similarly to the above-discussed case of the uniaxial compression simulation, one can evaluate, for assumed critical values of the shear strain κ_c and the effective strain-rate invariant $\bar{I}_c$, the corresponding value of the parameter c defining the shear strain-rate variation with time. Then, by employing (7.169), it is possible to express the ratio of the current strain-rate invariant $\bar{I}$ to the critical value $\bar{I}_c$ as a function of the current shear strain κ relative to the prescribed critical shear κ_c, and hence to evaluate the value of the fabric strength scaling factor $r(\bar{I})$.

The results of the simple shear simulations are plotted in Fig. 7.21, illustrating the variation of the shear viscosity $\mu_{13}/\mu_0 = S_{13}/(2\,\mu_0 D_{13})$ with increasing strain κ, depending on the values of the critical strain κ_c. Shown are the results obtained for cold ice; qualitatively very similar behaviour is predicted by the model for warm ice (Staroszczyk and Morland 2001; Staroszczyk 2004). The solid lines in the figure are labelled by the respective values of the critical shear strain κ_c, and the dashed lines illustrate the behaviour of ice when no dynamic recrystallization occurs. It is seen in the plots that the transition from the strongly anisotropic to isotropic response is less abrupt than in the case of uniaxial compression (see Fig. 7.20), though a surprising prediction of the model is the increase in the normalized shear viscosity beyond unity prior to the restoration of isotropy (the same feature is predicted by the model for warm ice). In some way this fabric strengthening feature resembles the hardening of initially isotropic ice at the beginning of shearing (also predicted by the micromechanical models presented in Chap. 6). However, it is likely that such a property predicted by the multiplicative form of the constitutive model does not reflect the real behaviour of the material. This prompted an alternative approach, in which the flow law is decomposed additively into isotropic and anisotropic parts, in a manner discussed in the previous Sect. 7.5, and the fabric strength scaling which describes the ice fabric weakening due to recrystallization is applied only to the anisotropic part of the constitutive law.

7.6.3 Additive Form of the Model

The previous modified multiplicative form of the constitutive equation considered in Sect. 7.6.1 is constructed by scaling the deformation tensor $\boldsymbol{B}$ and its invariants in the flow law (7.147), but leaving the response coefficients f and G unchanged compared to the original (that is, with no fabric weakening due to recrystallization) orthotropic viscous flow law (7.43) on p. 229. Now a different method is followed, in which an additive form of the flow law is used, analogous to that considered in Sect. 7.5. In this alternative approach, the ice fabric strength scaling is applied only to the response coefficients arising in the anisotropic part of the flow law, whereas the deformation tensor $\boldsymbol{B}$ and its invariants are not modified. Accordingly, we are concerned here with the constitutive law of the general form given by (7.108) on p. 250, subject to (7.110), and adopt the isotropic and anisotropic responses, $\boldsymbol{H}_I$ and $\boldsymbol{H}_A$ respectively, see (7.115), defined by the following representations:

$$\begin{aligned}\boldsymbol{H}_I(\boldsymbol{D}) &= 2\,\mu_0\,\boldsymbol{D},\\ \boldsymbol{H}_A(\boldsymbol{D},\boldsymbol{B},\boldsymbol{M}^{(r)}) &= \mu_0\Big\{\sum_{r=1}^{3}\tilde{f}(b_r)\big[\boldsymbol{M}^{(r)}\boldsymbol{D}+\boldsymbol{D}\boldsymbol{M}^{(r)}-\tfrac{2}{3}\mathrm{tr}\,(\boldsymbol{M}^{(r)}\boldsymbol{D})\boldsymbol{I}\big]\,+\\ &\quad+\,K^{-1}\tilde{G}(K)\big[\boldsymbol{B}\boldsymbol{D}+\boldsymbol{D}\boldsymbol{B}-\tfrac{2}{3}\mathrm{tr}\,(\boldsymbol{B}\boldsymbol{D})\boldsymbol{I}\big]\Big\},\end{aligned} \tag{7.170}$$

where $\tilde{f}(b_r)$ and $\tilde{G}(K)$ are new fabric response coefficients. The vanishing of the anisotropic part $\boldsymbol{H}_A$ when $\boldsymbol{B}=\boldsymbol{I}$, see $(7.110)_1$, provides the normalization condition for the functions $\tilde{f}$ and $\tilde{G}$, replacing $(7.39)_2$ on p. 229,

$$\tilde{f}(1)+\tfrac{1}{3}\tilde{G}(3)=0. \tag{7.171}$$

The flow law (7.170) determines the instantaneous directional viscosities, originally given by (7.41), in the form

$$\mu_{ij}=\tfrac{1}{2}\,\mu_0\Big\{2+\Big[\tilde{f}(b_i)+\tilde{f}(b_j)+(b_i+b_j)\,K^{-1}\,\tilde{G}(K)\Big]\Big\}, \tag{7.172}$$

and the viscosity relations (7.63)–(7.68) translate to the same conditions on the functions $\tilde{f}$ and $\tilde{G}$ as previously on f and G, with the new normalization (7.171). Hence, the equality (7.66) for the plane flow yields the relation

$$\tilde{G}(K) = -\frac{K}{b_1 - b_1^{-1}} \left[\tilde{f}(b_1) - \tilde{f}(b_1^{-1})\right], \quad K \geq 3, \tag{7.173}$$

which connects $\tilde{f}$ and $\tilde{G}$, with b_1 defined in terms of K by (7.70). In the limit, as $b_1 \to 1$ and $K \to 3$, the latter equation supplies

$$\tilde{f}(1) = \tilde{f}'(1), \tag{7.174}$$

which replaces (7.71) in the multiplicative formulation, while the limit condition for $b_1 \to \infty$ now becomes

$$\tilde{f}(0) - \tilde{f}(\infty) - \tilde{G}(\infty) = 0. \tag{7.175}$$

For the uniaxial unconfined compression deformation, assumed to occur along the x_3-axis, and defined by (7.17)–(7.19), the constitutive relation (7.170) yields the normalized axial viscosity μ_{33}/μ_0 expressed by

$$\frac{S_{33}}{2\mu_0 D_{33}} = 1 + \frac{1}{3}\left[\tilde{f}(b_1) + 2\tilde{f}(b_1^{-2}) + \frac{\tilde{G}(K)}{K}(b_1 + 2b_1^{-2})\right], \tag{7.176}$$

which, as $b_1 \to \infty$, and hence $K \sim 2b_1$, provides the limit relation

$$1 + \tfrac{1}{3}\tilde{f}(\infty) + \tfrac{2}{3}\tilde{f}(0) + \tfrac{1}{6}\tilde{G}(\infty) = E_a^{-1}. \tag{7.177}$$

In the simple shear flow taking place in the plane Ox_1x_3, and defined by relations (7.54), (7.55) and (7.58), the normalized instantaneous shear viscosity μ_{13}/μ_0 is given by

$$\frac{S_{13}}{2\mu_0 D_{13}} = 1 + \frac{1}{2}\left[\tilde{f}(b_1) + \tilde{f}(b_1^{-1}) + \frac{\tilde{G}(K)}{K}(2 + \kappa^2)\right], \tag{7.178}$$

which for an indefinitely large shear strain $\kappa \to \infty$, when $b_1 \sim \kappa^2$ and $K \sim \kappa^2$, yields the limit condition

$$1 + \tfrac{1}{2}\tilde{f}(\infty) + \tfrac{1}{2}\tilde{f}(0) + \tfrac{1}{2}\tilde{G}(\infty) = E_s^{-1}. \tag{7.179}$$

The three relations (7.175), (7.177) and (7.179) determine the limit values of the two response functions $\tilde{f}$ and $\tilde{G}$ in terms of the axial and shear enhancement factors, E_a and E_s respectively, as follows:

$$\begin{aligned}\tilde{f}(0) &= E_s^{-1} - 1,\\ \tilde{f}(\infty) &= 6E_a^{-1} - 5E_s^{-1} - 1,\\ \tilde{G}(\infty) &= 6\,(E_s^{-1} - E_a^{-1}),\end{aligned} \tag{7.180}$$

with the additional restriction (7.174) imposed on the choice of the fabric response function $\tilde{f}(b_r)$. The latter expressions are the analogues of the limit relations (7.73) on p. 235 for the response functions in the unmodified multiplicative viscous flow law.

We can now simply describe the mechanism of recrystallization-induced ice fabric weakening with increase of the effective strain-rate invariant $\bar{I}$ by applying the fabric strength scaling function $r(\bar{I})$ directly to the response coefficients in the anisotropic function $\boldsymbol{H}_A$. Therefore,

$$\begin{aligned}&\breve{f}(b) = r(\bar{I})\,\tilde{f}(b), \quad \breve{G}(K) = r\,(\bar{I})\tilde{G}(K),\\ &\breve{\boldsymbol{H}}_A(\boldsymbol{D}, \boldsymbol{B}, \boldsymbol{M}^{(r)}) = r\,(\bar{I})\boldsymbol{H}_A(\boldsymbol{D}, \boldsymbol{B}, \boldsymbol{M}^{(r)}),\end{aligned} \tag{7.181}$$

which do not introduce any modified forms of the deformation tensor $\boldsymbol{B}$ or its invariants, and replace $\tilde{f}(b)$, $\tilde{G}(K)$ and $\boldsymbol{H}_A(\boldsymbol{D}, \boldsymbol{B}, \boldsymbol{M}^{(r)})$, respectively, in the viscous law (7.170). The instantaneous directional viscosities (7.172) then become

$$\mu_{ij} = \tfrac{1}{2}\mu_0\left\{2 + r(\bar{I})\left[\tilde{f}(b_i) + \tilde{f}(b_j) + (b_i + b_j)\,K^{-1}\,\tilde{G}(K)\right]\right\}, \tag{7.182}$$

and the six viscosity equalities and inequalities (7.63)–(7.68) are not changed by the factor r in expression (7.182); a consistent $\tilde{f}(b)$ when $r = 1$ is consistent for all r. The axial and shear flow responses (7.176) and (7.178) depend, of course, on how r changes during a given strain-rate history.

7.6.4 Flow Simulations with the Additive Law

The modified additive form of the constitutive law (7.170) was used to simulate the flows of ice under sustained unconfined uniaxial compression and simple shearing, defined, as in Sect. 7.6.2, by relations (7.166) and (7.168), respectively. The same form (7.76) on p. 236 of the fabric response function was adopted again for $\tilde{f}(b_r)$, with $m = 1.5$, and the limit values determined now by (7.180). Also the same scaling function $r(\bar{I})$ plotted in Fig. 7.19, and the same values of the enhancement factors ($E_a = 1/3$ and $E_s = 5$ for cold ice and $E_a = 3$ and $E_s = 8$ for warm ice) defining the limit viscous properties of the material were adopted in the calculations, with the purpose to compare the predictions of the two, multiplicative and additive, formulations of the orthotropic flow law.

The results of simulations of the uniaxial compression, illustrating the variation of the normalized axial viscosity with increasing lateral stretch λ_1, as defined by (7.176),

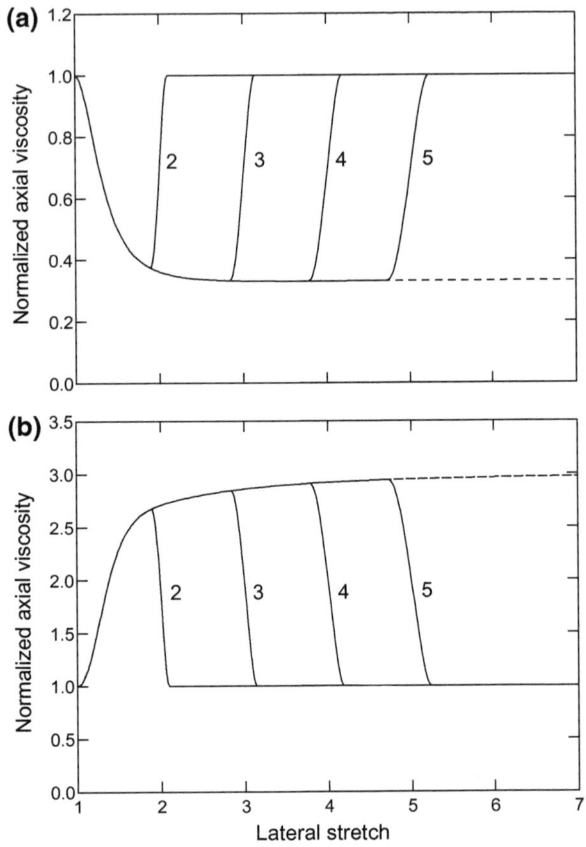

Fig. 7.22 Evolution of the normalized axial viscosity μ_{33}/μ_0 with the lateral stretch λ_1 in uniaxial compression for different values of the critical stretch λ_1^c, predicted by the additive form of the flow law: **a** warm ice, **b** cold ice. Reprinted from Staroszczyk and Morland (2001), Fig. 5, with permission of the Royal Society of London

are shown in Fig. 7.22, with the labels attached to the solid lines indicating again the respective values of the critical stretches λ_1^c, and the dashed lines showing the responses with no dynamic recrystallization taking place. By comparing the results plotted in the figure with those presented in Fig. 7.20 for cold ice one can note that the differences between the creep responses predicted by the two forms of the viscous flow law are practically insignificant. The same observation also applies to warm ice (Staroszczyk 2004).

The results of the simple shear flow simulations are illustrated in Fig. 7.23, showing the evolution of the shear viscosity with strain κ, as described by the relations (7.168). Comparison of the responses predicted by the two models (see Fig. 7.21 with the results for cold ice) shows that the additive form of the flow law yields results which significantly differ from those given by the multiplicative formulation. In particular, it is seen that the shear viscosities during the transition from the anisotropic to isotropic response now increase in a monotonic manner as the unit normalized viscosity is approached. This is qualitatively a different response than that predicted by the multiplicative form of the flow law considered in Sects. 7.6.1 and 7.6.2.

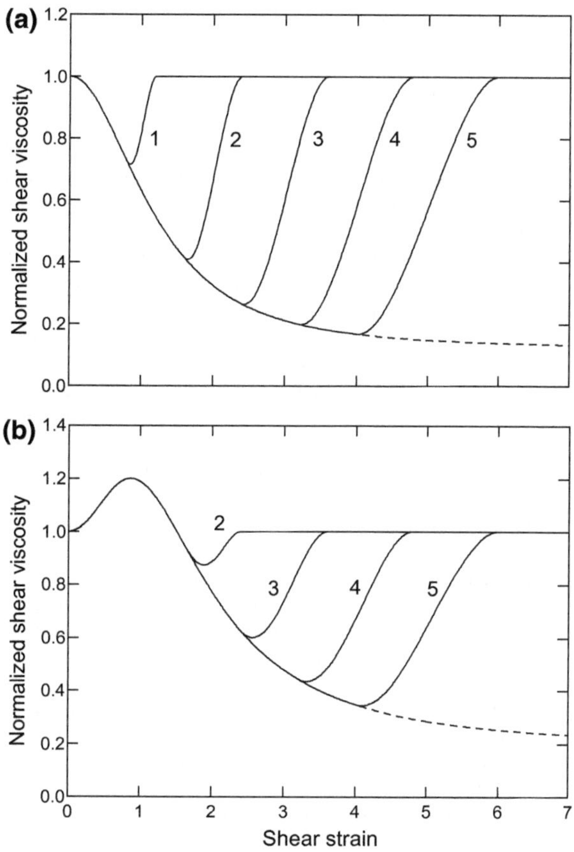

Fig. 7.23 Evolution of the normalized shear viscosity μ_{13}/μ_0 with the strain κ in simple shear for different values of the critical strain κ_c, predicted by the additive form of the flow law: **a** warm ice, **b** cold ice. Reprinted from Staroszczyk and Morland (2001), Fig. 6, with permission of the Royal Society of London

Although the behaviour illustrated in Fig. 7.23, with no increase in shear viscosity beyond unity prior to the isotropy restoration, seems to be physically more likely than the behaviour shown in Fig. 7.21, it is difficult to conclusively assess which of the two types of the viscous response is closer to the real material behaviour of polycrystalline ice, since no empirical data are available yet to compare the theory with experiment.

References

Budd WF, Jacka TH (1989) A review of ice rheology for ice sheet modelling. Cold Reg Sci Technol 16(2):107–144. https://doi.org/10.1016/0165-232X(89)90014-1

Chadwick P (1999) Continuum mechanics: concise theory and problems, 2nd edn. Dover, Mineola, New York

Ericksen JL, Rivlin RS (1954) Large elastic deformations of homogeneous anisotropic materials. J Ration Mech Anal 3:281–301

Gagliardini O, Meyssonnier J (1999) Analytical derivations for the behavior and fabric evolution of a linear orthotropic ice polycrystal. J Geophys Res 104(B8):17797–17809. https://doi.org/10.1029/1999JB900146

Hunter SC (1983) Mechanics of continuous media, 2nd edn. Ellis Horwood Ltd., Chichester

Jacka TH (1984) The time and strain required for development of minimum strain rates in ice. Cold Reg Sci Technol 8(3):261–268. https://doi.org/10.1016/0165-232X(84)90057-0

Jacka TH, Maccagnan M (1984) Ice crystallographic and strain rate changes with strain in compression and extension. Cold Reg Sci Technol 8(3):269–286. https://doi.org/10.1016/0165-232X(84)90058-2

Li J, Jacka TH, Budd WF (1996) Deformation rates in combined compression and shear for ice which is initially isotropic and after the development of strong anisotropy. Ann Glaciol 23:247–252

Liu IS (2002) Continuum mechanics. Springer, Berlin

Morland LW (1993) The flow of ice sheets and ice shelves. In: Hutter K (ed) Continuum mechanics in environmental sciences and geophysics. Springer, Wien, pp 403–466

Morland LW (2001) Influence of bed topography on steady plane ice sheet flow. In: Straughan B, Greve R, Ehrentraut H, Wang Y (eds) Continuum mechanics and applications in geophysics and the environment. Springer, Berlin, pp 276–304

Morland LW, Staroszczyk R (1998) Viscous response of polar ice with evolving fabric. Continuum Mech Thermodyn 10(3):135–152. https://doi.org/10.1007/s001610050086

Morland LW, Staroszczyk R (2003a) Strain-rate formulation of ice fabric evolution. Ann Glaciol 37:35–39

Morland LW, Staroszczyk R (2003b) Stress and strain-rate formulations for fabric evolution in polar ice. Contin Mech Thermodyn 15(1):55–71. https://doi.org/10.1007/s00161-002-0104-2

Morland LW, Staroszczyk R (2009) Ice viscosity enhancement in simple shear and uni-axial compression due to crystal rotation. Int J Eng Sci 47(11–12):1297–1304. https://doi.org/10.1016/j.ijengsci.2008.09.011

Rivlin RS (1955) Further remarks on the stress-deformation relations for isotropic materials. J Ration Mech Anal 4:681–701

Rivlin RS, Ericksen JL (1955) Stress-deformation relations for isotropic materials. J Ration Mech Anal 4:323–425

Smith GD, Morland LW (1981) Viscous relations for the steady creep of polycrystalline ice. Cold Reg Sci Technol 5(2):141–150

Smith GF (1994) Constitutive equations for anisotropic and isotropic materials. North-Holland, Amsterdam

Smith GF, Rivlin RS (1957) The anisotropic tensors. Q Appl Math 15:309–314

Spencer AJM (1980) Continuum mechanics. Longman, Harlow

Spencer AJM (1987a) Anisotropic invariants and additional results for invariant and tensor representations. In: Boehler JP (ed) Applications of tensor functions in solid mechanics. Springer, Wien, pp 171–186

Spencer AJM (1987b) Isotropic polynomial invariants and tensor functions. In: Boehler JP (ed) Applications of tensor functions in solid mechanics. Springer, Wien, pp 141–169

Staroszczyk R (2001) An orthotropic constitutive model for secondary creep of ice. Arch Mech 53(1):65–85

Staroszczyk R (2003) Plane ice sheet flow with evolving and recrystallising fabric. Ann Glaciol 37(1):247–251. https://doi.org/10.3189/172756403781815834

Staroszczyk R (2004) Constitutive modelling of creep induced anisotropy of ice. IBW PAN Publishing House, Gdańsk

Staroszczyk R, Gagliardini O (1999) Two orthotropic models for the strain-induced anisotropy of polar ice. J Glaciol 45(151):485–494

Staroszczyk R, Morland LW (1999) Orthotropic viscous model for ice. In: Hutter K, Wang Y, Beer H (eds) Advances in cold-region thermal engineering and sciences. Springer, Berlin, pp 249–258. https://doi.org/10.1007/BFb0104187

Staroszczyk R, Morland LW (2000a) Orthotropic viscous response of polar ice. J Eng Math 37(1–3):191–209. https://doi.org/10.1023/A:1004738429373

Staroszczyk R, Morland LW (2000b) Plane ice-sheet flow with evolving orthotropic fabric. Ann Glaciol 30:93–101

Staroszczyk R, Morland LW (2001) Strengthening and weakening of induced anisotropy in polar ice. Proc R Soc Lond A 457(2014):2419–2440. https://doi.org/10.1098/rspa.2001.0817

Treverrow A, Budd WF, Jacka TH, Warner RC (2012) The tertiary creep of polycrystalline ice: experimental evidence for stress-dependent levels of strain-rate enhancement. J Glaciol 58(208):301–314. https://doi.org/10.3189/2012JoG11J149

Truesdell C, Noll W (2004) The non-linear field theories of mechanics, 3rd edn. Springer, Berlin

Chapter 8
Polar Ice Sheet Flow Models

Polar ice sheet dynamics is concerned with the flow of large land-based ice masses on geophysical time scales. In essence, it deals with two fundamental problems. The first one is the calculation of the ice thickness (the ice free surface profile) and velocity and temperature fields which, for given bedrock topography and the distributions of the ice accumulation/ablation and temperature at the surface, maintain existing ice sheets in a steady state (that is, keep their geometry unchanged in time). The results obtained by solving this flow problem are necessary for the determination of the age of ice at different depths (to help to interpret ice core samples retrieved from polar ice caps), and are needed for the evaluation of the free surface ice velocities (these, for instance, will help to predict the future locations of Antarctic research stations or infrastructure which drift on the ice sheet surface). The second, and more difficult, problem is an unsteady one and deals with the evaluation of an ice sheet response to changes in the surface acumulation/ablation rates and temperature. The solution of this problem is important to those involved in the modelling of past or future global climate scenarios.

The ice sheet dynamics uses the methods of continuum mechanics to describe the motion of ice by a system of equations that express the laws of thermodynamics (mass, momentum and heat balances), the mechanical and thermal interactions of ice with the atmosphere and the bedrock, and the physical properties of the polycrystalline, strongly anisotropic, material. Since, for given climatic conditions, an ice sheet domain is not a priori known, we deal with a complex initial/boundary value problem on an unknown domain with a moving surface. Due to the intrinsic complexity, the solution of the complete system of the thermomechanically coupled equations describing the flow problem is difficult. For this reason, a number of simplifying assumptions were adopted in early ice sheet flow models to reduce the complexity of calculations. These included: (1) treatment of the ice as an incompressible body, (2) prescription of a temperature field to uncouple the energy balance from the mass and momentum balances, (3) reduction of geometrical dimensionality by taking advantage of symmetries in a flow field and solving either a plane or an axially symmetric problem, and (4) an assumption of the ice isotropy; for details refer to the papers by Hooke et al. (1979), Raymond (1983), Hodge (1985),

R. Staroszczyk, *Ice Mechanics for Geophysical and Civil Engineering Applications*, GeoPlanet: Earth and Planetary Sciences, https://doi.org/10.1007/978-3-030-03038-4_8

Hindmarsh et al. (1987), Hanson (1995), Hvidberg (1996), Morland (1997, 2000, 2001, 2009) Morland and Drăghicescu (1998), Drăghicescu (2001) and Cliffe and Morland (2000, 2001, 2002, 2004). The above assumptions were gradually relaxed in later developments, as our understanding of physical processes improved, more general theories were formulated, and the power of available computational hardware increased.

In order to solve the system of equations governing the flow of polar ice, two types of methods are in general use. The first type comprises analytic approaches, most often the methods of asymptotic expansions, and the other type includes discrete approaches, usually based on either finite-difference or finite-element methods. In the analytic methods based on the asymptotic expansions, the flow problem equations are solved in an approximate way by taking advantage of the characteristic aspect ratio of natural ice masses. The latter aspect ratio, a small dimensionless parameter ϵ, reflects varying conditions in lateral directions compared to the normal direction through an ice sheet thickness, and can be defined as the ratio of a thickness magnitude to a lateral ice sheet span magnitude, or the ratio of the stress and velocity gradients in the lateral direction of an ice sheet to those in the vertical direction. An alternative approach is defining the small parameter ϵ as a dimensionless viscosity parameter (Morland and Johnson 1980). By applying the method of asymptotic expansions in ϵ, and subsequent integration of the mass and momentum balance equations through an ice sheet depth, the number of space variables can be reduced by one, enabling thus a considerable simplification of the flow problem solution. Such a method is known in glaciology as the *Shallow Ice Approximation* (SIA), or the *Reduced Model*, and was originated by Fowler and Larson (1978), and was subsequently developed by Morland and Johnson (1980), Hutter (1981, 1983) and Morland (1984). The first application of the SIA to numerical modelling of ice sheet flows was due to Hutter et al. (1986); since then the method has been widely implemented, with various modifications, in many large-scale polar ice sheet models (Hindmarsh et al. 1987; Herterich 1988; Dahl-Jensen 1989; Huybrechts 1990; Fabre et al. 1995; Hindmarsh 2004). The Reduced Model, however, has a limitation: it is strictly valid only for bed slopes no greater than ϵ. For larger bed slope magnitudes, the theory was extended by Morland (2000, 2001) to constitute the *Enhanced Reduced Model*, uniformly valid for bed topographies with moderate slopes.

Although the SIA method has proved to be very effective in the large-scale ice sheet modelling, and soon became a standard method in theoretical glaciology, its application, in the original formulation, was restricted to the isotropic ice flows. Only few attempts were carried out to extend the SIA applicability to anisotropic ice. Mangeney and Califano (1998) investigated the flow problem in the ice divide region by adopting a transversely isotropic fabric representing that found in an ice core drilled within the Greenland Icecore Project (GRIP). The same type of anisotropic fabric was also used with the complete system of mechanical equations by Mangeney et al. (1996, 1997) to solve, by a discrete method, the problem of a steady-state flow of ice under isothermal conditions. In the latter papers, however, the empirically derived anisotropic fabrics were, in fact, functions of the ice depth only, since no anisotropic constitutive laws relating the fabric evolution to the flow field were incorporated in the models. Hence, the fabrics adopted in the above-cited papers were treated as *static*; that is, uncoupled from current flow fields. The full coupling between the flow and the *evolving* fabric, so that the flow field is not only a function of the anisotropic ice fabric, but also the fabric itself is a function of local field variables (current strains and strain-rates), was incorporated in finite-element models constructed by Staroszczyk and Morland (2000) and Staroszczyk (2003) (see Sect. 8.1 below), and later in SIA models developed by Morland and Staroszczyk (2006) and Staroszczyk (2006) (see Sect. 8.2).

It turns out that the inclusion of evolving ice anisotropy in the analysis significantly complicates the flow problem solution compared to the standard SIA method for isotropic ice, since a system of additional differential equations of the hyperbolic type, required to describe the evolution of the material properties along ice particle paths, has to be solved on top of the common mass and momentum balance equations. This might explain why no further progress in the development of SIA models for anisotropic ice has been observed since the publication of the above-cited papers (Morland and Staroszczyk 2006; Staroszczyk 2006), and it seems that in recent years the interest is primarily focused on the construction of discrete models for polar ice sheet flows. The examples of such models are those presented by Gillet-Chaulet et al. (2005), Greve and Blatter (2009), Ma et al. (2010), Gillet-Chaulet and Hindmarsh (2011), Seddik et al. (2012), Bargmann et al. (2012) and Pan et al. (2013).

In this Chapter both types of the afore-discussed ice sheet flow models are presented. In Sect. 8.1, a steady plane flow problem is considered, and its solution is based on a finite-element method. Section 8.2, in turn, deals with a steady axially-symmetric ice sheet flow problem, the solution of which is constructed as a leading-order approximation to the results obtained by applying the SIA theory. Both models incorporate the mechanism of dynamic recrystallization of ice. The results given by the two models illustrate the effects of the evolving macroscopic anisotropy of ice on the velocity fields in polar glaciers and their size and shape (the free surface profiles) under prescribed surface accumulation/ablation and basal melting rates. More details regarding the assumptions adopted in the models, the solution methods applied and the results obtained are provided in the following text.

In the ice sheet flow models presented in this chapter, the viscous behaviour of polycrystalline ice is described by the phenomenological constitutive laws formulated in Chap. 7. The direct incorporation into the ice sheet models of the micro-mechanical constitutive relations developed in Chap. 6 is not feasible yet because of an enormous number of variables involved (the necessity of tracking the behaviour of hundreds of ice grains separately at every material point of a numerical model consisting of hundreds of thousands of discrete nodes). In order to circumvent these difficulties of the computational nature, one can follow an approximate method proposed by Gillet-Chaulet et al. (2005). The idea of this method is to correlate the micro-mechanical and macroscopic responses in such a way that the microscopic response of ice is expressed by a small set of geometric parameters describing the micro-structure of the polycrystalline aggregate. The latter parameters are then used to evaluate macroscopic directional viscosities by applying an optimization

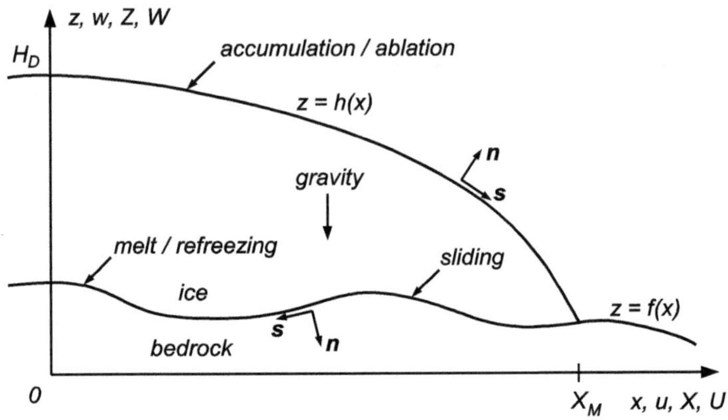

Fig. 8.1 Ice sheet geometry, coordinate axes and notations

technique to minimize the approximation errors. This operation is performed only once at the start of simulations, and in subsequent computations all required values of macroscopic viscosities are calculated by interpolating the earlier tabulated results.

8.1 Plane Ice Sheet Flow

In this section, a plane-strain, gravity-driven, steady flow of an ice sheet moving over a rigid bedrock is considered. The ice sheet is subjected to ice accumulation or ablation at its free surface, and to melting or refreezing at its base. For simplicity, it is assumed that the temperature field within the body of ice does not change in time, but can change with the depth of ice. The ice sheet cross-section is sketched in Fig. 8.1. The free surface of the glacier is traction-free, with the stress measured relative to the atmospheric pressure assumed to be constant (independent of the free surface elevation). At the glacier base either no-slip (when ice is frozen to the bed) or sliding can take place. All heat fluxes across the ice sheet boundaries are neglected in this analysis.

8.1.1 Flow Problem Formulation

The flow problem is solved in spatial rectangular Cartesian coordinates $Oxyz$, see Fig. 8.1, with the horizontal x-axis in the direction of flow to the right of the ice sheet divide located at $x = 0$, and the vertical z-axis directed upwards. The corresponding material coordinate axes, X and Z, are also shown in the figure. In the problem considered, there is no dependence on the transverse coordinate y. The geometry of the ice sheet is defined by the free surface elevation $z = h(x)$ and the bed elevation

$z = f(x)$. The ice sheet ends at a margin X_M where the ice thickness becomes zero, $h = f$. The ice accumulation rate, which is a normal mass flux per unit area, is denoted by q. It is assumed that q is positive (accumulation) in central regions at higher elevations, and negative (ablation) at lower elevations near the ice sheet margins. In general, q depends on both elevation h and location x. Similarly, the normal mass flux across the bed $f(x)$ is denoted by the melt rate b, which is assumed positive when ice melts, and negative when ice refreezes. The ice velocity vector $\boldsymbol{v}$ is defined by the components u, v and w in the x, y and z directions respectively, with the transverse component $v = 0$ in a plane flow. In terms of the velocity components u and w, the non-vanishing components of the strain-rate tensor $\boldsymbol{D}$ are given by

$$D_{xx} = \frac{\partial u}{\partial x}, \quad D_{zz} = \frac{\partial w}{\partial z}, \quad D_{xz} = \frac{1}{2}\left(\frac{\partial u}{\partial z} + \frac{\partial w}{\partial x}\right). \tag{8.1}$$

The mass balance equation is here the ice incompressibility condition given by $\operatorname{tr} \boldsymbol{D} = 0$, which in components reads

$$\frac{\partial u}{\partial x} + \frac{\partial w}{\partial z} = 0. \tag{8.2}$$

The Cauchy stress tensor $\boldsymbol{\sigma}$ has non-vanishing components σ_{xx}, σ_{yy}, σ_{zz} and σ_{xz}, which due to the decomposition (6.12) on p. 174 are defined in terms of the components of the deviatoric stress $\boldsymbol{S}$ and the mean pressure p by

$$\sigma_{xx} = S_{xx} - p, \quad \sigma_{yy} = S_{yy} - p, \quad \sigma_{zz} = S_{zz} - p, \quad \sigma_{xz} = S_{xz}. \tag{8.3}$$

The horizontal and vertical momentum balances, in the absence of inertia forces and due to negligibly small Reynolds numbers in typical polar ice sheet flows, are expressed by the equations of equilibrium under gravity

$$\frac{\partial S_{xx}}{\partial x} + \frac{\partial S_{xz}}{\partial z} - \frac{\partial p}{\partial x} = 0, \tag{8.4}$$

$$\frac{\partial S_{xz}}{\partial x} + \frac{\partial S_{zz}}{\partial z} - \frac{\partial p}{\partial z} = \varrho g, \tag{8.5}$$

where ϱ is the ice density and g is the gravitational acceleration.

The mass and momentum balance equations (8.2), (8.4) and (8.5) are subject to boundary conditions at the free surface and the base of an ice sheet, expressing the interactions of a glacier with the atmosphere and the underlying bedrock. Let us define the unit outward normal and tangent vectors $\boldsymbol{n}$ and $\boldsymbol{s}$ in a right-hand sense, as shown in Fig. 8.1. Then, at the free surface $h(x)$, the position of which is in general unknown, the unit vectors have components given by

$$\boldsymbol{n} = \Delta_h^{-1}[-h'(x),\ 0,\ 1], \quad \boldsymbol{s} = \Delta_h^{-1}[1,\ 0,\ h'(x)], \tag{8.6}$$

where

$$\Delta_h = \{1 + [h'(x)]^2\}^{1/2}, \tag{8.7}$$

with the primes denoting spatial derivatives. The zero traction condition at $z = h(x)$ is conveniently expressed in terms of vanishing normal and tangential tractions, $t_n = \boldsymbol{n} \cdot \boldsymbol{\sigma n}$ and $t_s = \boldsymbol{s} \cdot \boldsymbol{\sigma n}$, in the Oxz plane. Hence,

$$z = h(x): \quad \Delta_h^2 t_n = -\Delta_h^2 p + [h'(x)]^2 S_{xx} + S_{zz} - 2h'(x) S_{xz} = 0, \tag{8.8}$$

$$z = h(x): \quad \Delta_h^2 t_s = h'(x)(S_{zz} - S_{xx}) + \{1 - [h'(x)]^2\} S_{xz} = 0. \tag{8.9}$$

The prescription of the mass flux across the free surface due to ice accumulation or ablation yields the kinematic condition in the form

$$z = h(x): \quad h'(x)u - w = \Delta_h q, \tag{8.10}$$

where q denotes the ice accumulation rate.

At the prescribed bed $z = f(x)$, the unit vectors $\boldsymbol{n}$ and $\boldsymbol{s}$ have components given by

$$\boldsymbol{n} = \Delta_f^{-1}[f'(x), 0, -1], \quad \boldsymbol{s} = \Delta_f^{-1}[-1, 0, -f'(x)], \tag{8.11}$$

where

$$\Delta_f = \{1 + [f'(x)]^2\}^{1/2}. \tag{8.12}$$

With (8.11) and (8.12), normal and tangential tractions at the prescribed bed $z = f(x)$, t_n and t_s, respectively, are expressed by

$$z = f(x): \quad \Delta_f^2 t_n = -\Delta_f^2 p + [f'(x)]^2 S_{xx} + S_{zz} - 2f'(x) S_{xz}, \tag{8.13}$$

$$z = f(x): \quad \Delta_f^2 t_s = f'(x)(S_{zz} - S_{xx}) + \{1 - [f'(x)]^2\} S_{xz} 0, \tag{8.14}$$

and normal and tangential velocities at the bed, v_n and v_s, respectively, are

$$z = f(x): \quad \Delta_f v_n = f'(x)u - w, \quad \Delta_f v_s = -u - f'(x)w. \tag{8.15}$$

The kinematic condition defining the normal basal mass flux due to ice melt (drainage) or refreezing is given by

$$z = f(x): \quad f'(x)u - w = \Delta_f b, \tag{8.16}$$

with b denoting the basal melt rate.

At the glacier bed either no-slip or sliding can occur. In the former case of no sliding, the ice particle velocity component which is tangential to the surface $f(x)$ is zero; that is, $v_s = 0$. This, in view of $(8.15)_2$, provides the relation

$$u + f'(x)w = 0. \tag{8.17}$$

In the case of basal sliding, the tangential traction t_s at $z = f(x)$ is related to the tangential velocity v_s and the normal pressure $p = -t_n$ by a sliding law. A linear form of the latter law is adopted

$$z = f(x): \quad t_s = \lambda t_n v_s, \tag{8.18}$$

where λ is a constant friction coefficient. The proportionality of t_s to t_n ensures that, as an ice sheet margin X_M is approached and the pressure decreases to zero, the free surface slope $h'(X_M)$ at the margin is bounded (Morland and Johnson 1980). Otherwise, in the case of no-slip basal conditions, the slope at the margin would be unbounded.

The mass conservation balance (8.2) and the equilibrium equations (8.4) and (8.5), subject to relevant initial and boundary conditions, have to be supplemented by a flow law describing the viscous creep of anisotropic ice. For the constitutive description of the material, the phenomenological constitutive equations formulated in Chap. 7 are used, which can be generalized by the form

$$\boldsymbol{S} = 2\mu_0 \boldsymbol{H}(\boldsymbol{D}, \boldsymbol{B}, \boldsymbol{M}^{(r)}) \quad (r = 1, 2, 3). \tag{8.19}$$

In the above relation, the isotropic ice viscosity μ_0 is assumed to be a function of temperature and stress/strain-rate invariant, as described by Eqs. (7.44)–(7.48) in Sect. 7.3. The class of anisotropic flow laws (8.19), apart from the common dependence of the deviatoric stress $\boldsymbol{S}$ on the strain-rate $\boldsymbol{D}$, also incorporates the dependence on the strain measure $\boldsymbol{B}$, the left Cauchy-Green tensor. Since $\boldsymbol{B}$, as defined by (7.9) on p. 223, depends on the deformation gradient $\boldsymbol{F}$, we need to follow the evolution of $\boldsymbol{F}$ along ice particle paths, as ice deforms and develops anisotropy during its passage from the free surface to depth in an ice sheet. In a plane flow considered here, the deformation gradient has five non-zero components: F_{xx}, $F_{yy} = 1$, F_{zz}, F_{xz} and F_{zx}. By (7.9), the non-zero components of the tensor $\boldsymbol{B}$ are expressed by

$$\begin{aligned} B_{xx} &= F_{xx}^2 + F_{xz}^2, \quad B_{yy} = 1, \quad B_{zz} = F_{zx}^2 + F_{zz}^2, \\ B_{xz} &= B_{zx} = F_{xx}F_{zx} + F_{xz}F_{zz}. \end{aligned} \tag{8.20}$$

The evolution of the components F_{ij} is described by the kinematic relation (7.12), involving the velocity gradient $\boldsymbol{L}$. The non-zero components of the latter, defined by (7.13), are

$$L_{xx} = \frac{\partial u}{\partial x}, \quad L_{zz} = \frac{\partial w}{\partial z}, \quad L_{xz} = \frac{\partial u}{\partial z}, \quad L_{zx} = \frac{\partial w}{\partial x}. \tag{8.21}$$

Accordingly, (7.12) provides four first-order differential equations for non-trivial components of $\boldsymbol{F}$ as follows:

$$\frac{\partial F_{ij}}{\partial t} + v_k \frac{\partial F_{ij}}{\partial x_k} = F_{kj} \frac{\partial v_i}{\partial x_k} \quad (i, j, k = 1, 3), \tag{8.22}$$

with the equivalence $x_1 = x$, $x_3 = z$, $v_1 = u$, $v_3 = w$. Due to the ice incompressibility constraint $\det \boldsymbol{F} = 1$, only three evolution equations (8.22) are independent. The latter equations must be solved together with the mass balance relation (8.2) and the equilibrium equations (8.4) and (8.5), and are subject to the condition $\boldsymbol{F} = \boldsymbol{I}$ at the part of the free surface where the ice accumulation ($q > 0$) occurs, and where, initially isotropic, individual ice elements enter the ice sheet body.

8.1.2 Scaled Equations

In order to solve the mass conservation and equilibrium equations (8.2), (8.4) and (8.5), combined with the adopted constitutive law (8.19), two-step scaling is performed. First, physical dimensions are eliminated from the equations by using typical magnitudes of quantities involved, and, in the second step, the horizontal coordinates and velocities are stretched in such a way that both, lateral and vertical, coordinates and velocity components become order unity variables. This enables proper estimation of relative magnitudes of all terms entering the flow problem equations, so that those terms which are less important than other can be eliminated from the analysis to simplify the calculations. Hence, we adopt characteristic magnitudes: h^*, a typical ice thickness, used as a length scale, and v^*, a typical accumulation rate, used as a velocity unit. These two characteristic quantities determine other scaling parameters: a stress unit $\tau^* = \varrho g h^*$, a strain-rate unit $D^* = v^*/h^*$, a time scale $t^* = h^*/v^* = 1/D^*$, and a viscosity unit $\mu^* = \tau^* h^*/v^*$. By applying the adopted scales, we introduce dimensionless variables, indicated by a superposed bar, defined by

$$
\begin{gathered}
(\bar{x}, \bar{z}) = (x, z)/h^*, \quad (\bar{u}, \bar{w}) = (u, w)/v^*, \quad (\bar{\boldsymbol{S}}, \bar{p}) = (\boldsymbol{S}, p)/\tau^*, \\
(\bar{\boldsymbol{L}}, \bar{\boldsymbol{D}}) = (\boldsymbol{L}, \boldsymbol{D})/D^*, \quad \bar{t} = t/t^*, \quad \bar{\mu}_0 = \mu_0/\mu^*.
\end{gathered}
\tag{8.23}
$$

Then, by means of the small aspect ratio parameter ϵ, we stretch the lateral coordinate $\bar{x}$ and the velocity $\bar{u}$, while leaving the vertical counterparts unchanged, to obtain

$$
X = \epsilon \bar{x}, \quad Z = \bar{z}, \quad U = \epsilon \bar{u}, \quad W = \bar{w}. \tag{8.24}
$$

This results in X, Z, U and W all being order unity. We also introduce the normalized free surface and bed profiles, $H(X) = h/h^*$ and $F(X) = f/h^*$ respectively, together with their corresponding slopes $H'(X) = \Gamma(X)$ and $F'(X) = \beta(X)$, all being order unity variables as well. In terms of the normalized stresses and strain-rates, the constitutive law (8.19) takes the dimensionless form

$$
\bar{\boldsymbol{S}} = 2\bar{\mu}_0 \boldsymbol{H}(\bar{\boldsymbol{D}}, \boldsymbol{B}, \boldsymbol{M}^{(r)}). \tag{8.25}
$$

The components of the dimensionless stress $\bar{\mathbf{S}}$ can be rescaled by using the small parameter ϵ to give

$$\bar{S}_{xx} = \epsilon\Sigma_{xx}, \quad \bar{S}_{zz} = \epsilon\Sigma_{zz}, \quad \bar{S}_{xz} = \epsilon\Sigma_{xz}, \quad \bar{p} = P, \tag{8.26}$$

so that Σ_{ij} are (at most) order unity components of a normalized deviatoric stress tensor $\boldsymbol{\Sigma}$, and P is order unity normalized pressure.

Application of the above scalings transforms the mass balance relation (8.2) to

$$\frac{\partial U}{\partial X} + \frac{\partial W}{\partial Z} = 0, \tag{8.27}$$

while the equilibrium equations (8.4) and (8.5) become

$$\epsilon\,\frac{\partial \Sigma_{xx}}{\partial X} + \frac{\partial \Sigma_{xz}}{\partial Z} - \frac{\partial P}{\partial X} = 0, \tag{8.28}$$

$$\epsilon^2\,\frac{\partial \Sigma_{xz}}{\partial X} + \epsilon\,\frac{\partial \Sigma_{zz}}{\partial Z} - \frac{\partial P}{\partial Z} = 1, \tag{8.29}$$

where all the derivatives involved are order unity. In normalized stretched variables, the zero free surface tractions (8.8) and (8.9) are expressed by

$$Z = H(X): \quad -\Delta_h^2 P + \epsilon\Sigma_{zz} - 2\epsilon^2\Gamma\Sigma_{xz} + \epsilon^3\Gamma^2\Sigma_{xx} = 0, \tag{8.30}$$

$$Z = H(X): \quad (1 - \epsilon^2\Gamma^2)\Sigma_{xz} + \epsilon\Gamma(\Sigma_{zz} - \Sigma_{xx}) = 0, \tag{8.31}$$

and the free surface kinematic condition (8.10) takes the form

$$Z = H(X): \quad \Gamma U - W = \Delta_h Q, \tag{8.32}$$

where $Q = q/v^*$ is a normalized ice accumulation rate. Similarly, the scaled relations for the basal normal and tangential tractions, (8.13) and (8.14), are given by

$$Z = F(X): \quad \Delta_f^2 T_n = -\Delta_f^2 P + \epsilon\Sigma_{zz} - 2\epsilon^2\beta\Sigma_{xz} + \epsilon^3\beta^2\Sigma_{xx}, \tag{8.33}$$

$$Z = F(X): \quad \Delta_f^2 T_s = (1 - \epsilon^2\beta^2)\Sigma_{xz} + \epsilon\beta(\Sigma_{zz} - \Sigma_{xx}), \tag{8.34}$$

where $T_n = t_n/\tau^*$ and $T_s = \epsilon^{-1}t_s/\tau^*$ are scaled basal tractions, both being order unity variables. The normal and tangential basal velocities, given by (8.15), are

$$Z = F(X): \quad \Delta_f V_n = \beta U - W, \quad \Delta_f V_s = -U - \epsilon^2\beta W, \tag{8.35}$$

where $V_n = v_n/v^*$ and $V_s = \epsilon v_s/v^*$ denote order unity components of the basal velocity vector. The kinematic condition at the bed, (8.16), becomes

$$Z = F(X): \quad \beta U - W = \Delta_f B, \tag{8.36}$$

where $B = b/v^*$ is a normalized basal melt rate. The expressions for Δ_h and Δ_f, see (8.7) and (8.12), now have the forms

$$\Delta_h = \left(1 + \epsilon^2 \Gamma^2\right)^{1/2}, \quad \Delta_f = \left(1 + \epsilon^2 \beta^2\right)^{1/2}. \tag{8.37}$$

Further, the sliding law, in physical variables defined by (8.18), in the normalized dimensionless form is expressed by

$$T_s = \Lambda\, T_n V_s\,, \tag{8.38}$$

where $\Lambda = \epsilon^{-2} v^* \lambda$ is an order unity or greater normalized basal friction coefficient, the basal tractions T_n and T_s are defined by (8.33) and (8.34), and the basal velocity V_s is given by $(8.35)_2$.

Finally, the constitutive law (8.25) needs its arguments to be expressed in the scaled variables either, which requires the strain-rate components (8.1) to be defined in terms of the stretched coordinates and velocities (8.24). Hence, we have

$$\bar{D}_{xx} = \frac{\partial U}{\partial X}, \quad \bar{D}_{zz} = \frac{\partial W}{\partial Z}, \quad \bar{D}_{xz} = \frac{1}{2}\left(\epsilon^{-1}\frac{\partial U}{\partial Z} + \epsilon \frac{\partial W}{\partial X}\right), \tag{8.39}$$

showing that the dominant strain-rate component, of order ϵ^{-1}, is $\bar{D}_{xz}$, with the other two components, $\bar{D}_{xx}$ and $\bar{D}_{zz}$, being order unity. In addition, also the velocity gradient components (8.21) need to be expressed in the scaled variables in order to solve the deformation gradient evolution equations (8.22) along ice particle paths. These components are given by

$$\bar{L}_{xx} = \frac{\partial U}{\partial X}, \quad \bar{L}_{zz} = \frac{\partial W}{\partial Z}, \quad \bar{L}_{xz} = \epsilon^{-1}\frac{\partial U}{\partial Z}, \quad \bar{L}_{zx} = \epsilon \frac{\partial W}{\partial X}. \tag{8.40}$$

The deformation gradient components, entering the evolution equations (8.22), are expressed in the stretched coordinates as

$$F_{xx} = \frac{\partial X}{\partial X^*}, \quad F_{zz} = \frac{\partial Z}{\partial Z^*}, \quad F_{xz} = \epsilon^{-1}\frac{\partial X}{\partial Z^*}, \quad F_{zx} = \epsilon \frac{\partial Z}{\partial X^*}, \tag{8.41}$$

where X^* and Z^* are the stretched normalized reference coordinates, related to their physical counterparts by the scalings analogous to (8.23) and (8.24).

Since the dominant normalized strain-rate, $\bar{D}_{xz}$, is of order ϵ^{-1}, and the corresponding normalized stress $\bar{S}_{xz}$ is of order ϵ (this can be seen from Eq. $(8.26)_3$, in which Σ_{xz} is order unity), it follows from the constitutive law (8.25) (Staroszczyk 2004) that the viscosity $\bar{\mu}_0$ is of order ϵ^2. This prompts rescaling of the ice viscosity by introducing an order unity isotropic ice viscosity $\tilde{\mu}_0$ defined by

$$\bar{\mu}_0 = \epsilon^2 \tilde{\mu}_0\,. \tag{8.42}$$

Such a viscosity renormalization corresponds equivalently to treating ϵ^2 as a dimensionless viscosity (Morland and Johnson 1980; Morland 1984). The requirement that $\tilde{\mu}_0$ is of order unity, on account of the scalings (8.23) and Eqs. (7.44)–(7.48) in Sect. 7.3 expressing the physical viscosity μ_0 in terms of temperature and the strain-rate/stress invariants, yields a relation which defines the small parameter ϵ as

$$\epsilon = \frac{1}{h^*}\left(\frac{\sigma_0 v^*}{\varrho g D_0}\right)^{1/2} . \tag{8.43}$$

Choosing typical magnitudes of the ice sheet thickness as $h^* = 2000$ m and the ice accumulation rate as $v^* = 1\,\mathrm{m\,yr^{-1}} = 3.17 \times 10^{-8}\,\mathrm{m\,s^{-1}}$ (the unit 'yr' denotes a year), relation (8.43) gives $\epsilon = 0.00167 \sim 1/600$ (the units σ_0 and D_0 are defined in Sect. 7.3, and the ice density $\rho = 917\,\mathrm{kg\,m^{-3}}$).

The ice sheet flow governing equations (8.27)–(8.29), combined with the deformation gradient evolution equations (8.22), describe the problem in terms of the set of dimensionless quantities and the small parameter $\epsilon \ll 1$. In this section, concerned with the plane flow problem, the governing equations are solved in their full forms by applying a discrete approach, based on the finite-element method. In the following Sect. 8.2, dealing with the axially-symmetric ice sheet flow, the solution is constructed as a leading-order approximation, derived by neglecting in the flow equations all the terms which are of order ϵ or smaller compared to unity. The ensuing leading-order equations are solved by first integrating them through the ice thickness to eliminate one spatial coordinate, and then by calculating the ice sheet free surface profile and the velocity and stress fields by applying a numerical method for integration of a system of differential equations.

Two forms of the constitutive laws are used here in the plane ice sheet flow models: (1) the multiplicative form (7.43) on p. 229 and (2) the additive form (7.170) on p. 267 employed to describe the mechanism of dynamic recrystallization of ice. By expressing the flow law (7.43) in the normalized variables, as prescribed by the general representation (8.25), one obtains the following relations for the deviatoric stresses given in terms of the velocities:

$$\begin{aligned}
\Sigma_{xx} &= \tilde{\mu}_0 \left[a_1 \epsilon \frac{\partial U}{\partial X} + a_3 \left(\frac{\partial U}{\partial Z} + \epsilon^2 \frac{\partial W}{\partial X} \right) \right], \\
\Sigma_{zz} &= \tilde{\mu}_0 \left[a_2 \epsilon \frac{\partial W}{\partial Z} + a_3 \left(\frac{\partial U}{\partial Z} + \epsilon^2 \frac{\partial W}{\partial X} \right) \right], \\
\Sigma_{xz} &= \tilde{\mu}_0 \, a_4 \left(\frac{\partial U}{\partial Z} + \epsilon^2 \frac{\partial W}{\partial X} \right),
\end{aligned} \tag{8.44}$$

where the coefficients depending on the fabric response functions $f(b)$ and $G(K)$ are given by

$$
\begin{aligned}
a_1 &= \frac{2}{3}\Big[f(b_1)\left(1+M_{11}^{(1)}\right)+f(b_3)\left(1+M_{11}^{(3)}\right)+K^{-1}G(K)\,(2B_{xx}+B_{zz})\Big],\\
a_2 &= \frac{2}{3}\Big[f(b_1)\left(1+M_{33}^{(1)}\right)+f(b_3)\left(1+M_{33}^{(3)}\right)+K^{-1}G(K)\,(B_{xx}+2B_{zz})\Big],\\
a_3 &= \frac{1}{3}\Big[f(b_1)M_{13}^{(1)}+f(b_3)M_{13}^{(3)}+K^{-1}G(K)B_{xz}\Big],\\
a_4 &= \frac{1}{2}\big[f(b_1)+f(b_3)+K^{-1}G(K)\big(B_{xx}+B_{zz}\big)\big].
\end{aligned}
$$

In the initial undeformed state, when $\boldsymbol{B}=\boldsymbol{I}$, so that $b_1=b_2=b_3=1$ and $K=3$, on account of the normalization relation $(7.39)_2$, the above four coefficients become

$$
a_1=a_2=2,\quad a_3=0,\quad a_4=1, \tag{8.45}
$$

and relations (8.44) reduce to the isotropic ice viscous flow relations

$$
\Sigma_{xx}=2\tilde{\mu}_0\epsilon\,\frac{\partial U}{\partial X},\quad \Sigma_{zz}=2\tilde{\mu}_0\epsilon\,\frac{\partial W}{\partial Z},\quad \Sigma_{xz}=\tilde{\mu}_0\left(\frac{\partial U}{\partial Z}+\epsilon^2\,\frac{\partial W}{\partial X}\right), \tag{8.46}
$$

confirming a result known from the SIA theory that, to the leading order, the normalized axial deviatoric stresses Σ_{xx} and Σ_{zz} are order ϵ quantities, while the normalized shear stress Σ_{xz} is order unity.

Similarly, the additive constitutive equations (7.170), when expressed in the normalized variables, give the following deviatoric stress components:

$$
\begin{aligned}
\Sigma_{xx} &= \tilde{\mu}_0\left[(2+c_1)\,\epsilon\,\frac{\partial U}{\partial X}+c_3\left(\frac{\partial U}{\partial Z}+\epsilon^2\,\frac{\partial W}{\partial X}\right)\right],\\
\Sigma_{zz} &= \tilde{\mu}_0\left[(2+c_2)\,\epsilon\,\frac{\partial W}{\partial Z}+c_3\left(\frac{\partial U}{\partial Z}+\epsilon^2\,\frac{\partial W}{\partial X}\right)\right],\\
\Sigma_{xz} &= \tilde{\mu}_0\,(1+c_4)\left(\frac{\partial U}{\partial Z}+\epsilon^2\,\frac{\partial W}{\partial X}\right),
\end{aligned} \tag{8.47}
$$

where the coefficients depending on the fabric response functions $\tilde{f}(b)$ and $\tilde{G}(K)$ are defined by

$$
\begin{aligned}
c_1 &= \frac{2}{3}r(\bar{I})\Big[\tilde{f}(b_1)\left(1+M_{11}^{(1)}\right)+\tilde{f}(b_3)\left(1+M_{11}^{(3)}\right)+K^{-1}\tilde{G}(K)\,(2B_{xx}+B_{zz})\Big],\\
c_2 &= \frac{2}{3}r(\bar{I})\Big[\tilde{f}(b_1)\left(1+M_{33}^{(1)}\right)+\tilde{f}(b_3)\left(1+M_{33}^{(3)}\right)+K^{-1}\tilde{G}(K)\,(B_{xx}+2B_{zz})\Big],\\
c_3 &= \frac{1}{3}r(\bar{I})\Big[\tilde{f}(b_1)M_{13}^{(1)}+\tilde{f}(b_3)M_{13}^{(3)}+K^{-1}\tilde{G}(K)B_{xz}\Big],\\
c_4 &= \frac{1}{2}r(\bar{I})\Big[\tilde{f}(b_1)+\tilde{f}(b_3)+K^{-1}\tilde{G}(K)\big(B_{xx}+B_{zz}\big)\Big].
\end{aligned}
$$

Recall that $r(\bar{I})$ is the ice fabric strength scaling factor defined by equation (7.153) on p. 261, see also Fig. 7.19. In the initial undeformed state, given by $\boldsymbol{B} = \boldsymbol{I}$, all four coefficients c_i become zero, regardless of the value of the scaling factor $r(\bar{I})$. These coefficients are also zero for $r = 0$; that is, for the effective strain-rate invariant $\bar{I}$ exceeding its upper critical level $\bar{I}_{cu}$, after the dynamic recrystallization process has been completed. In either case, formulae (8.47) reduce to the isotropic viscous fluid flow relations (8.46) again.

8.1.3 Finite-Element Model

The flow problem defined by the ice incompressibility condition (8.27) and the equilibrium equations (8.28) and (8.29), with the deviatoric stress tensor components given by either (8.44) for the multiplicative, or (8.47) for the additive form of the constitutive law, is solved for the normalized velocities U and W and pressure P by employing a finite-element method (FEM). A weighted residual, or Galerkin, approach is adopted (Zienkiewicz et al. 2005b) in which the problem equations are satisfied in an integral mean sense. The plane domain bounded by the ice sheet free surface $H(X)$ and the ice–bedrock interface $F(X)$ is discretized by a mesh of triangular finite elements. Since in the ensuing equilibrium equations the spatial derivatives of the velocity functions U and W are by one order higher than the derivatives of the pressure function P, for numerical reasons, the velocity field is approximated by applying polynomial interpolation (shape) functions which are by one order higher than those for the pressure field. Accordingly, six-node elements are adopted, with three nodes at the vertices and three nodes at the mid-side points, see Fig. 8.2. The unknown values of the discrete velocities, U_j and W_j, are defined at all six nodal points, while the discrete values of the pressure field, P_k, are defined only at the three vertices. Therefore, in each finite element there are 15 discrete parameters to be calculated. In terms of the nodal parameters, the continuous functions U, W and P are approximated by the representations

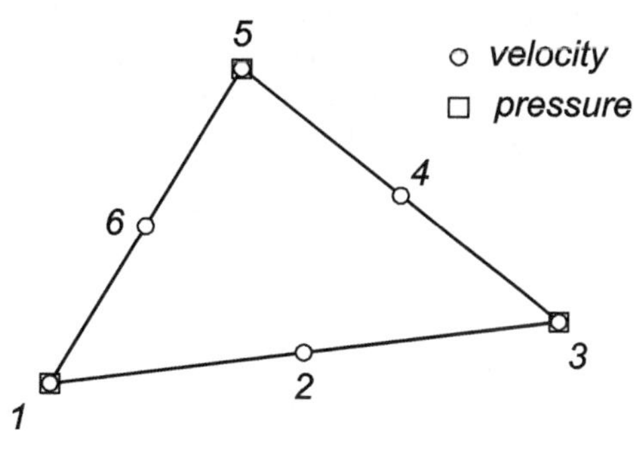

Fig. 8.2 A triangular finite element with six nodes

$$\begin{aligned} U(X,Z) &= \Phi^v_j(X,Z)\,U_j\,, \\ W(X,Z) &= \Phi^v_j(X,Z)\,W_j\,, \quad (j=1,\ldots,6), \\ P(X,Z) &= \Phi^p_k(X,Z)\,P_k\,, \quad (k=1,3,5), \end{aligned} \tag{8.48}$$

where the summation convention over repeated indices applies. The shape (interpolation) functions Φ^v_j and Φ^p_k are assumed to be different for the velocity and pressure fields: the velocities are approximated by bi-quadratic shape functions, whereas the pressure field is approximated by bi-linear shape functions. Standard polynomial forms of these functions (Zienkiewicz et al. 2005b) are adopted in the model.

In the weighted residual method, the problem governing equations are solved in their weak forms. Hence, the mass balance and equilibrium equations are first multiplied by a set of continuous and sufficiently smooth weighting functions, which in the Galerkin method are identical to the element shape functions, here Φ^v_j and Φ^p_k. The resulting relations are then integrated over the whole ice sheet domain, with the aim to minimize the error of the approximate solution. In the process, to reduce the order of spatial derivatives, Green's theorem is used. This transforms the problem to the solution of a set of algebraic equations, which in a matrix form is expressed by

$$\boldsymbol{K}\boldsymbol{w} = \boldsymbol{f}\,, \tag{8.49}$$

where the vector $\boldsymbol{w}$ includes the unknown values of the velocities U_l and W_l and the pressures P_l at all nodal points l of the discrete system. For a single element, with 15 degrees of freedom, the element nodal parameters are assumed to be arranged in the following order

$$\boldsymbol{w}^e = (P_1, U_1, W_1, U_2, W_2, P_3, U_3, W_3, U_4, W_4, P_5, U_5, W_5, U_6, W_6)^T. \tag{8.50}$$

The global matrix $\boldsymbol{K}$ and the forcing vector $\boldsymbol{f}$ are assembled from respective single element matrices and vectors, $\boldsymbol{K}^e$ and $\boldsymbol{f}^e$ respectively, in a manner typical of the finite-element method. The element matrices $\boldsymbol{K}^e$, each 15×15 in size, are, in turn, composed of 9 submatrices $\boldsymbol{k}^{rs}$ $(r, s = 1, 2, 3)$, each of dimensions 5×5. Similarly, the element vectors $\boldsymbol{f}^e$, of the same structure as the vector $\boldsymbol{w}^e$ in (8.50), are made up of three vectors $\boldsymbol{f}^r$, each of length 5. With the ordering of the nodal parameters defined by (8.50), the non-vanishing entries in the components matrices $\boldsymbol{k}^{rs}$ and vectors $\boldsymbol{f}^r$, for the multiplicative constitutive law (7.43) and hence the stresses expressed by (8.44), are defined by the integrals

$$k^{rs}_{1,2n} = \int_A \Phi^p_{2r-1}\frac{\partial \Phi^v_j}{\partial X}\,\mathrm{d}A, \quad k^{rs}_{1,2n+1} = \int_A \Phi^p_{2r-1}\frac{\partial \Phi^v_j}{\partial Z}\,\mathrm{d}A,$$

$$k^{rs}_{2m,1} = -\int_A \frac{\partial \Phi^v_i}{\partial X}\Phi^p_{2s-1}\,\mathrm{d}A, \quad k^{rs}_{2m+1,1} = -\int_A \frac{\partial \Phi^v_i}{\partial Z}\Phi^p_{2s-1}\,\mathrm{d}A,$$

$$
\begin{aligned}
k^{rs}_{2m,\,2n} &= \tilde{\mu}_0 \int\limits_A \left(\epsilon^2 a_1 \frac{\partial \Phi^v_i}{\partial X}\frac{\partial \Phi^v_j}{\partial X} + \epsilon a_3 \frac{\partial \Phi^v_i}{\partial X}\frac{\partial \Phi^v_j}{\partial Z} + a_4 \frac{\partial \Phi^v_i}{\partial Z}\frac{\partial \Phi^v_j}{\partial Z} \right) \mathrm{d}A, \\
k^{rs}_{2m,\,2n+1} &= \tilde{\mu}_0 \int\limits_A \left(\epsilon^3 a_3 \frac{\partial \Phi^v_i}{\partial X}\frac{\partial \Phi^v_j}{\partial X} + \epsilon^2 a_4 \frac{\partial \Phi^v_i}{\partial Z}\frac{\partial \Phi^v_j}{\partial X} \right) \mathrm{d}A, \qquad (8.51)\\
k^{rs}_{2m+1,\,2n} &= \tilde{\mu}_0 \int\limits_A \left(\epsilon a_3 \frac{\partial \Phi^v_i}{\partial Z}\frac{\partial \Phi^v_j}{\partial Z} + \epsilon^2 a_4 \frac{\partial \Phi^v_i}{\partial X}\frac{\partial \Phi^v_j}{\partial Z} \right) \mathrm{d}A, \\
k^{rs}_{2m+1,\,2n+1} &= \tilde{\mu}_0 \int\limits_A \left(\epsilon^2 a_2 \frac{\partial \Phi^v_i}{\partial Z}\frac{\partial \Phi^v_j}{\partial Z} + \epsilon^3 a_3 \frac{\partial \Phi^v_i}{\partial Z}\frac{\partial \Phi^v_j}{\partial X} + \epsilon^4 a_4 \frac{\partial \Phi^v_i}{\partial X}\frac{\partial \Phi^v_j}{\partial X} \right) \mathrm{d}A, \\
f^r_{2m+1} &= -\int\limits_A \Phi^v_i \,\mathrm{d}A,
\end{aligned}
$$

where

$$
r, s = 1, 2, 3; \quad m, n = 1, 2; \quad i = 2(r-1) + m, \quad j = 2(s-1) + n,
$$

and A denotes the plane domain of integration within a given triangular element. The analogous relations for the additive constitutive law (7.170), for which the stresses are given by (8.47), are obtained by replacing the coefficients a_1, a_2, a_3 and a_4 in (8.51) by their counterparts $(c_1 + 2)$, $(c_2 + 2)$, c_3 and $(c_4 + 1)$, respectively.

The surface integrals prescribed by (8.51) are evaluated numerically by applying the Gauss-Legendre quadrature with seven sampling points within a triangular element (Zienkiewicz et al. 2005b). Simultaneously with the finite-element equations for U, W and P, four first-order differential equations (8.22) describing the evolution of the deformation gradient $\boldsymbol{F}$ are solved to determine the anisotropic fabric associated with the current deformation field. The continuous functions F_{xx}, F_{xz}, F_{zx} and F_{zz} are approximated within finite elements in the same way as the pressures field P. That is, bi-linear interpolation of discrete values given at the three vertex nodal points is applied, so that the variation of the deformation gradient components is described by relations analogous to $(8.48)_3$.

The finite-element model described above was used to simulate plane, steady in time flow of anisotropic ice over a horizontal bed. To find a steady solution, an iterative algorithm was used. In this algorithm, the calculations are started by assuming that ice is isotropic throughout an ice sheet; that is, $\boldsymbol{F} = \boldsymbol{I}$ everywhere. As an initial isotropic ice flow solution, the velocities and pressures predicted by the SIA method are used. Starting from this initial state (which is slightly unbalanced due to the approximate nature of the SIA solution), new ice velocities are calculated by solving the FEM equations, and these results are used to determine the velocity gradient components L_{ij}. The latter are necessary to evaluate new deformation gradient components F_{ij} by solving the evolution equation (8.22). The updated deformation gradients are then used to calculate the values and directions of the principal stretches λ_1 and λ_3

($\lambda_2 = 1$) defining the current ice fabric, which, in turn, enables the evaluation of the response coefficients in the constitutive laws. After completing the above steps, a new iteration is started, in which the FEM equations are solved to provide new velocities that are used again to update the fabric and the response coefficients, etc. The whole iteration process as described above is continued until a stationary solution is obtained. In our simulations, depending on the choice of parameters defining the limit strength of ice anisotropy (the specific values of the enhancement factors), the stationary flow was usually reached for the values of the dimensionless time $\bar{t} = t/t^*$ ranging from about 0.2 to about 0.3.

The ice incompressibility constraint (8.27), expressing the mass conservation balance, requires special numerical treatment, since the absence of a pressure term in (8.27), in combination with equilibrium equations (8.28) and (8.29), gives rise to an ill-conditioned system of algebraic equations that are solved in the finite-element method. A number of approaches have been developed and used in practice to resolve the above numerical problem associated with the incompressibility of the medium. It appears that the first such an approach was proposed by Chorin (1967), and is known as the pseudo-incompressibility method. In this method, a compressibility term in the form of a pressure time-derivative is added to relation (8.27), and when a steady state is reached, this artificial term vanishes. In an alternative approach, a so-called pressure correction method (Hirsch 1992), an iterative procedure between the velocity and pressure fields is applied, in which the momentum balance equations are solved together with a Poisson equation for the pressure correction term that is derived by taking the divergence of the momentum equations. In incompressible elasticity and plasticity two other methods are often used (Zienkiewicz et al. 2005b). One is the standard penalty method, in which a pressure term times some small penalty parameter is inserted in the incompressibility relation. The other method, which can be viewed as an extension of the penalty method, consists in subtracting from both sides of the incompressibility relation (8.27) a pressure term times some small parameter. The resulting system of the momentum and modified mass conservation equations is then solved by iterations, in which the initially adopted pressure is gradually updated until the stationary solution is found. It seems that nowadays the most common approach to solving the problems involving incompressible materials is a projection method, known in computational fluid mechanics as a fractional step method (Zienkiewicz et al. 2005a). In this method, at each computational time step, the velocity field is first calculated without enforcing incompressibility, and then the intermediate velocity field is projected onto a divergent-free space to satisfy the incompressibility constraint (Chorin 1968); the latter requires the solution of a Poisson equation for the pressure field. It is likely that the projection method was first introduced to the field of theoretical glaciology by Mangeney et al. (1996). In the finite-element model presented here, two methods have been applied: the pseudo-incompressibility method and the penalty method.

8.1.4 Numerical Flow Simulations

The finite-element model described in Sect. 8.1.3 is a general-purpose model that can be used to solve numerically a wide class of problems involving the viscous flow of a large polar ice sheet, provided that the ice deformation can be treated as occurring in plane flow conditions. Hence, steady or unsteady flows can be simulated in which the anisotropy of ice evolves, with possible non-linear effects, such as the dependence of ice viscosity on current strain-rate or stress, incorporated in the model. Here, however, the model is applied to solve a flow problem of a simple, idealized geometry, in order to demonstrate how the mechanism of induced anisotropy affects the flow of ice (in particular, the velocities within an ice sheet) compared to the case of isotropic ice flow.

Accordingly, a steady flow of an ice sheet with a prescribed geometry is considered, for which the velocity depth profiles and the free surface accumulation rates necessary to maintain the steady-state flow are calculated. It is assumed for simplicity that the ice flows over a flat horizontal bed $F(X) = 0$, and the free surface elevation $H(X)$ is given by the equation

$$H + X^2 = 1. \tag{8.52}$$

Thus, the ice divide is at $X = 0$ and the margin is at $X = 1$, so that the lateral span of the glacier, L, is equal to unity in the stretched normalized coordinates. It is also assumed that the ice sheet free surface profile is symmetric about the plane $X = 0$, and therefore the considerations are restricted to the region $0 \leq X \leq 1$. The simulations were carried out for the small aspect ratio parameter $\epsilon = 10^{-2}$. At the bed $Z = 0$ no-slip conditions are assumed, with zero melt rate ($B = 0$). These, together with the flow symmetry conditions at the divide $X = 0$, are expressed by

$$Z = 0: \quad U = W = 0; \quad X = 0: \quad U = 0, \quad \frac{\partial W}{\partial Z} = 0. \tag{8.53}$$

All the results discussed below were obtained by running a discrete model consisting of 6414 six-node triangular elements shown in Fig. 8.2, with a total number of 29 376 degrees of freedom in the discrete system.

As first, the results of simulations of an isothermal ice flow, with the viscous response of ice defined by the multiplicative constitutive relation (7.43) on p. 229, are presented. These results were obtained for the constant normalized viscosity $\tilde{\mu}_0 = 1$, with no dependence on temperature and strain-rate or stress invariants. The fabric response functions were adopted in the form defined by (7.76) on p. 236, with the parameter $m = 2$.

Figure 8.3 illustrates how the overall flow rate of an ice sheet depends on the limit viscous properties of anisotropic ice, defined by the magnitudes of the axial and shear enhancement factors, E_a and E_s, respectively. The plots show the variation of the horizontal and vertical velocities at the free surface, U_s and W_s, and the accumulation rate Q (given by (8.32)) required to maintain the ice sheet geometry prescribed by (8.52). The symbols A and S denote here the limit viscosity factors

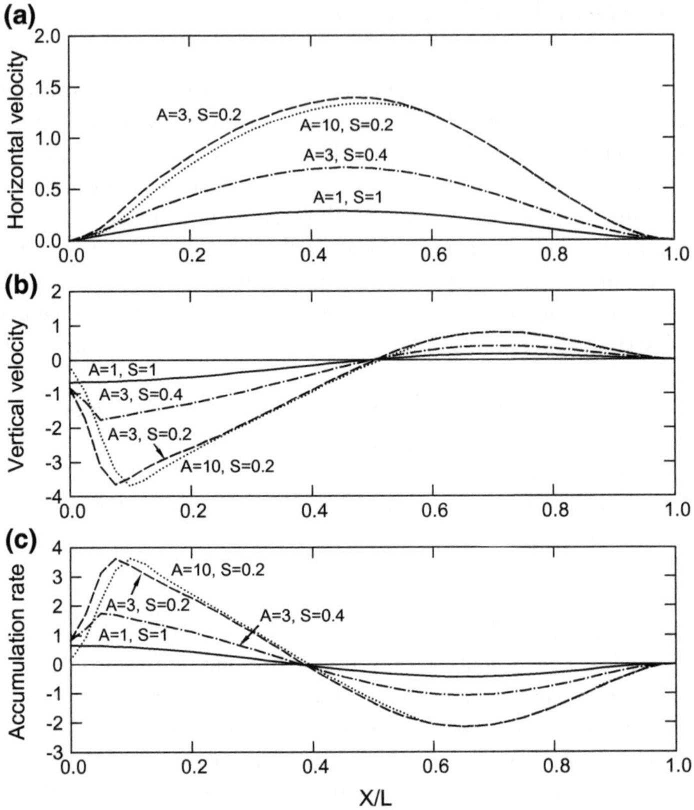

Fig. 8.3 **a** Free surface horizontal velocities U_s, **b** vertical velocities W_s, and **c** the accumulation rates Q, for different combinations of the limit viscosity factors $A = 1/E_a$ and $S = 1/E_s$. Reprinted from Staroszczyk and Morland (2000), Fig. 4, with permission of the International Glaciological Society

for compression and shear respectively, equal to the reciprocals of the respective enhancement factors E_a and E_s; that is, $A = 1/E_a$ and $S = 1/E_s$. It can be noted that the effect on the global flow of the parameter describing the behaviour of ice in compression is very limited. Comparison of the results obtained for $A = 3$ and $S = 0.2$ (dashed lines) with the results for $A = 10$ and the same S (dotted lines) shows that the differences between the free surface velocities are small and essentially confined to the zone of positive ice accumulation Q. On the contrary, it is seen in the plots that the significance of the shear enhancement is crucial, as the change in its value considerably changes the global ice sheet flow-rate. Comparison of the results obtained for the values $A = 3$ and $S = 0.2$ ($E_s = 5$), dashed lines, with those for $A = 3$ and $S = 0.4$ ($E_s = 2.5$), dashed-dotted lines, indicates that both horizontal and vertical velocities, over practically the whole span of the ice sheet, are in the former case about twice as large as in the latter case. That is, the global flow-rate

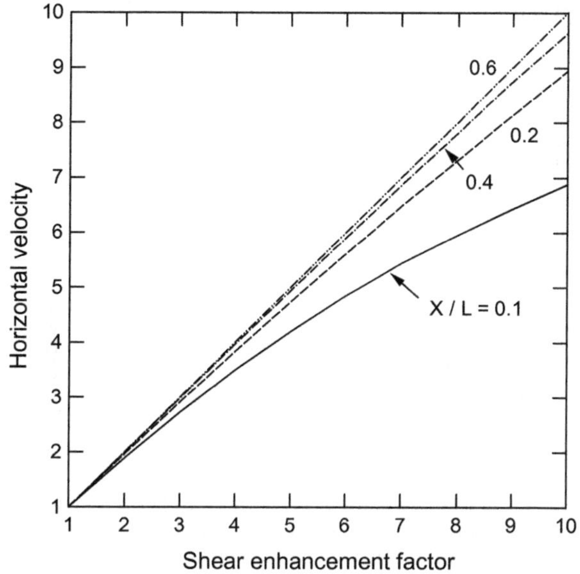

Fig. 8.4 Variation of the free surface horizontal velocity U_s with increasing shear enhancement factor E_s at different locations X/L (for the axial enhancement factor $E_a = 1/3$). Reprinted from Staroszczyk and Morland (2000), Fig. 5, with permission of the International Glaciological Society

of ice is approximately proportional to the magnitude of the shear enhancement factor E_s.

The observations from the previous figure are confirmed by the results shown in Fig. 8.4, illustrating how the free surface horizontal velocity, U_s, obtained for anisotropic ice ($E_s > 1$) increases with respect to the corresponding velocity for isotropic ice ($E_s = 1$); the results were obtained for the axial enhancement factor $E_a = 1/3$. The horizontal velocity ratio is plotted against the value of the enhancement factor E_s for different locations X/L indicated in the figure. It can be seen that at locations distant from the ice divide $X = 0$, curves for $X/L = 0.4$ and $X/L = 0.6$, the increase in the flow velocity U_s is almost exactly proportional to the magnitude of the shear enhancement factor. Only near the ice divide, curves for $X/L = 0.1$ and $X/L = 0.2$, is the increase slightly smaller. Such a feature of the ice sheet viscous behaviour can be explained by the fact that in the near-divide region both axial and shear deviatoric stresses are of comparable magnitudes, whereas in the regions far from the divide the shear stresses increase steadily (they are approximately proportional to the free surface slope), so that the viscous flow of ice becomes increasingly dominated by the shear stresses.

The plots in Fig. 8.5 display depth profiles of the horizontal and vertical velocities at different locations X/L, indicated by the respective labels in the figure. Compared are the results calculated for anisotropic ice, defined by the parameters $E_a = 1/3$ and $E_s = 5$ pertinent to cold ice, and for isotropic ice, $E_a = E_s = 1$. The velocities are normalized by respective values at the free surface $Z = H$; that is, the ratios $U(X, Z)/U_s(X)$ and $W(X, Z)/W_s(X)$ are plotted against the normalized elevation Z/H. One can see that the depth profiles of the horizontal velocities (Fig. 8.5a) obtained for anisotropic ice vary with X, in contrast to the isotropic ice flow for which

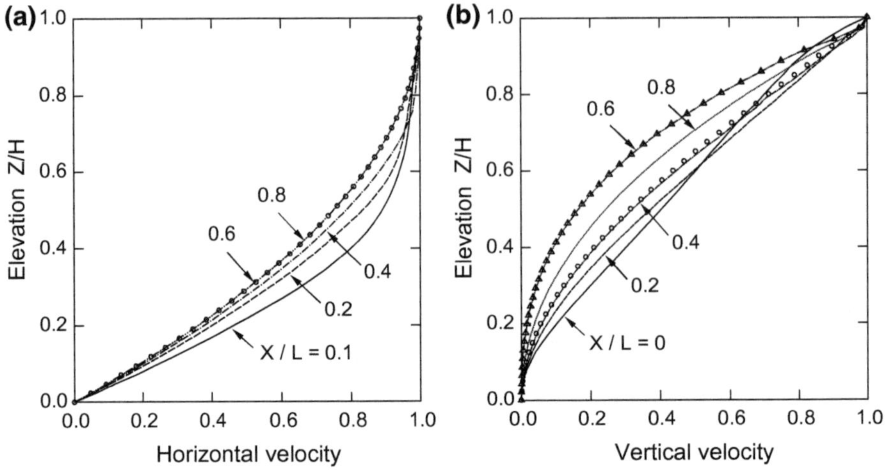

Fig. 8.5 **a** Depth profiles of the normalized horizontal and **b** vertical velocities at different locations X/L for anisotropic ice (lines) and isotropic ice (symbols). Reprinted from Staroszczyk and Morland (2000), Fig. 6, with permission of the International Glaciological Society

the normalized velocity profile (indicated by circles) is common for all locations X/L. The latter velocity pattern for isotropic ice flow differs slightly from those predicted by the models in which Glen's creep flow law (3.15) on p. 39, with the exponent $n = 3$, was used (Dahl-Jensen 1989; Hvidberg 1996). This discrepancy between the results is due to the assumption of the constant (independent of stress) isotropic ice viscosity adopted here. The plots in Fig. 8.5a show that the most significant differences between the horizontal velocities calculated for anisotropic and isotropic ice occur in the near-divide region, whereas for $X/L \gtrsim 0.6$ the respective profiles for both cases coincide. As regards the vertical velocity profiles presented in Fig. 8.5b, a more complicated pattern is observed, in which the depth profiles for not only anisotropic, but also isotropic ice, vary with X/L; for the sake of clarity only the profiles for the divide $X/L = 0$ (circles) and $X/L = 0.6$ (triangles) are plotted for the isotropic case. The largest discrepancies between the velocities for both types of ice occur, as in the case of the horizontal velocities, in the vicinity of the ice divide at $X = 0$; again, the corresponding profiles for anisotropic and isotropic ice practically coincide at locations distant from the divide. A similar property, in which the effect of the ice anisotropy is restricted only to the near-divide region and is negligibly small elsewhere, was also observed with regard to the depth profiles of the deviatoric stress components (Staroszczyk and Morland 2000).

Next, the results of simulations of a polythermal ice sheet flow, with a prescribed temperature field $T(X, Z)$, are presented. Again, the free surface profile (8.52) was adopted in the simulations. Compared to the above case of an isothermal ice flow, the mechanism of dynamic recrystallization is also considered in order to investigate its effect on the overall flow-rate of an ice sheet. Thus, the additive constitutive law (7.170) on p. 267 is applied to describe the creep behaviour of ice. To prescribe

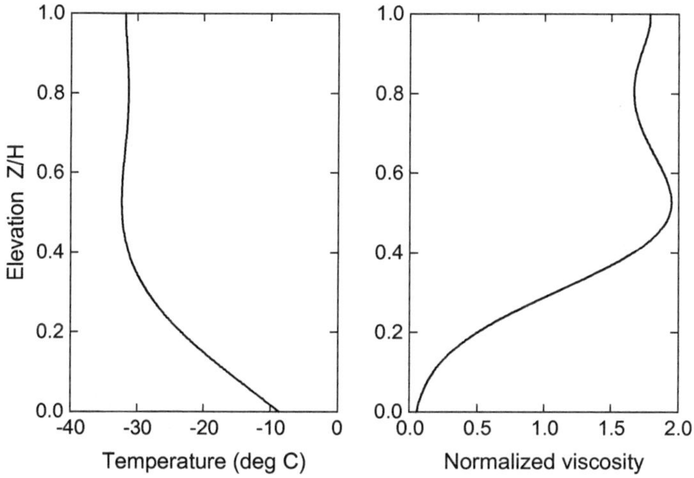

Fig. 8.6 Depth profiles of temperature T and isotropic ice viscosity $\tilde{\mu}_0$

the temperature field in an ice sheet, a temperature depth profile determined for an ice core drilled in Central Greenland (Gundestrup et al. 1993) was used. For simplicity, it was assumed that the temperature distribution was independent of the horizontal coordinate X, and the same temperature depth profile, when expressed in terms of the local dimensionless elevation Z/H, was applied throughout the whole ice sheet. The isotropic ice viscosity dependence on temperature was expressed in terms of the rate factor $a(T)$, as defined by Eqs. (3.18)–(3.20) on p. 40. The adopted depth profile of temperature T and the calculated depth profile of the viscosity $\tilde{\mu}_0$ are illustrated in Fig. 8.6. The viscosity is presented in a normalized form $\tilde{\mu}_0(T)/\tilde{\mu}_0(T_{av})$, where $T_{av} = -27.9$ °C is an average temperature along the vertical profile; that is, $\tilde{\mu}_0(T_{av}) = 1$. For the adopted temperature variation, the ratio of the isotropic ice viscosity at the free surface $Z = H$ to that at the bed $Z = 0$ is equal to about 33.4.

The magnitudes of the critical effective strain-rate invariant $\bar{I}_c$, a free parameter in the adopted constitutive model which defines the onset of dynamic recrystallization, was chosen in such a way that the maximum thicknesses (occurring at $X/L \sim 0.6$ for the assumed ice sheet geometry) of the recrystallized ice are equal, respectively, to about 0.025, 0.05, and 0.075 of the normalized ice depth $H(X)$. The other free parameter of the constitutive model (7.170), δ, which determines the rate of the recrystallization process, was assumed as 0.2.

Figure 8.7 illustrates the depth profiles of the horizontal and vertical velocity components, U and W, at different locations X/L. The labels (1), (2) and (3) in the plots refer to the above three values 0.025, 0.05 and 0.075 of the normalized ice depth. The velocities are normalized by the respective free surface values U_s and W_s, and are plotted against the normalized elevation Z/H. It is seen in the plots that in the vicinity of the ice divide (plots for $X/L = 0.2$) and near the ice sheet margin (plots for $X/L = 0.8$) the flow patterns are little affected by the mechanism of dynamic recrystallization. In particular, this observation applies to the normalized

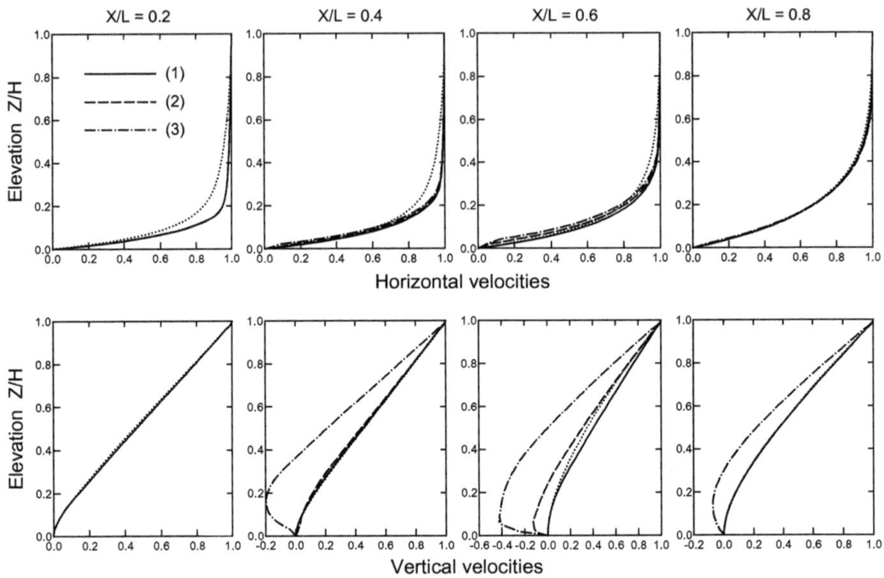

Fig. 8.7 Depth profiles of the normalized horizontal and vertical velocities at different locations X/L for different values of the critical strain-rate invariant $\bar{I}_c$. The dotted lines show the results for isotropic ice with no dynamic recrystallization involved. Reprinted from Staroszczyk (2003), Fig. 2, with permission of the International Glaciological Society

horizontal velocities U, the depth profiles of which, even in the middle part of the ice sheet ($X/L \sim 0.5$), are found to be little sensitive to the process of recrystallization (keep in mind, however, that the absolute velocity magnitudes do differ from each other). On the contrary, the normalized vertical velocity profiles in the middle part of the glacier are strongly influenced by the presence of the underlying recrystallized (isotropic) ice near the base. Specifically, significant changes in the flow pattern are seen in the lower part of the ice sheet ($Z/H \lesssim 1/3$).

Finally, Fig. 8.8 presents the variation of the shear strain-rates for different flow regimes at two locations X/L in the middle part of the ice sheet, and shows the values of $\epsilon \bar{D}_{xz}$ as functions of the relative elevation Z/H. Now the label (1) indicates the isotropic ice response, while the labels (2), (3) and (4) refer, in turn, to the values 0.025, 0.05 and 0.075 of the normalized ice thickness at which recrystallization starts. One can see that the magnitudes of the strain-rates undergo abrupt changes in the thin transition layer separating the recrystallized ice from the overlying strongly anisotropic ice. We can also note that with increasing thickness of the bottom layer of the recrystallized ice the shear strain-rates in this layer decrease and approach the values occurring in the isotropic ice flow with no dynamic recrystallization occurring (see the dashed-dotted lines in the plots).

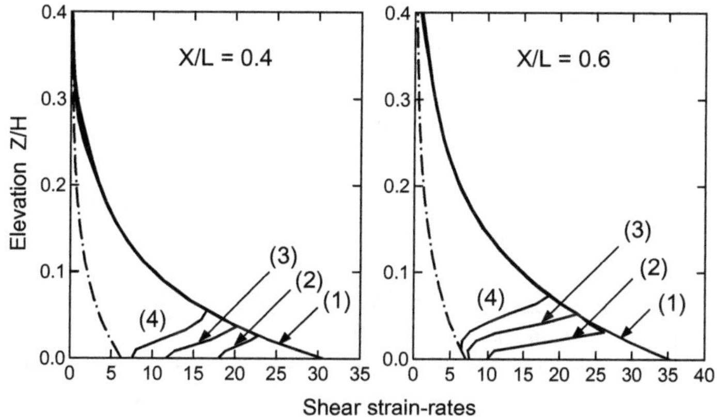

Fig. 8.8 Depth profiles of the normalized shear strain-rate $\epsilon \bar{D}_{xz}$ at two locations X/L for different values of the critical strain-rate invariant $\bar{I}_c$ (solid lines). The dashed-dotted lines show the results for isotropic ice with no dynamic recrystallization involved. Reprinted from Staroszczyk (2003), Fig. 3, with permission of the International Glaciological Society

8.2 Radially-Symmetric Ice Sheet Flow

In the previous Sect. 8.1, the problem of a plane ice sheet flow is analysed. In this section, a related problem, involving the flow of ice with evolving anisotropic fabric is considered, in which the ice sheet geometry is axially symmetric about the vertical axis passing through the ice divide. Hence, the problem of a gravity-driven, steady flow of polar ice in a radially-symmetric configuration is solved. The ice sheet is assumed to be subjected to accumulation or ablation at its free surface, and to melting or refreezing at its base. The top surface of the glacier is treated as traction-free, and at its base the glacier can either slide or be frozen to the underlying bedrock. Again, for the sake of simplicity, all heat fluxes in the ice sheet and across its boundaries are neglected, and a prescribed, constant in time, temperature distribution in ice is adopted. The solution of the flow problem is constructed as a leading-order approximation derived from asymptotic expansions in a small parameter ϵ which reflects the small ratio of the stress and velocity gradient components in the lateral direction of the ice sheet to those in the ice thickness direction.

8.2.1 Flow Problem Formulation

The radially-symmetric flow problem is solved in cylindrical polar coordinates (r, θ, z), with the r-axis on the horizontal plane, and the vertical z-axis directed upwards. It is assumed that the ice sheet geometry and all associated flow variables are radially symmetric about the axis Oz $(r = 0)$, so that they are independent of the polar angle θ and, hence, are functions of only r and z. The ice sheet cross-section, with the coordinate axes and other relevant notations, is shown in Fig. 8.9. The ice sheet geometry is defined by the free surface elevation $z = h(r)$ and the bed elevation

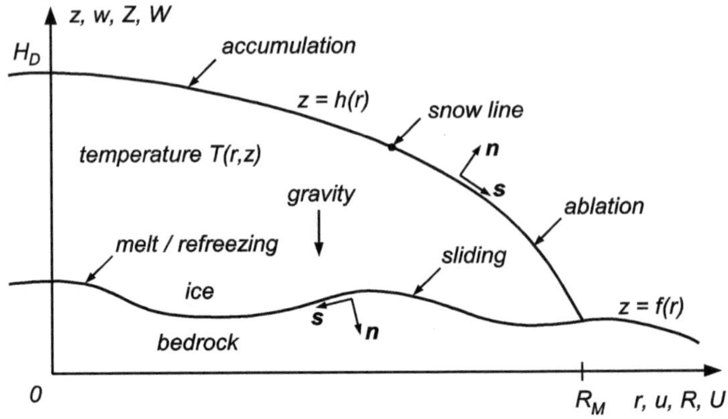

Fig. 8.9 Axi-symmetric ice sheet geometry and coordinate axes

$z = f(r)$. The ice sheet margin is at $r = R_M$ and the free surface elevation at the ice divide at $r = 0$ is $z = H_D$. At the free surface $h(r)$ and at the bed surface $f(r)$ there are ice mass fluxes defined by the accumulation rate q and the melt rate b; their meaning is explained in Sect. 8.1.1.

The velocity field $\boldsymbol{v}$ is described by the radial and vertical components $u(r, z)$ and $w(r, z)$, respectively. The four non-vanishing components of the strain-rate tensor $\boldsymbol{D}$ are expressed in terms of the ice velocities by

$$D_{rr} = \frac{\partial u}{\partial r}, \quad D_{\theta\theta} = \frac{u}{r}, \quad D_{zz} = \frac{\partial w}{\partial z}, \quad D_{rz} = \frac{1}{2}\left(\frac{\partial u}{\partial z} + \frac{\partial w}{\partial r}\right). \tag{8.54}$$

The deviatoric stress tensor $\boldsymbol{S}$ has also four independent non-zero components, corresponding to the strain-rate components defined by (8.54). The mass balance equation, given by the ice incompressibility condition $\mathrm{tr}\,\boldsymbol{D} = 0$, is expressed by

$$\frac{\partial u}{\partial r} + \frac{u}{r} + \frac{\partial w}{\partial z} = 0. \tag{8.55}$$

The horizontal radial and vertical momentum balances, in the absence of inertia forces in the extremely slow flow considered here, are given by the relations of equilibrium under gravity

$$\frac{\partial S_{rr}}{\partial r} + \frac{S_{rr} - S_{\theta\theta}}{r} + \frac{\partial S_{rz}}{\partial z} - \frac{\partial p}{\partial r} = 0, \tag{8.56}$$

$$\frac{\partial S_{rz}}{\partial r} + \frac{S_{rz}}{r} + \frac{\partial S_{zz}}{\partial z} - \frac{\partial p}{\partial z} - \varrho g = 0, \tag{8.57}$$

and the circumferential balance is identically satisfied because of the radial symmetry of the problem.

The mass and momentum balances expressed by Eqs. (8.55)–(8.57) are solved with the boundary conditions which are analogous to those formulated in Sect. 8.1.1, since essentially the same physical problem is considered, only the geometry of the flow domain is now different, and hence different coordinates in the horizontal plane are involved in the description. The unit normal and tangent vectors, $\boldsymbol{n}$ and $\boldsymbol{s}$ respectively, have components defined at the free surface $z = h(r)$ by (8.6) and (8.7) on p. 280, and at the bed $z = f(r)$ by (8.11) and (8.12), provided that x is replaced by r as the argument of the functions h and f. Therefore, the boundary conditions for the radially-symmetric flow can formally be expressed by those for the plane flow presented in Sect. 8.1.1 by replacing S_{xx} by S_{rr} and S_{xz} by S_{rz} in respective relations. Hence, the zero traction conditions at the free surface are given by the equations analogous to (8.8) and (8.9), the expressions for the basal normal and tangential traction components, t_n and t_s, are analogous to (8.13) and (8.14), the kinematic conditions at the surfaces $h(r)$ and $f(r)$ are expressed by (8.10) and (8.16), respectively, and the velocity components v_n and v_s at the bed are given by (8.15). Further, the sliding law in its linear form is defined again by (8.18).

The constitutive law of the general form given by (8.19) requires the components of the Cauchy-Green deformation tensor $\boldsymbol{B}$ expressed in the cylindrical polar coordinates. In these coordinates, the symmetric tensor $\boldsymbol{B}$ has four non-vanishing components involved in the ice fabric evolution relations (recall that in the plane flow only three components, B_{xx}, B_{zz} and B_{rz}, were required, since the fourth one, B_{yy}, was identically unity). These four non-vanishing components of $\boldsymbol{B}$ are determined by the following five non-zero components of the deformation gradient tensor $\boldsymbol{F}$

$$\begin{aligned} B_{rr} &= F_{rr}^2 + F_{rz}^2, \quad B_{\theta\theta} = F_{\theta\theta}^2, \quad B_{zz} = F_{zr}^2 + F_{zz}^2, \\ B_{rz} &= B_{zr} = F_{rr}F_{zr} + F_{rz}F_{zz}, \end{aligned} \tag{8.58}$$

with the components of $\boldsymbol{F}$ defined by relations

$$F_{rr} = \frac{\partial r}{\partial r^*}, \quad F_{\theta\theta} = \frac{r}{r^*}, \quad F_{zz} = \frac{\partial z}{\partial z^*}, \quad F_{rz} = \frac{\partial r}{\partial z^*}, \quad F_{zr} = \frac{\partial z}{\partial r^*}, \tag{8.59}$$

where r^* and z^* denote the particle reference (material) coordinates. There are now also five non-trivial components of the velocity gradient tensor $\boldsymbol{L}$,

$$L_{rr} = \frac{\partial u}{\partial r}, \quad L_{\theta\theta} = \frac{u}{r}, \quad L_{zz} = \frac{\partial w}{\partial z}, \quad L_{rz} = \frac{\partial u}{\partial z}, \quad L_{zr} = \frac{\partial w}{\partial r}, \tag{8.60}$$

so (8.22), describing the deformation gradient evolution, is equivalent to five first-order differential equations; however, only four of them are independent due to the ice incompressibility condition $\det \boldsymbol{F} = 1$.

8.2.2 Scaled Equations

To solve the mass conservation and equilibrium equations (8.55)–(8.57), together with the constitutive equation of the form (8.19) and the fabric evolution equations (8.22), the same method as that already applied in Sect. 8.1.2 is followed. Accordingly, the equations are first scaled by using the characteristic magnitudes of the variables involved, as described by relations (8.23), with x and $\bar{x}$ now being replaced by r and $\bar{r}$, respectively. The horizontal coordinates and velocities are then stretched by using the small parameter ϵ. Hence, by analogy to (8.24), we now have

$$R = \epsilon\bar{r}, \quad Z = \bar{z}, \quad U = \epsilon\bar{u}, \quad W = \bar{w}, \tag{8.61}$$

where R, Z, U and W are all order unity quantities. The small parameter ϵ is defined in terms of the typical physical magnitudes by relation (8.43). The scalings result in the dimensionless strain-rate and velocity gradient components, defined by (8.54) and (8.60), to be expressed by

$$\bar{D}_{rr} = \frac{\partial U}{\partial R}, \quad \bar{D}_{\theta\theta} = \frac{U}{R}, \quad \bar{D}_{zz} = \frac{\partial W}{\partial Z}, \quad \bar{D}_{rz} = \frac{1}{2}\left(\epsilon^{-1}\frac{\partial U}{\partial Z} + \epsilon\frac{\partial W}{\partial R}\right), \tag{8.62}$$

$$\bar{L}_{rr} = \frac{\partial U}{\partial R}, \quad \bar{L}_{\theta\theta} = \frac{U}{R}, \quad \bar{L}_{zz} = \frac{\partial W}{\partial Z}, \quad \bar{L}_{rz} = \epsilon^{-1}\frac{\partial U}{\partial Z}, \quad \bar{L}_{zr} = \epsilon\frac{\partial W}{\partial R}, \tag{8.63}$$

showing that the dominant components, both of order ϵ^{-1}, are $\bar{D}_{rz}$ and $\bar{L}_{rz}$. Similarly to (8.26), the stress tensor components are scaled by

$$\bar{S}_{rr} = \epsilon\Sigma_{rr}, \quad \bar{S}_{\theta\theta} = \epsilon\Sigma_{\theta\theta}, \quad \bar{S}_{zz} = \epsilon\Sigma_{zz}, \quad \bar{S}_{rz} = \epsilon\Sigma_{rz}, \quad \bar{p} = P, \tag{8.64}$$

where Σ_{ij} and P are all order unity variables. The isotropic ice viscosity is scaled again by using (8.42), so that we deal with an order unity viscosity $\tilde{\mu}_0$ in the scaled constitutive law.

On application of the above scalings, the mass conservation balance (8.55) becomes

$$\frac{\partial U}{\partial R} + \frac{U}{R} + \frac{\partial W}{\partial Z} = 0, \tag{8.65}$$

and the momentum balance equations, (8.56) and (8.57), take the forms

$$\epsilon\left(\frac{\partial\Sigma_{rr}}{\partial R} + \frac{\Sigma_{rr} - \Sigma_{\theta\theta}}{R}\right) + \frac{\partial\Sigma_{rz}}{\partial Z} - \frac{\partial P}{\partial R} = 0, \tag{8.66}$$

$$\epsilon^2\left(\frac{\partial\Sigma_{rz}}{\partial R} + \frac{\Sigma_{rz}}{R}\right) + \epsilon\frac{\partial\Sigma_{zz}}{\partial Z} - \frac{\partial P}{\partial Z} = 1. \tag{8.67}$$

In the normalized stretched coordinates (R, Z), the free surface elevation is described by $H(R) = h/h^*$ and the bed topography by $F(R) = f/h^*$, and the respective slopes are $\Gamma(R) = H'(R)$ and $\beta(R) = F'(R)$. In terms of the latter dimensionless variables, the expressions for Δ_h and Δ_f are given by (8.37). All the boundary conditions at the surfaces $H(R)$ and $F(R)$ have the forms analogous to those for the plane flow and given in Sect. 8.1.2. The expressions for the traction conditions can be obtained by replacing the stress components Σ_{xx} and Σ_{xz} in the plane flow by Σ_{rr} and Σ_{rz}, respectively, in the radially-symmetric flow; the stress component $\Sigma_{\theta\theta}$ does not enter the boundary conditions. The kinematic conditions are formally expressed in the same forms in both plane and radially-symmetric flows (bearing in mind that now the coordinate R replaces X as the argument of the functions involved). For reference, all the boundary conditions necessary for solving the system of Eqs. (8.65)–(8.67) are given below. Accordingly, the zero traction conditions at the free surface $Z = H(R)$, when expressed in the normalized stretched coordinates, are defined by

$$Z = H(R): \quad -\Delta_h^2 P + \epsilon \Sigma_{zz} - 2\epsilon^2 \Gamma \Sigma_{rz} + \epsilon^3 \Gamma^2 \Sigma_{rr} = 0, \tag{8.68}$$

$$Z = H(R): \quad (1 - \epsilon^2 \Gamma^2) \Sigma_{rz} + \epsilon \Gamma (\Sigma_{zz} - \Sigma_{rr}) = 0, \tag{8.69}$$

and the kinematic condition at the free surface is

$$Z = H(R): \quad \Gamma(R) U - W = \Delta_h Q, \tag{8.70}$$

where $Q(R)$ is a normalized accumulation rate. Similarly, the scaled relations for the basal normal and tangential tractions, T_n and T_s, are defined by

$$Z = F(R): \quad \Delta_f^2 T_n = -\Delta_f^2 P + \epsilon \Sigma_{zz} - 2\epsilon^2 \beta \Sigma_{rz} + \epsilon^3 \beta^2 \Sigma_{rr}, \tag{8.71}$$

$$Z = F(R): \quad \Delta_f^2 T_s = (1 - \epsilon^2 \beta^2) \Sigma_{rz} + \epsilon \beta (\Sigma_{zz} - \Sigma_{rr}). \tag{8.72}$$

Further, the normal and tangential components of the basal velocity, V_n and V_s, are

$$Z = F(R): \quad \Delta_f V_n = \beta(R) U - W, \quad \Delta_f V_s = -U - \epsilon^2 \beta(R) W, \tag{8.73}$$

and the kinematic condition at the bed is expressed by

$$Z = F(R): \quad \beta(R) U - W = \Delta_f B, \tag{8.74}$$

where $B(R)$ is a normalized basal melt rate. Finally, the linear sliding law, when expressed in the normalized dimensionless form, becomes

$$T_s = \Lambda \, T_n V_s \,, \tag{8.75}$$

where Λ is an order unity or greater normalized basal friction coefficient.

The orthotropic constitutive law, in the general form given by (8.19) and in the normalized form by (8.25), after applying the scalings (8.64) and (8.42), becomes

$$\boldsymbol{\Sigma} = 2\tilde{\mu}_0\epsilon \boldsymbol{H}(\bar{\boldsymbol{D}}, \boldsymbol{B}, \boldsymbol{M}^{(r)}) \quad (r = 1, 2, 3). \tag{8.76}$$

In this section we confine attention to the additive forms of the viscous flow law, in which case the tensor function $\boldsymbol{H}(\bar{\boldsymbol{D}}, \boldsymbol{B}, \boldsymbol{M}^{(r)})$ is given by

$$\begin{aligned}\boldsymbol{H}(\bar{\boldsymbol{D}}, \boldsymbol{B}, \boldsymbol{M}^{(r)}) = \bar{\boldsymbol{D}} + \sum_{r=1}^{3} f(\xi_r)\left[\boldsymbol{M}^{(r)}\bar{\boldsymbol{D}} + \bar{\boldsymbol{D}}\boldsymbol{M}^{(r)} - \tfrac{2}{3}\operatorname{tr}(\boldsymbol{M}^{(r)}\bar{\boldsymbol{D}})\boldsymbol{I}\right] + \\ + g(\zeta)\left[\boldsymbol{B}\bar{\boldsymbol{D}} + \bar{\boldsymbol{D}}\boldsymbol{B} - \tfrac{2}{3}\operatorname{tr}(\boldsymbol{B}\bar{\boldsymbol{D}})\boldsymbol{I}\right].\end{aligned} \tag{8.77}$$

The above form generalizes the additive constitutive equations considered in Sects. 7.5 and 7.6.3. Therefore, the response functions in (8.77), f and g, represent either the functions $\bar{f}$ and $\bar{g}$ used in (7.115) on p. 251, or the functions $\tilde{f}$ and $\tilde{g}$ used in (7.170) on p. 267. Similarly, the function arguments, the invariants ξ_r and ζ, defined by (7.111), are replaced by the invariants b_r and K, respectively, when the functions $\tilde{f}$ and $\tilde{g}$ are applied.

The viscous flow law (8.76) with the tensor function $\boldsymbol{H}$ given by (8.77) determines the following deviatoric stress tensor components:

$$\begin{aligned}
\Sigma_{rr} &= 2\tilde{\mu}_0\epsilon\left[(1+2c_1)\bar{D}_{rr} - c_2\bar{D}_{\theta\theta} - c_3\bar{D}_{zz} + c_4\bar{D}_{rz}\right],\\
\Sigma_{\theta\theta} &= 2\tilde{\mu}_0\epsilon\left[-c_1\bar{D}_{rr} + (1+2c_2)\bar{D}_{\theta\theta} - c_3\bar{D}_{zz} - 2c_4\bar{D}_{rz}\right],\\
\Sigma_{zz} &= 2\tilde{\mu}_0\epsilon\left[-c_1\bar{D}_{rr} - c_2\bar{D}_{\theta\theta} + (1+2c_3)\bar{D}_{zz} + c_4\bar{D}_{rz}\right],\\
\Sigma_{rz} &= 2\tilde{\mu}_0\epsilon\left[c_4(\bar{D}_{rr} + \bar{D}_{zz}) + (1+c_5)\bar{D}_{rz}\right],
\end{aligned} \tag{8.78}$$

with the coefficients c_i $(i = 1, \ldots, 5)$ defined by

$$\begin{aligned}
c_1 &= \tfrac{1}{3}\left[f(\xi_1)M^{(1)}_{rr} + f(\xi_3)M^{(3)}_{rr} + g(\zeta)B_{rr}\right],\\
c_2 &= \tfrac{1}{3}\left[f(\xi_2) + g(\zeta)B_{\theta\theta}\right],\\
c_3 &= \tfrac{1}{3}\left[f(\xi_1)M^{(1)}_{zz} + f(\xi_3)M^{(3)}_{zz} + g(\zeta)B_{zz}\right],\\
c_4 &= \tfrac{1}{3}\left[f(\xi_1)M^{(1)}_{rz} + f(\xi_3)M^{(3)}_{rz} + g(\zeta)B_{rz}\right],\\
c_5 &= \tfrac{1}{2}\left[f(\xi_1) + f(\xi_3) + g(\zeta)(B_{rr} + B_{zz})\right].
\end{aligned} \tag{8.79}$$

It can be proved that in the initial undeformed state when $\boldsymbol{B} = \boldsymbol{I}$, so that $b_1 = b_2 = b_3 = 1$ and $K = 3$, implying $\xi_1 = \xi_2 = \xi_3 = 0$ and $\zeta = 0$, all the coefficients c_i $(i = 1, \ldots, 5)$ vanish, irrespective of which constitutive law formulation, (7.115) or (7.170), is employed. As a result, relations (8.78) reduce to four equations $\Sigma_{ij} = 2\tilde{\mu}_0\epsilon\bar{D}_{ij}$, meaning that the adopted flow law describes the viscous behaviour of an isotropic fluid.

Finally, the components of the deformation gradient $\boldsymbol{F}$, given by (8.59), are expressed in the normalized stretched coordinates by

$$F_{rr} = \frac{\partial R}{\partial R^*}, \quad F_{\theta\theta} = \frac{R}{R^*}, \quad F_{zz} = \frac{\partial Z}{\partial Z^*},$$
$$F_{rz} = \epsilon^{-1}\frac{\partial R}{\partial Z^*}, \quad F_{zr} = \epsilon\,\frac{\partial Z}{\partial R^*}, \tag{8.80}$$

where R^* and Z^* are the stretched normalized reference coordinates, related to the physical coordinates r^* and z^* in the same way as the corresponding spatial coordinates are, that is, by $R^* = \epsilon r^*/h^*$ and $Z^* = z^*/h^*$. Relations (8.80) show that the components of $\boldsymbol{F}$ are of different orders, with F_{rz} of the largest magnitude of order ϵ^{-1}, and F_{zr} of the smallest magnitude of order ϵ. A similar structure, in terms of typical normalized magnitudes, is shown by the velocity gradient tensor $\bar{\boldsymbol{L}}$, see (8.63). The components of the tensors $\boldsymbol{F}$ and $\bar{\boldsymbol{L}}$ can be rescaled once again (Staroszczyk 2004) so that they all become order unity quantities, but the details are omitted here. Inspection of relations (8.58) for the components of the deformation tensor $\boldsymbol{B}$ shows that the largest component, of order ϵ^{-2}, is B_{rr}, and the smallest, of order ϵ^2, is B_{zz} (the components $B_{rz} = B_{zr}$ are of order ϵ^{-1}).

8.2.3 Leading-Order Equations and Solutions

In the previous Sect. 8.2.2, the flow problem differential equations and boundary conditions are derived in the form of asymptotic expansions in the small parameter $\epsilon \ll 1$. These equations are now solved in an approximate way by constructing a leading-order solution, instead of attempting to solve them in their full forms. For this purpose, all terms in the full equations which are of order ϵ or smaller compared to unity are first neglected, and then the simplified equations are solved to yield the leading-order approximations to the exact solutions of the axi-symmetric flow problem considered. The asymptotic solutions are constructed under the standard SIA assumption that the bed slopes $f'(r)$ are of order ϵ or less; that is, the normalized slopes $\beta = F'(R)$ are of order unity or less (Morland and Johnson 1980; Hutter 1983; Morland 1984). The situations in which the bed slopes are of larger magnitudes were treated by the method of asymptotic expansions, for isotropic ice, by Morland (2000, 2001) and Schoof (2003), and for anisotropic ice by Morland and Staroszczyk (2006) and Staroszczyk (2006).

Accordingly, to simplify the relevant equations in Sect. 8.2.2, we omit all terms with factors ϵ, ϵ^2 and ϵ^3, by considering them to be negligibly small. The mass conservation balance (8.65), which does not involve ϵ, is solved in its full form. To leading order, the horizontal and vertical equilibrium equations, (8.66) and (8.67), become

$$\frac{\partial P}{\partial R} = \frac{\partial \Sigma_{rz}}{\partial Z}, \quad \frac{\partial P}{\partial Z} = -1. \tag{8.81}$$

Expressions (8.37) on p. 284, due to the assumption that the normalized free surface and bed slopes, Γ and β respectively, do not exceed unity, yield

$$\Delta_h = 1, \quad \Delta_f = 1. \tag{8.82}$$

Therefore, the zero traction conditions at the free surface, given by (8.68) and (8.69), become, to leading order,

$$Z = H(R): \quad P = 0, \quad \Sigma_{rz} = 0, \tag{8.83}$$

and the free surface kinematic condition (8.70) takes the form

$$Z = H(R): \quad \Gamma U - W = Q. \tag{8.84}$$

At the bed $Z = F(R)$, the normal and tangential traction components, (8.71) and (8.72), are defined, to leading order, by

$$Z = F(R): \quad T_n = -P, \quad T_s = \Sigma_{rz}, \tag{8.85}$$

the ice velocity components, (8.73), are now

$$Z = F(R): \quad V_n = \beta U - W, \quad V_s = -U, \tag{8.86}$$

and the basal kinematic condition, (8.74), is

$$Z = F(R): \quad \beta U - W = B. \tag{8.87}$$

With the basal tractions and velocities given by (8.85) and (8.86), the sliding law (8.75) becomes

$$Z = F(R): \quad \Sigma_{rz} = \Lambda P U. \tag{8.88}$$

The normalized deviatoric stress components are prescribed by relations (8.78) and (8.79). The coefficients c_i $(i = 1, \dots, 5)$ describe changes in the dimensionless viscosities relative to unity as ice fabric evolves from its initially isotropic to strongly anisotropic state. Experimental evidence indicates that the polar ice viscosities do not change by more than a factor of about 10. On the other hand, as shown by relations (8.62), the shear strain-rate component $\bar{D}_{rz}$ is by a factor of $\epsilon^{-1} \sim 600 \gg 10$ larger than each of the remaining strain-rate components. Since $\bar{D}_{rz}$ enters each of the four relations (8.78), this implies that for a strongly anisotropic fabric all four tensor components Σ_{ij} are of the same order of magnitude, governed by the magnitude of $\bar{D}_{rz}$. Therefore, the leading-order approximations to the deviatoric stresses are given by

$$\Sigma_{rr} = \Sigma_{zz} = -\frac{1}{2}\Sigma_{\theta\theta} = \tilde{\mu}_0 \, c_{rr} \, \frac{\partial U}{\partial Z}, \quad \Sigma_{rz} = \tilde{\mu}_0 \, c_{rz} \, \frac{\partial U}{\partial Z}, \tag{8.89}$$

where

$$c_{rr}(R,Z) = \tfrac{1}{3}\left[f(\xi_1)M_{rz}^{(1)} + f(\xi_3)M_{rz}^{(3)} + g(\zeta)B_{rz}\right], \tag{8.90}$$
$$c_{rz}(R,Z) = 1 + \tfrac{1}{2}\left[f(\xi_1) + f(\xi_3) + g(\zeta)(B_{rr} + B_{zz})\right]. \tag{8.91}$$

In order to solve the above leading-order equations, the method first developed by Morland (1997, 2000) is followed. The new features in this analysis compared to the latter papers are: (1) the inclusion of the ice anisotropy in the flow problem, (2) no stream function is introduced to construct the solution, and (3) a different method is applied to solve the ensuing differential equation for the free surface elevation function $H(R)$. Accordingly, we start from integration of equilibrium equations (8.81) with the boundary conditions (8.83) at the free surface, which gives the leading-order approximations for pressure and stress fields

$$P(R,Z) = H(R) - Z, \quad \Sigma_{rz}(R,Z) = -\Gamma(R)[H(R) - Z]. \tag{8.92}$$

These two relations express the pressure P_b and shear stress Σ_{rzb} at the base $Z = F(R)$ as

$$P_b(R) = \Delta(R), \quad \Sigma_{rzb}(R) = -\Gamma(R)\Delta(R), \quad \Delta(R) = H(R) - F(R), \tag{8.93}$$

where $\Delta(R)$ is the ice thickness. With the latter basal stress components, the sliding law (8.88) determines the leading-order horizontal velocity at the base, U_b, as

$$U_b(R) = -\frac{\Gamma(R)}{\Lambda(R)}, \tag{8.94}$$

where the friction coefficient Λ is assumed to depend only on the position R; that is, it does not depend on stress and temperature. The kinematic condition at the bed, (8.87), expresses then the basal vertical velocity W_b as

$$W_b(R) = \beta(R)U_b(R) - B(R), \tag{8.95}$$

where the basal melt-rate B is assumed to depend only on R. The basal velocity components (8.94) and (8.95) are used as boundary conditions in integration of the flow equations over the depth Z to determine the velocity field inside the ice sheet. Hence, the shear stress Σ_{rz} is first eliminated from $(8.92)_2$ by using the last of the constitutive relations (8.89), and then, by integrating the resulting equation, one obtains

$$U(R,Z) = U_b(R) + G_1(R,Z), \tag{8.96}$$

where the function G_1 is defined by

$$G_1(R,Z) = -\Gamma(R)\int_F^Z \frac{H(R) - Z'}{\tilde{\mu}_0(R,Z')c_{rz}(R,Z')}\,\mathrm{d}Z'. \tag{8.97}$$

The prime denotes a running integration variable. With the horizontal velocity component $U(R, Z)$ given by (8.96), the vertical component $W(R, Z)$ is calculated by solving the incompressibility equation (8.65). Thus,

$$W(R,Z) = W_b(R) - \frac{1}{R}\,U_b(R)[Z - F(R)] - \frac{\mathrm{d}U_b}{\mathrm{d}R}\,[Z - F(R)] + \\ - \frac{1}{R}\,G_2(R,Z) - \frac{\partial G_2}{\partial R}(R,Z), \tag{8.98}$$

where the function G_2 is given by

$$G_2(R,Z) = -\Gamma(R)\int_F^Z \frac{(Z - Z')(H - Z')}{\tilde{\mu}_0(R,Z')c_{rz}(R,Z')}\,\mathrm{d}Z'. \tag{8.99}$$

The above leading-order relations express the stress and velocity components in terms of the free surface elevation function $H(R)$, which is yet unknown, and the prescribed bed elevation function $F(R)$, together with their derivatives $\Gamma(R)$ and $\beta(R)$. In order to find $H(R)$, use is made of the kinematic conditions (8.84) and (8.87), defined at the free surface and bed, respectively. Hence, we first difference both kinematic equations to obtain the relation

$$\Gamma U(R,H) - \beta U_b(R) - W(R,H) + W_b(R) = Q^*[R, H(R)], \tag{8.100}$$

where $Q^* = Q - B$, and then substitute into it expressions for U and W, (8.96) and (8.98). This yields the ordinary differential equation

$$\frac{\mathrm{d}}{\mathrm{d}R}\{RU_b(R)\Delta(R) - R\Gamma(R)I(R)\} = RQ^*[R, H(R)], \tag{8.101}$$

where $I(R)$ is defined by

$$I(R) = \int_{F(R)}^{H(R)} \frac{(H - Z')^2}{\tilde{\mu}_0(R,Z')\,c_{rz}(R,Z')}\,\mathrm{d}Z'. \tag{8.102}$$

Integration of (8.101) from the ice divide $R = 0$ to the margin $R = R_M$, due to $U(0) = 0$ and $I(R_M) = 0$, gives the relation

$$\int_0^{R_M} RQ^*[R, H(R)]\,\mathrm{d}R = 0, \tag{8.103}$$

which means that there is no net flux of mass into the ice sheet. In other words, the total mass of ice remains constant, which is consistent with the steady flow assumption.

The differential equation (8.101) is second-order in the unknown function $H(R)$, since it involves the term $\Gamma(R) = H'(R)$ in the curly braces on the left-hand side. The equation is solved by transforming it into an equivalent system of two first-order equations. For this purpose, let us denote the expression in the braces in (8.101) by

$$K(R) = -R\Gamma(R)\left\{\Lambda^{-1}(R)\,\Delta(R) + I(R)\right\}, \tag{8.104}$$

where relation (8.94) for U_b has been substituted. Hence, (8.101) becomes a first-order equation

$$\frac{\mathrm{d}K}{\mathrm{d}R} = RQ^*[R, H(R)]. \tag{8.105}$$

The above equation is solved over R ranging from zero to the margin at R_M. Since at $R = R_M$ we have $\Delta = 0$, hence $I(R) = 0$ and $K = 0$, the boundary conditions for (8.105) are given by

$$K(0) = 0, \quad K(R_M) = 0. \tag{8.106}$$

Relation (8.104) expresses $\Gamma(R) = H'(R)$ as

$$\frac{\mathrm{d}H}{\mathrm{d}R} = -\frac{K(R)}{R\left\{\Lambda^{-1}(R)\,\Delta(R) + I(R)\right\}}, \tag{8.107}$$

which is the second first-order equation to be solved simultaneously with (8.105). The boundary conditions for (8.107) are defined by

$$H(0) = H_D, \quad H(R_M) = F(R_M), \tag{8.108}$$

where H_D is the ice divide elevation, and R_M defines the point at which $H = F$, that is, $\Delta = 0$; both H_D and R_M are unknown quantities that need to be calculated as part of the solution.

Note that the dependence of the leading-order solution on the anisotropic properties of ice is described by the two functions c_{rr} and c_{rz}, defined by (8.90) and (8.91). The function c_{rz} explicitly enters the denominators of the integrands in (8.97), (8.99) and (8.102), whereas the function c_{rr} is involved implicitly through the invariant tr $\boldsymbol{S}^2$ (and hence tr $\boldsymbol{\Sigma}^2$) on which the magnitude of the ice viscosity depends, see relations (8.89) for the dimensionless stress tensor components.

The two first-order parabolic differential equations, (8.105) and (8.107), for the free surface profile $H(R)$, with the boundary conditions (8.106) and (8.108) prescribed at a priori unknown positions of the ice divide and margin, form a *two-point boundary-value problem*. A shooting method has been applied to solve the two equations. First, for given distributions of the fabric functions c_{rz} and c_{rr}, trial values of H_D and R_M are assumed, and then, starting from the endpoints $R = 0$ and $R = R_M$ and moving inwards, numerical integration is carried out by using a Runge-Kutta-Fehlberg scheme with an adaptive step size. The routine is repeated until the elevation H and the value of the function K are matched at a chosen interior

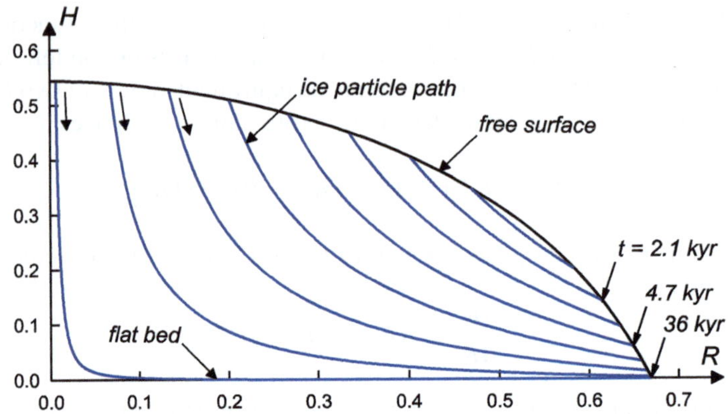

Fig. 8.10 Ice particle paths in an ice sheet flowing on a flat horizontal bed, with no basal melting. The times (in thousands of years) indicate the age of ice leaving the ice sheet due to ablation

fitting point. The validity and accuracy of the solution is verified by employing the integral property (8.103). Having found the free surface profile $H(R)$, the flow field variables (the velocities and their gradients) are evaluated and the spatial distributions of the functions c_{rz} and c_{rr} are updated, before starting calculations to find the next approximation for the free surface elevation function H. Such iterations between the ice fabric and the flow fields are repeated until a convergent solution is obtained; the first iteration is started by assuming an isotropic fabric, for which $c_{rz} = 1$ and $c_{rr} = 0$. Simultaneously, in each iteration, the system of five hyperbolic partial differential equations defined by (8.22) on p. 281 is solved to follow the evolution of the fabric along ice particle paths (streamlines). The paths start at the free surface in the accumulation zone ($Q > 0$), where $\boldsymbol{F} = \boldsymbol{I}$, pass through the interior of the ice sheet, and end either at the free surface in the ablation zone ($Q < 0$), or at the bed if basal melting occurs ($B > 0$). Typical ice particle paths in an ice sheet creeping over a flat bed, with no ice melting at the base, are illustrated in Fig. 8.10. For selected paths shown is the corresponding age of ice (given in thousands of years, kyr) when it leaves the ice sheet due to ablation. The parameters describing the flow of ice are provided in the following Sect. 8.2.4.

8.2.4 Numerical Ice Sheet Flow Simulations

The leading order equations derived in the previous section were solved numerically to simulate the steady flow of ice in a number of simple, idealized configurations. The main objective was to explore the effects of ice anisotropy on the ice sheet free surface geometry and the velocity variation across its depth. A range of model functions proposed by Morland (1997) was used in the simulations to prescribe the ice accumulation and temperature spatial distributions, and to define the bed topography.

This allowed the validity and accuracy of the numerical method applied here to be verified by comparing its results with the isotropic ice solutions obtained earlier by Morland (1997) and Cliffe and Morland (2000, 2001, 2002), before starting the investigation of the anisotropic ice flows.

The two differential equations (8.105) and (8.107) were integrated by using 500 or 1000 basic integral intervals along R to calculate the free surface profile $H(R)$. The integration of the equations was started with trial values of H_D and R_M, and the solutions for the functions $K(R)$ and $H(R)$ were matched with an accuracy of 10^{-6} at a fitting point located in the middle of a current integration range; that is, at $R = R_M/2$. In the matching procedure, a number of numerical routines from the book by Press et al. (2001) were employed. 100 points along the vertical were used to perform all depth integrations, and up to 5000 intervals were needed to calculate, from the evolution equations (8.22), the deformation gradient components along the longest characteristics (streamlines). The iterations between successive solutions for the flow and the fabric fields were continued until the maximum difference between the free surface profiles in two consecutive iterations was less than 10^{-5}. Typically, this required about 50 iterations.

The results presented below were obtained by adopting the characteristic magnitudes of the ice thickness $h^* = 2000$ m and the accumulation rate (or the vertical ice velocity) $v^* = 1\,\mathrm{m\,yr}^{-1}$. These values determine the small aspect ratio parameter $\epsilon = 0.00167 \sim 1/600$, and hence the length unit in the radial direction is equal to 1200 km, and the horizontal velocity unit is 600 m yr^{-1}. Further, the associated time unit is equal to 2000 yr and the strain-rate unit is the reciprocal of the latter value. To test the numerical ice flow model, the simulations were first run for two alternative constitutive laws in their additive forms, considered in Sects. 7.5 and 7.6.3 and expressed by Eqs. (7.115) and (7.170), respectively. The results showed that the two constitutive models predict very similar behaviour of an ice sheet, with the lateral spans R_M and the divide elevations H_D differing by no more than 2×10^{-3}. The results presented below were obtained by applying the flow law in the form given by (7.170).

As first, the viscous flow of a polythermal ice sheet over a flat bed $F(R)$=0 is considered. At the glacier base, the ice sliding conditions are adopted, with the basal friction coefficient $\Lambda = 10$. The ice sheet is subject to the free surface accumulation, the rate of which is described by the following idealized distribution function proposed by Morland (1997):

$$Q = Q_\infty - (Q_\infty - Q_0)\exp(-H/H^*), \tag{8.109}$$

where $Q_\infty > 0$ and $Q_0 < 0$ define the accumulation rates at $H \to \infty$ and $H = 0$, respectively, and H^* is a decay height. The snow line altitude, at which $Q = 0$, is then given by

$$H_e = H^* \ln(1 - Q_0/Q_\infty). \tag{8.110}$$

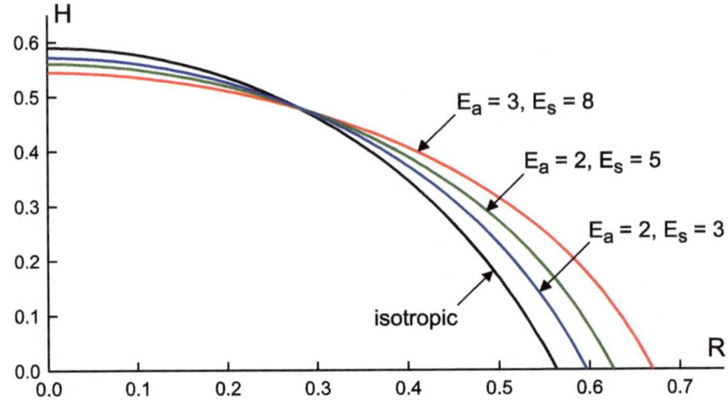

Fig. 8.11 Free surface profiles $H(R)$ for isotropic and anisotropic ice with different combinations of enhancement factors E_a and E_s. Adapted from Staroszczyk (2006), Fig.2, with permission of the Polish Academy of Sciences

In the simulations, the values $Q_\infty = 0.5$, $Q_0 = -1$ and $H^* = 0.25$ were used, implying $H_e = 0.275$, corresponding to $h = 550$ m. The adopted temperature distribution function, also due to Morland (1997), is expressed by

$$\bar{T}_1 = -\tfrac{4}{5}H + \tfrac{1}{2}\left\{H - Z - \tfrac{1}{4}\left[\Delta^2(H - Z) - \tfrac{1}{2}\Delta(H - Z)^2\right]\right\}. \qquad (8.111)$$

Recall that $\bar{T}$ is a normalized temperature defined by (3.19) on p. 40:

$$\bar{T} = (T - T_m)/\Delta_T, \quad \Delta_T = 20 \text{ K}, \qquad (8.112)$$

where $T_m = 273.15$ K is the ice melting temperature at the atmospheric pressure.

Figure 8.11 illustrates the effect of ice anisotropy on the free surface profiles $H(R)$. No basal melting is assumed, so $B = 0$ and $Q^* = Q$. Compared are the results for isotropic and anisotropic ice, characterized by different combinations of the axial and shear enhancement factors. The pair $E_a = 3$ and $E_s = 8$ describes the limit viscous properties of so-called warm ice.

The results plotted in the figure show that the influence of the ice anisotropy is more pronounced on the ice sheet extent R_M, rather than on the ice thickness H_D at the divide $R = 0$. For the most anisotropic ice illustrated in the figure (red line, warm ice), the sheet span increases by about 19% compared to the isotropic ice, while the ice divide thickness decreases by about 8%.

The results displayed in Fig. 8.12 show depth profiles of the scaled horizontal and vertical velocities U and W at $R = R_M/2$ (the middle of the ice sheet span), for the same ice sheet geometry and the viscous properties of anisotropic ice as those illustrated in Fig. 8.11. It can be observed that the ice anisotropy significantly affects the horizontal velocities across an ice sheet (Fig. 8.12a), while its effect on the vertical velocity components is relatively small, with very similar depth profiles for different

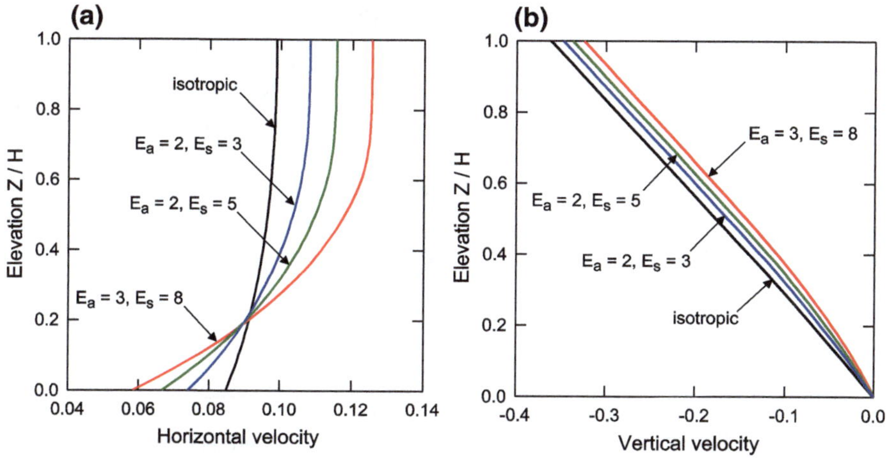

Fig. 8.12 **a** Horizontal and **b** vertical velocity depth profiles at $R = R_M/2$ for isotropic and anisotropic ice with different combinations of enhancement factors. Fig. (a) adapted from Staroszczyk (2006), Fig.3a, with permission of the Polish Academy of Sciences

combinations of the enhancement factors. An interesting feature is the decrease of the basal horizontal velocity U_b with increasing anisotropy of ice (increasing value of E_s); that is, with increasing ease of ice shearing, so that the fastest flowing ice near the bed is the isotropic (the 'stiffest') ice. On the other hand, as one can expect, the fastest flowing ice near the free surface is the most anisotropic (the 'softest') one.

The plots in Fig. 8.13 present the depth profiles of the horizontal and vertical velocities at different locations R/R_M in a flow of ice of the properties defined by the parameters $E_a = 3$ and $E_s = 8$ (representing warm ice). It is seen again that the vertical velocities (Fig. 8.13b) do not change much along the ice sheet (to some extent, this may be the consequence of the adopted accumulation distribution function (8.109)). The horizontal velocities, in turn, vary considerably down the ice sheet, growing steadily with increasing distance from the ice divide. The increasing slopes of the respective velocity curves near the bed reflect increasing shear strain-rates near the glacier base, facilitated by the anisotropic ice fabric development (decrease in shear viscosities) in near-base regions.

Figure 8.14 illustrates the effects of basal ice melting and the temperature distribution on the ice sheet geometry (the free surface profile $H(R)$, the lateral span R_M and the divide height H_D). Presented are the free surface profiles for the flows of warm ice ($E_a = 3$ and $E_s = 8$) for three different melt rates, $B = 0$ (no basal melt), $B = 0.05$ and $B = 0.10$, all for the temperature distribution T_1 given by (8.111). In addition, also the surface profile $H(R)$ for no basal melting conditions, but with a different temperature field, denoted by T_2, is displayed. The latter temperature distribution is the one in which ice is assumed to have, at a given R, constant temperature across the ice depth, which is an average of the corresponding surface and base temperatures prescribed by (8.111). Hence, T_2 defines the surface temperature

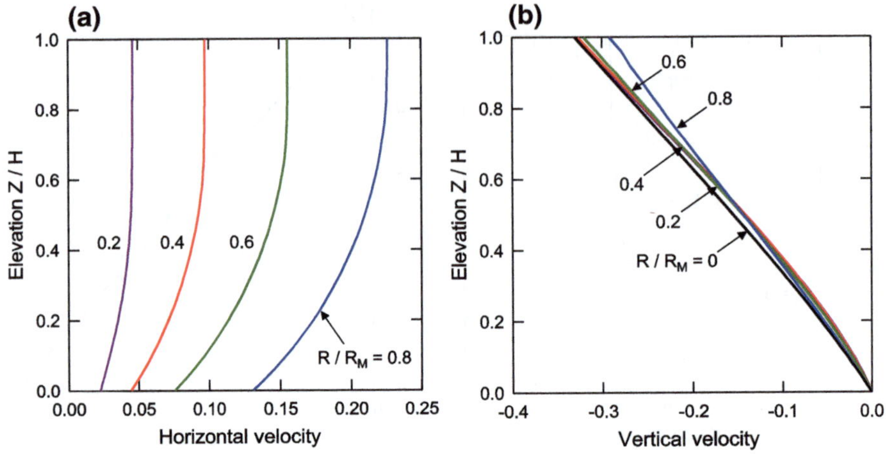

Fig. 8.13 **a** Horizontal and **b** vertical velocity depth profiles at different locations R/R_M for anisotropic ice defined by enhancement factors $E_a = 3$ and $E_s = 8$ (warm ice). Fig. (a) adapted from Staroszczyk (2006), Fig.3b, with permission of the Polish Academy of Sciences

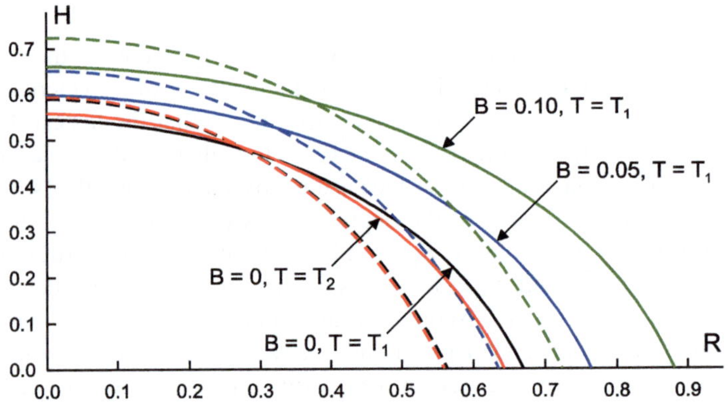

Fig. 8.14 Free surface profiles $H(R)$ for flows with different basal melt rates B and temperature distributions T, for anisotropic ice defined by enhancement factors $E_a = 3$ and $E_s = 8$. Corresponding isotropic ice profiles are plotted by dashed lines in the matching colours. Adapted from Staroszczyk (2006), Fig.5, with permission of the Polish Academy of Sciences

which is higher, and the base temperature which is lower, than the respective temperatures given by T_1. In the figure, the profiles for the anisotropic ice (solid lines) are compared with the corresponding results for the isotropic ice (dashed lines in the matching colours). One can note that the occurrence of basal melting significantly increases (for a given surface accumulation rate distribution) the total volume of ice in steady flow, increasing both the lateral span R_M and the divide thickness H_D by approximately the same rates when compared to the no basal melt flow. The comparison of the free surface profiles for the anisotropic and isotropic ice indicates that the anisotropy effects, reflected by flattening of the glacier, are similar in the three cases of B investigated. Further, it appears that the presence of basal melting does not modify considerably the overall behaviour of the anisotropic ice sheet (that is, the shapes of the profiles $H(R)$ are very similar, only their size differs). The plots for no-melt conditions ($B = 0$) and the two different temperature distributions, T_1 and T_2, show that the influence of temperature is moderate, leading to a relative decrease in the radial extent R_M by about 4% for the distribution T_2; that is, when the ice at the glacier base is colder, and hence more viscous, than it is in the case of the temperature field T_1.

The next three figures illustrate the effect of the mechanism of dynamic recrystallization on the free surface shape and the velocity depth profiles across an ice sheet. The results presented below were obtained for the ice rheological parameters pertaining to warm ice ($E_a = 3$ and $E_s = 8$), and again the surface ice accumulation distribution Q, given by relation (8.109) with $Q_\infty = 0.5$, $Q_0 = -1$ and $H^* = 0.25$, and the temperature field $\bar{T}_1$ defined by (8.111), were used in the simulations. No basal melting ($B = 0$) was assumed, and the basal friction coefficient was $\Lambda = 10$.

The effects of the ice anisotropy and the mechanism of migration recrystallization on the free surface profile $H(R)$ are illustrated in Fig. 8.15. By comparing the surface profile for the isotropic ice with those for the anisotropic ice, it can be observed that the anisotropy influences more the ice sheet span, defined by the margin radius R_M, than the ice divide thickness H_D at $R = 0$. For the anisotropic ice with no recrystallization taking place (the blue line in the plot), the relative increase in the ice sheet horizontal extent is about 28%, versus the ice divide thickness decreasing by about 12%. The occurrence of the migration recrystallization process, resulting in the restoration of the isotropy in the near-bed layer of the ice which makes it 'stiffer', leads to shortening and steepening of the free surface profile (the red line). Compared to the non-recrystallizing ice, the ice margin radius R_M decreases by about 7%, while the divide elevation increases by nearly 4%. In physical units, against the isotropic ice, the margin moves by about 191 km in the case of non-recrystallizing anisotropic ice, and by about 127 km in the case of recrystallizing ice. The respective changes in the ice divide thickness are -131 m for non-recrystallizing, and -90 m for recrystallizing ice.

The plots in Fig. 8.16 compare the horizontal velocity depth profiles for the isotropic and anisotropic ice flows, with and without the occurrence of the dynamic recrystallization process. The same input parameters were used as those in the previous figure. Figure 8.16a shows the velocity distributions at the mid-span location $R = R_M/2$. It is seen that the recrystallization mechanism has a moderate effect on the velocity field U. In the normalized units, the difference between the free surface horizontal velocities of non-recrystallizing and recrystallizing ice does not exceed 0.01. In physical units, the free surface velocities for non-recrystallizing and recrystallizing ice, compared to those of the isotropic ice, are greater by about 16 and 21 m yr^{-1}, respectively. The plots in Fig. 8.16b compare the horizontal velocity depth profiles for non-recrystallizing (the solid lines) and recrystallizing (the dashed lines) anisotropic ice at different locations R/R_M along the lateral direction. One can note that the influence of the recrystallization mechanism on the velocity field is

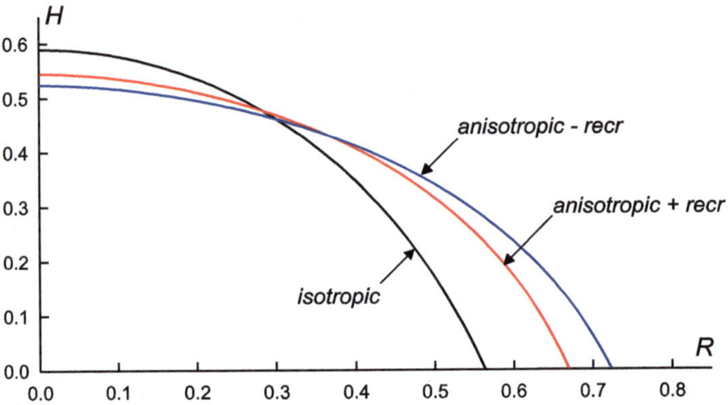

Fig. 8.15 Free surface profiles $H(R)$ for isotropic ice, and for anisotropic ice with and without dynamic recrystallization occurring

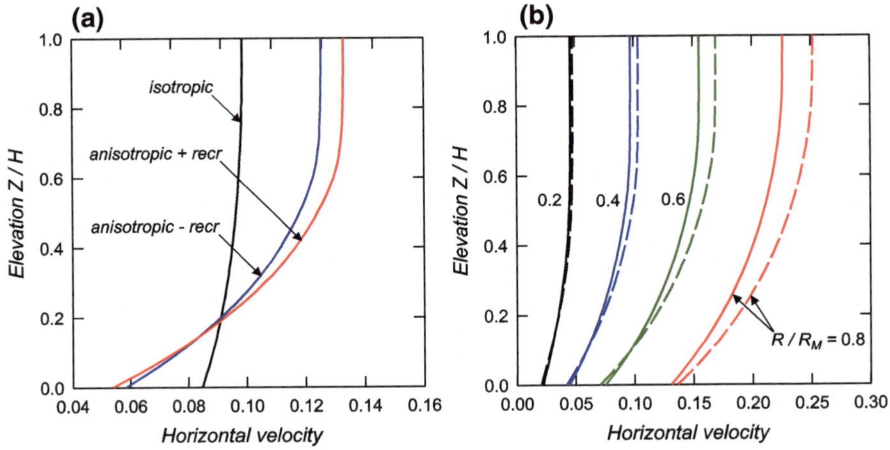

Fig. 8.16 Horizontal velocity depth profiles: **a** at $R = R_M/2$ for isotropic and anisotropic non-recrystallizing and recrystallizing ice; **b** at different locations R/R_M for non-recrystallizing and recrystallizing anisotropic ice defined by enhancement factors $E_a = 3$ and $E_s = 8$ (warm ice)

small in the central part of the glacier ($R/R_M \lesssim 0.4$), but in the region close to the glacier margin ($R/R_M \gtrsim 0.8$) its role becomes more significant.

Figure 8.17 presents the effect of the dynamic recrystallization process on the free surface profile $H(R)$ at different basal melt rates. The cases of $B = 0$ (no melting), $B = 0.05$ and $B = 0.10$ are illustrated. It can be seen that an increase in the basal melt rates, with the adopted free surface accumulation/ablation rates Q (depending on the elevation H), increases both the lateral span R_M and the divide thickness H_D by about the same amount. The comparison of the corresponding curves for recrystallizing ice (the solid lines) and non-recrystallizing ice (the dashed lines) shows that the effect

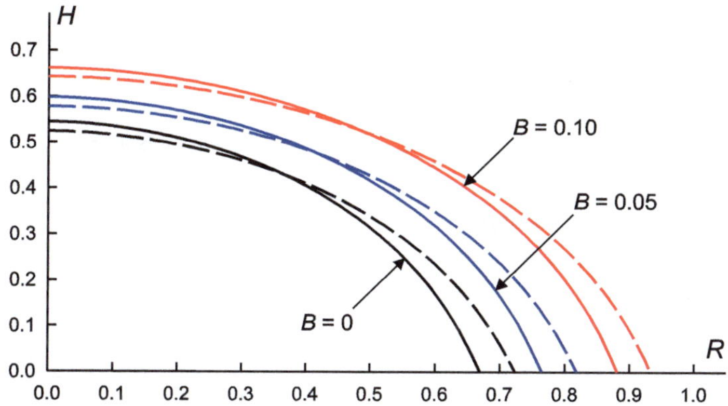

Fig. 8.17 Free surface profiles $H(R)$ for flows with different basal melt rates B, for recrystallizing (solid lines) and non-recrystallizing (dashed lines) anisotropic ice

of the recrystallization process on the free surface profile depends relatively little on the basal melting rates.

Finally, the flow of ice on uneven bed is considered to see how the bed topography affects the span and thickness of ice. Two simple bed forms are adopted for illustrations: a single hump of a moderate slope, and a single hollow, being the mirror image of the slope (see Figs. 8.19 and 8.20). It was assumed in the simulations that an ice sheet was subject to the elevation-dependent accumulation/ablation described by (8.109), with the parameters $Q_\infty = 0.5$ and $H^* = 0.25$ of the same values as those adopted for the flow problems with flat beds, illustrated above. However, the parameter Q_0, defining the accumulation rate at $H = 0$, was assumed to be equal to -6, which describes much larger ablation near an ice sheet margin than that in the above-illustrated cases for $Q_0 = -1$. The adopted set of parameters implies the equilibrium height (snowline), where $Q = 0$, equal to $H = 0.641$, corresponding to $h = 1282$ m (compared to $h = 550$ m in the flow problems considered earlier). The temperature field in the ice sheet was described again by relation (8.111). The coefficient of the basal friction was $\Lambda = 25$ (compared to $\Lambda = 10$ in the previous simulations), and no basal melting ($B = 0$) case was simulated.

In most of the ice sheet flow cases illustrated above, the limit viscous properties of ice (defined by the enhancement factors E_a and E_s) were assumed to be those of warm ice ($E_a = 3$ and $E_s = 8$). Now also the behaviour of cold ice, defined by the enhancement factors $E_a = 1/3$ and $E_s = 5$, is illustrated. It has turned out (Morland and Staroszczyk 2006) that the numerical modelling of ice sheet flows involving cold ice is much more difficult than that involving warm ice.

The results of simulations carried out for the input parameters described above are presented in Figs. 8.18, 8.19 and 8.20. For reference, the first of these figures illustrates the case of flow on a flat bed, for isotropic, and for anisotropic warm and cold ice. It is immediately seen that the assumption of a much larger ablation rate Q_0 at the ice margin level results in much larger values of the lateral span R_M and the

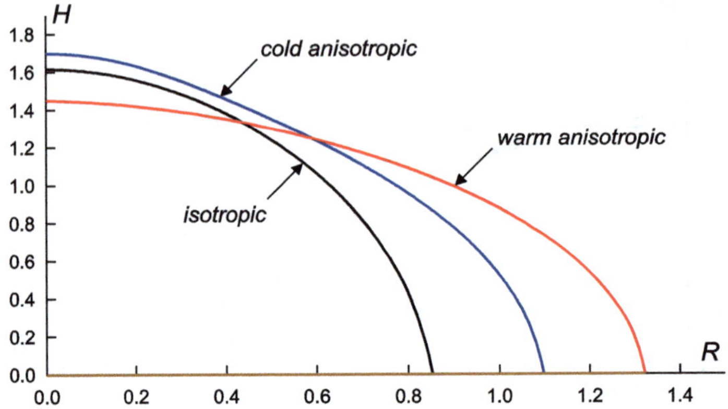

Fig. 8.18 Free surface profiles $H(R)$ for isotropic, and for anisotropic cold and warm ice flowing on the flat bed. Reprinted from Morland and Staroszczyk (2006), Fig. 3, with permission of the International Glaciological Society

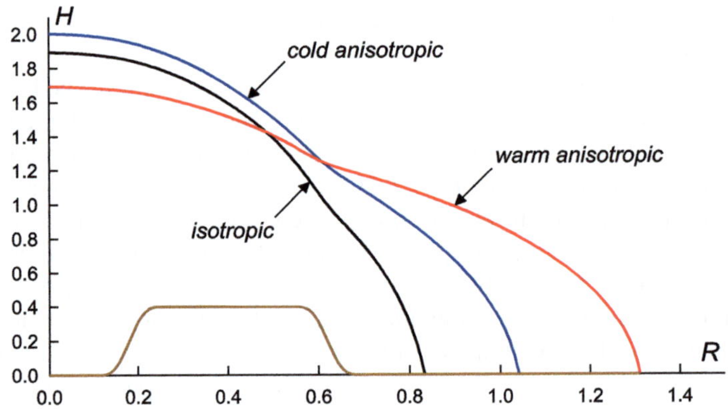

Fig. 8.19 Free surface profiles $H(R)$ for isotropic, and for anisotropic cold and warm ice flowing on the illustrated hump bed. Reprinted from Morland and Staroszczyk (2006), Fig. 4, with permission of the International Glaciological Society

divide height H_D than those shown in Figs. 8.11 and 8.15. It can be also observed that the free surface profiles for the cases of cold and warm ice are quite different: the cold ice profile is steeper, with a larger divide height H_D and a smaller lateral span R_M than the corresponding values for warm ice.

Figure 8.19 illustrates the free surface profiles for the ice sheet flow over a single hump centred at $R = 0.4$, of a span $40h^*$, a height $0.4h^*$, and a slope 0.01. One can see that, compared to the flat bed case presented in Fig. 8.18, the corresponding ice sheet spans are now a little smaller than those in the previous flow configuration, while the ice divide elevations are significantly larger, by about 17–18%, depending on the type of ice. One can also note the pronounced local changes in the free surface slopes reflecting the effect of the underlying hump on the ice flow.

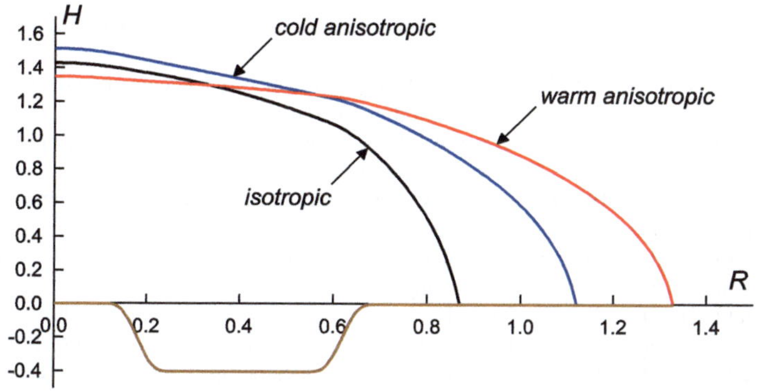

Fig. 8.20 Free surface profiles $H(R)$ for isotropic, and for anisotropic cold and warm ice flowing on the illustrated basin bed. Reprinted from Morland and Staroszczyk (2006), Fig. 5, with permission of the International Glaciological Society

The last Fig. 8.20 shows the free surface profiles for the flow over a hollow (basin), of the geometry being a mirror image of the hump (that is, the spans and slopes are the same, but the hollow depth elevation is the negative hump height elevation). Compared to the flat bed case (see Fig. 8.18), the corresponding lateral spans are again similar, but the divide heights are considerably smaller, changing from about 7% (cold ice) to about 12% (isotropic ice). Again, there is some effect of the bed topography visible on the free surface slopes, but this effect is less pronounced than in the case of flow over a hump.

References

Bargmann S, Seddik H, Greve R (2012) Computational modeling of flow-induced anisotropy of polar ice for the EDML deep drilling site, Antarctica: the effect of rotation recrystallization and grain boundary migration. Int J Numer Anal Meth Geomech 36(7):892–917. https://doi.org/10.1012/nag.1034

Chorin AJ (1967) A numerical method for solving incompressible viscous flow problems. J Comp Phys 2(1):12–26

Chorin AJ (1968) Numerical solution of the Navier-Stokes equations. Math Comp 22(104):745–762

Cliffe KA, Morland LW (2000) Full and reduced model solutions of steady axi-symmetric ice sheet flow over small and large bed topography slopes. Continuum Mech Thermodyn 12(3):195–216

Cliffe KA, Morland LW (2001) A thermo-mechanically coupled test case for axi-symmetric ice sheet flow. Continuum Mech Thermodyn 13(2):135–148. https://doi.org/10.1007/s001610100047

Cliffe KA, Morland LW (2002) Full and reduced model solutions of steady axi-symmetric ice sheet flow over bed topography with moderate slopes. Continuum Mech Thermodyn 14(2):149–164. https://doi.org/10.1007/s001610100059

Cliffe KA, Morland LW (2004) Full and reduced model solutions of unsteady axi-symmetric ice sheet flow over a flat bed. Continuum Mech Thermodyn 16(5):481–494. https://doi.org/10.1007/s00161-004-0175-3

Dahl-Jensen D (1989) Steady thermomechanical flow along two-dimensional flow lines in large grounded ice sheets. J Geophys Res 94(B8):10355–10362
Drăghicescu A (2001) Steady plane nonlinearly viscous flow of ice sheets on beds with moderate slope topography. Continuum Mech Thermodyn 13(6):421–438
Fabre A, Letreguilly A, Ritz C, Mangeney A (1995) Greenland under changing climate: sensitivity experiments with a new three-dimensional ice-sheet model. Ann Glaciol 21:1–7
Fowler AC, Larson DA (1978) On the flow of polythermal glaciers. I. Model and preliminary analysis. Proc R Soc Lond A 363(1713):217–242
Gillet-Chaulet F, Gagliardini O, Meyssonnier J, Montagnat M, Castelnau O (2005) A user-friendly anisotropic flow law for ice-sheet modelling. J Glaciol 51(172):3–14
Gillet-Chaulet F, Hindmarsh RCA (2011) Flow at ice-divide triple junctions: 1. Three-dimensional full-Stokes modeling. J Geophys Res 116(F02023). https://doi.org/10.1029/2009JF001611
Greve R, Blatter H (2009) Dynamics of ice sheets and glaciers. Springer, Berlin, Heidelberg
Gundestrup NS, Dahl-Jensen D, Johnsen SJ, Rossi A (1993) Bore-hole survey at dome GRIP 1991. Cold Reg Sci Technol 21(4):399–402
Hanson B (1995) A fully three-dimensional finite-element model applied to velocities on Storglaciären. Sweden J Glaciol 41(137):91–102
Herterich K (1988) A three-dimensional model of the Antarctic ice sheet. Ann Glaciol 11:32–35
Hindmarsh RCA (2004) A numerical comparison of approximations to the Stokes equations used in ice sheet and glacier modeling. J Geophys Res 109(F1):F01012. https://doi.org/10.1029/2003JF000065
Hindmarsh RCA, Morland LW, Boulton GS, Hutter K (1987) The unsteady plane flow of ice-sheets, a parabolic problem with two moving boundaries. Geophys Astrophys Fluid Dyn 39(3):183–225
Hirsch C (1992) Numerical computation of internal and external flows, vol 2. Wiley, Chichester
Hodge SM (1985) Two-dimensional, time-dependent modeling of an arbitrarily shaped ice mass with the finite element technique. J Glaciol 31(109):350–359
Hooke RL, Raymond CF, Hotchkiss RL, Gustafson RJ (1979) Calculations of velocity and temperature in a polar glacier using the finite-element method. J Glaciol 24(90):131–145
Hutter K (1981) The effect of longitudinal strain on the shear stress of an ice sheet: in defence of using stretched coordinates. J Glaciol 27(95):39–56
Hutter K. (1983) Theoretical glaciology. Material science of ice and the mechanics of glaciers and ice sheets. Reidel, Dordrecht
Hutter K, Yakowitz S, Szidarovszky F (1986) A numerical study of plane ice sheet flow. J Glaciol 32(111):139–160
Huybrechts P (1990) A 3-D model for the Antarctic ice sheet: a sensitivity study on the glacial-interglacial contrast. Climate Dyn 5(2):79–92
Hvidberg CS (1996) Steady-state thermomechanical modelling of ice flow near the centre of large ice sheets with the finite-element technique. Ann Glaciol 23:116–123
Ma Y, Gagliardini O, Ritz C, Gillet-Chaulet F, Durand G, Montagnat M (2010) Enhancement factors for grounded ice and ice shelves inferred from an anisotropic ice-flow model. J Glaciol 56(199):805–812
Mangeney A, Califano F (1998) The shallow ice approximation for anisotropic ice: formulation and limits. J Geophys Res 103(B1):691–705
Mangeney A, Califano F, Castelnau O (1996) Isothermal flow of an anisotropic ice sheet in the vicinity of an ice divide. J Geophys Res 101(B12):28189–28204
Mangeney A, Califano F, Hutter K (1997) A numerical study of anisotropic, low Reynolds number, free surface flow for ice sheet modeling. J Geophys Res 102(B10):22749–22764
Morland LW (1984) Thermomechanical balances of ice sheet flows. Geophys Astrophys Fluid Dyn 29:237–266
Morland LW (1997) Radially symmetric ice sheet flow. Phil Trans R Soc Lond A 355:1873–1904
Morland LW (2000) Steady plane isothermal linearly viscous flow of ice sheets on beds with moderate-slope topography. Proc R Soc Lond A 456(1999):1711–1739

Morland LW (2001) Influence of bed topography on steady plane ice sheet flow. In: Straughan B, Greve R, Ehrentraut H, Wang Y (eds) Continuum mechanics and applications in geophysics and the environment. Springer, Berlin, pp 276–304

Morland LW (2009) A three-dimensional ice-sheet flow solution. J Glaciol 55(191):473–480. https://doi.org/10.3189/002214309788816588

Morland LW, Drăghicescu A (1998) Steady plane isothermal linearly viscous flow of ice sheets on beds with large slope topography. Environmetrics 9:459–492

Morland LW, Johnson IR (1980) Steady motion of ice sheets. J Glaciol 25(92):229–246

Morland LW, Staroszczyk R (2006) Steady radial ice sheet flow with fabric evolution. J Glaciol 52(177):267–280. https://doi.org/10.3189/172756506781828719

Pan W, Tartakovsky AM, Monaghan JJ (2013) Smoothed particle hydrodynamics non-newtonian model for ice-sheet and ice-shelf dynamics. J Comput Phys 242:828–842. https://doi.org/10.1016/j.jcp.2012.10.027

Press WH, Teukolsky SA, Vettering WT, Flannery BP (2001) Numerical recipes in Fortran 77, vol. 1, 2nd edn. Cambridge University Press, Cambridge

Raymond CF (1983) Deformation in the vicinity of ice divides. J Glaciol 29(103):357–373

Schoof C (2003) The effect of basal topography on ice sheet dynamics. Continuum Mech Thermodyn 15(3):295–307

Seddik H, Greve R, Zwinger T, Gillet-Chaulet F, Gagliardini O (2012) Simulations of the Greenland ice sheet 100 years into the future with the full Stokes model Elmer/Ice J Glaciol 58(209):427–440. https://doi.org/10.3189/2012JoG11J177

Staroszczyk R (2003) Plane ice sheet flow with evolving and recrystallising fabric. Ann Glaciol 37(1):247–251. https://doi.org/10.3189/172756403781815834

Staroszczyk R (2004) Constitutive modelling of creep induced anisotropy of ice. IBW PAN Publishing House, Gdańsk

Staroszczyk R (2006) Axi-symmetric ice sheet flow with evolving anisotropic fabric. Bull Pol Ac Tech 54(4):419–428

Staroszczyk R, Morland LW (2000) Plane ice-sheet flow with evolving orthotropic fabric. Ann Glaciol 30:93–101

Zienkiewicz OC, Taylor RL, Nithiarasu P (2005a) The finite element method for fluid dynamics, 6th edn. Elsevier Butterworth-Heinemann, Amsterdam

Zienkiewicz OC, Taylor RL, Zhu JZ (2005b) The finite element method: its basis and fundamentals, 6th edn. Elsevier Butterworth-Heinemann, Amsterdam

Appendix A
Physical Parameters Relevant to Ice

Parameter	Value	Unit
Albedo of bare sea ice	$\sim$0.5	–
Albedo of snow-covered sea ice	$\sim$0.9	–
Bulk modulus of isotropic ice at 263 K	7.94	GPa
Compressive fracture strength of ice	$\sim$5	MPa
Creep activation energy for ice at 263 K	6.7×10^4	$\mathrm{J\,mol^{-1}}$
Density of bubble-free ice	916.7	$\mathrm{kg\,m^{-3}}$
Fracture toughness of ice of grain size 1 mm	0.10	$\mathrm{MPa\,m^{1/2}}$
Freezing temperature of water of salinity 35 ppt	271.24	K
Latent heat of ice melting	3.34×10^5	$\mathrm{J\,kg^{-1}}$
Poisson's ratio for isotropic polycrystalline ice	0.31	–
Shear modulus of isotropic ice at 263 K	3.45	GPa
Specific heat of air at sea level at 273.15 K	1.00×10^3	$\mathrm{J\,kg^{-1}\,K^{-1}}$
Specific heat of air-free water	4.18×10^3	$\mathrm{J\,kg^{-1}\,K^{-1}}$
Specific heat of ice at 263 K	2.11×10^3	$\mathrm{J\,kg^{-1}\,K^{-1}}$
Tensile fracture strength of ice	$\sim$1	MPa
Thermal conductivity of air at 273 K	0.024	$\mathrm{W\,m^{-1}\,K^{-1}}$
Thermal conductivity of ice at 272 K	2.22	$\mathrm{W\,m^{-1}\,K^{-1}}$
Thermal conductivity of water at 273 K	0.58	$\mathrm{W\,m^{-1}\,K^{-1}}$
Thermal diffusivity coefficient for ice	1.15×10^{-6}	$\mathrm{m^2\,s^{-1}}$
Thermal expansion coefficient of ice at 263 K	5.2×10^{-5}	$\mathrm{K^{-1}}$
Universal gas constant	8.314	$\mathrm{J\,K^{-1}\,mol^{-1}}$
Viscosity of isotropic ice at 263 K at 0.1 MPa	$\sim 1 \times 10^{14}$	Pa s
Water triple point pressure	612	Pa
Water triple point temperature	273.16	K
Young's modulus of isotropic ice at 263 K	9.05	GPa

R. Staroszczyk, *Ice Mechanics for Geophysical and Civil Engineering Applications*,
GeoPlanet: Earth and Planetary Sciences, https://doi.org/10.1007/978-3-030-03038-4

Appendix B
Tensor Representations for Constitutive Laws

Admissible constitutive relations must satisfy the *principle of objectivity* (or *principle of isotropy of space*, or *principle of frame-indifference*) (Boehler 1987; Spencer 1987a, b; Chadwick 1999; Liu 2002; Truesdell and Noll 2004), according to which the observed response of a material is the same for any pair of observers. In mathematical terms, an observer is interpreted as a frame of reference, so the principle of objectivity requires constitutive laws to be invariant under changes of frame that preserve the essential structure of space and time (so that the distance of any pair of points, the orientation of space, and the time lapse between any two events, all remain unaltered). An observer frame transformation connecting position $\boldsymbol{x}$ and time t in one reference frame to position $\boldsymbol{x}^*$ and time t^* in another frame, which satisfies the above invariance properties, can be expressed by a time-dependent rigid transformation, also referred to as the *Euclidean transformation*:

$$\boldsymbol{x}^* = \boldsymbol{Q}(t)\boldsymbol{x} + \boldsymbol{c}(t), \quad t^* = t - a, \tag{B.1}$$

where $\boldsymbol{Q}$ is a proper orthogonal tensor, $\boldsymbol{c}$ is a vector-valued function of time, and a is a constant. The principle of frame-indifference must be distinguished from the usual transformation law for tensor components in different coordinate frames, since the latter law does not involve time (and, therefore, is less restrictive than the principle of tensor frame-invariance).

A time-dependent scalar $\phi(\boldsymbol{x}, t)$, vector $\boldsymbol{u}(\boldsymbol{x}, t)$, and a second-order tensor $\boldsymbol{T}(\boldsymbol{x}, t)$ are said to be, respectively, an *objective* (or *frame-indifferent*) scalar, vector or tensor, if, when subjected to a change of reference frame given by the Euclidean transformation (B.1), they obey the relations

$$\begin{aligned} \phi^*(\boldsymbol{x}^*, t^*) &= \phi(\boldsymbol{x}, t), \\ \boldsymbol{u}^*(\boldsymbol{x}^*, t^*) &= \boldsymbol{Q}(t)\, \boldsymbol{u}(\boldsymbol{x}, t), \\ \boldsymbol{T}^*(\boldsymbol{x}^*, t^*) &= \boldsymbol{Q}(t)\, \boldsymbol{T}(\boldsymbol{x}, t)\, \boldsymbol{Q}^T(t), \end{aligned} \tag{B.2}$$

R. Staroszczyk, *Ice Mechanics for Geophysical and Civil Engineering Applications*, GeoPlanet: Earth and Planetary Sciences, https://doi.org/10.1007/978-3-030-03038-4

where the starred entities denote the counterparts of ϕ, $\boldsymbol{u}$ and $\boldsymbol{T}$ when transformed by applying (B.1). It can be proved that objective fields include Cauchy stress, the second Piola-Kirchhoff stress, left stretch tensor, left Cauchy-Green deformation tensor, strain-rate tensor, spatial gradient of an objective scalar field, spatial divergence of an objective vector field and spatial divergence of an objective tensor field. Non-objective quantities, that is those which do not satisfy (B.2), are the first Piola-Kirchhoff stress, referential deformation gradient, right stretch tensor, right Cauchy-Green deformation tensor, spatial velocity gradient, spin tensor, rotation tensor and referential gradient of an objective scalar field (Chadwick 1999; Liu 2002).

Assume that the behaviour of an anisotropic material is described by a constitutive equation which relates a second-order tensor $\boldsymbol{T}$ to a set of second-order tensors $\boldsymbol{A}_1$, $\boldsymbol{A}_2, \ldots, \boldsymbol{A}_m$ and $\boldsymbol{M}_1, \boldsymbol{M}_2, \ldots, \boldsymbol{M}_n$ by means of a tensor-valued function $\boldsymbol{H}$:

$$\boldsymbol{T} = \boldsymbol{H}(\boldsymbol{A}_1, \boldsymbol{A}_2, \ldots, \boldsymbol{A}_m; \boldsymbol{M}_1, \boldsymbol{M}_2, \ldots, \boldsymbol{M}_n), \tag{B.3}$$

where the tensors $\boldsymbol{A}_i$ represent mechanical arguments (such as stress, strain, strain-rate, *etc.*) on which the material behaviour depends, and the tensors $\boldsymbol{M}_j$ are a set of n so-called *structure tensors*, which define symmetry properties of the internal structure of the material.

A non-linear frame-indifferent constitutive equation involving the tensors $\boldsymbol{A}_i$ and $\boldsymbol{M}_j$, includes a set of tensor generators, which are formed from the tensors $\boldsymbol{A}_i$, $\boldsymbol{M}_j$ and their inner products, and a set of scalar-valued response coefficients α_k, which are functions of scalar invariants of the tensors $\boldsymbol{A}_i$, $\boldsymbol{M}_j$ and their inner products. For any particular type of the material symmetry, an *irreducible* (or *canonical*) form of a constitutive law can be derived (Boehler 1987), for which none of the tensor generators, and none of the invariant arguments of the response coefficients, can be expressed as a single-valued function of the others. Below, the irreducible forms of the constitutive laws for isotropic, transversely isotropic and orthotropic materials are presented, with the restriction that these laws relate one symmetric second-order tensor $\boldsymbol{T}$ to two other symmetric second-order tensors $\boldsymbol{A}_1$ and $\boldsymbol{A}_2$ (besides a relevant set of structure tensors $\boldsymbol{M}_j$). The constitutive laws involving only one tensor as a mechanical argument, $\boldsymbol{A}_1$ say, can be obtained by omitting in the respective relations all the terms containing $\boldsymbol{A}_2$.

Isotropic Medium

The only structure tensor $\boldsymbol{M}$ for an isotropic material is the unit tensor $\boldsymbol{I}$; therefore, there is no need to use any structure tensor in the constitutive description. The general frame-indifferent tensor representation for a constitutive law for an isotropic medium involves eight tensor generators, and is expressed by

$$\begin{aligned}\boldsymbol{T} = {} & \alpha_1 \boldsymbol{I} + \alpha_2 \boldsymbol{A}_1 + \alpha_3 \boldsymbol{A}_2 + \alpha_4 \boldsymbol{A}_1^2 + \alpha_5 \boldsymbol{A}_2^2 + \alpha_6(\boldsymbol{A}_1\boldsymbol{A}_2 + \boldsymbol{A}_2\boldsymbol{A}_1) + \\ & + \alpha_7(\boldsymbol{A}_1^2\boldsymbol{A}_2 + \boldsymbol{A}_2\boldsymbol{A}_1^2) + \alpha_8(\boldsymbol{A}_1\boldsymbol{A}_2^2 + \boldsymbol{A}_2^2\boldsymbol{A}_1).\end{aligned} \tag{B.4}$$

The scalar-valued response coefficients α_k $(k = 1, \ldots, 8)$ entering (B.4) are functions of ten invariants of the tensors $\boldsymbol{A}_1$ and $\boldsymbol{A}_2$:

$$\alpha_k = \alpha_k(\operatorname{tr} A_1,\ \operatorname{tr} A_1^2,\ \operatorname{tr} A_1^3,\ \operatorname{tr} A_2,\ \operatorname{tr} A_2^2,\ \operatorname{tr} A_2^3,\\ \operatorname{tr} A_1 A_2,\ \operatorname{tr} A_1^2 A_2,\ \operatorname{tr} A_1 A_2^2,\ \operatorname{tr} A_1^2 A_2^2). \tag{B.5}$$

Transversely Isotropic Medium

A transversely isotropic body has one privileged material direction, which is the axis of the material rotational symmetry. This symmetry property is described by a single structure tensor $\boldsymbol{M}$, given by

$$\boldsymbol{M} = \boldsymbol{e} \otimes \boldsymbol{e}, \tag{B.6}$$

where $\boldsymbol{e}$ is the unit vector along the axis of rotational symmetry. The general objective representation for a transversely isotropic material incorporates 13 tensor generators,

$$\begin{aligned} \boldsymbol{T} = {} & \alpha_1 \boldsymbol{I} + \alpha_2 \boldsymbol{M} + \alpha_3 \boldsymbol{A}_1 + \alpha_4 \boldsymbol{A}_2 + \alpha_5(\boldsymbol{M}\boldsymbol{A}_1 + \boldsymbol{A}_1\boldsymbol{M}) + \\ & + \alpha_6(\boldsymbol{M}\boldsymbol{A}_2 + \boldsymbol{A}_2\boldsymbol{M}) + \alpha_7 \boldsymbol{A}_1^2 + \alpha_8 \boldsymbol{A}_2^2 + \alpha_9(\boldsymbol{M}\boldsymbol{A}_1^2 + \boldsymbol{A}_1^2\boldsymbol{M}) + \\ & + \alpha_{10}(\boldsymbol{M}\boldsymbol{A}_2^2 + \boldsymbol{A}_2^2\boldsymbol{M}) + \alpha_{11}(\boldsymbol{A}_1\boldsymbol{A}_2 + \boldsymbol{A}_2\boldsymbol{A}_1) + \\ & + \alpha_{12}(\boldsymbol{A}_1^2\boldsymbol{A}_2 + \boldsymbol{A}_2\boldsymbol{A}_1^2) + \alpha_{13}(\boldsymbol{A}_1\boldsymbol{A}_2^2 + \boldsymbol{A}_2^2\boldsymbol{A}_1), \end{aligned} \tag{B.7}$$

and the response coefficients α_k ($k = 1, \ldots, 13$) are functions of 14 scalar invariants:

$$\alpha_k = \alpha_k(\operatorname{tr} A_1,\ \operatorname{tr} A_1^2,\ \operatorname{tr} A_1^3,\ \operatorname{tr} A_2,\ \operatorname{tr} A_2^2,\ \operatorname{tr} A_2^3,\ \operatorname{tr} A_1 A_2,\ \operatorname{tr} A_1^2 A_2,\\ \operatorname{tr} A_1 A_2^2,\ \operatorname{tr} M A_1,\ \operatorname{tr} M A_2,\ \operatorname{tr} M A_1^2,\ \operatorname{tr} M A_2^2,\ \operatorname{tr} M A_1 A_2). \tag{B.8}$$

Orthotropic Medium

An orthotropic body exhibits reflectional symmetries with respect to three mutually orthogonal planes, the planes of orthotropic symmetry. These symmetry properties are described by three structure tensors (but only two of them are independent), given by

$$\boldsymbol{M}_r = \boldsymbol{e}_r \otimes \boldsymbol{e}_r \quad (r = 1, 2, 3), \tag{B.9}$$

where $\boldsymbol{e}_r$ are the unit vectors along the axes of orthotropic symmetry. The general frame-indifferent constitutive representation for an orthotropic medium includes 12 tensor generators, and is defined by

$$\begin{aligned} \boldsymbol{T} = {} & \alpha_1 \boldsymbol{M}_1 + \alpha_2 \boldsymbol{M}_2 + \alpha_3 \boldsymbol{M}_3 + \alpha_4(\boldsymbol{M}_1\boldsymbol{A}_1 + \boldsymbol{A}_1\boldsymbol{M}_1) + \\ & + \alpha_5(\boldsymbol{M}_2\boldsymbol{A}_1 + \boldsymbol{A}_1\boldsymbol{M}_2) + \alpha_6(\boldsymbol{M}_3\boldsymbol{A}_1 + \boldsymbol{A}_1\boldsymbol{M}_3) + \\ & + \alpha_7(\boldsymbol{M}_1\boldsymbol{A}_2 + \boldsymbol{A}_2\boldsymbol{M}_1) + \alpha_8(\boldsymbol{M}_2\boldsymbol{A}_2 + \boldsymbol{A}_2\boldsymbol{M}_2) + \\ & + \alpha_9(\boldsymbol{M}_3\boldsymbol{A}_2 + \boldsymbol{A}_2\boldsymbol{M}_3) + \alpha_{10}\boldsymbol{A}_1^2 + \alpha_{11}\boldsymbol{A}_2^2 + \alpha_{12}(\boldsymbol{A}_1\boldsymbol{A}_2 + \boldsymbol{A}_2\boldsymbol{A}_1), \end{aligned} \tag{B.10}$$

where the response coefficients α_k ($k = 1, \ldots, 12$) are functions of 19 invariants:

$$\begin{aligned}\alpha_k = \alpha_k(\,&\mathrm{tr}\, \boldsymbol{M}_1\boldsymbol{A}_1,\ \mathrm{tr}\, \boldsymbol{M}_2\boldsymbol{A}_1,\ \mathrm{tr}\, \boldsymbol{M}_3\boldsymbol{A}_1,\ \mathrm{tr}\, \boldsymbol{M}_1\boldsymbol{A}_2,\ \mathrm{tr}\, \boldsymbol{M}_2\boldsymbol{A}_2,\ \mathrm{tr}\, \boldsymbol{M}_3\boldsymbol{A}_2,\\ &\mathrm{tr}\, \boldsymbol{M}_1\boldsymbol{A}_1^2,\ \mathrm{tr}\, \boldsymbol{M}_2\boldsymbol{A}_1^2,\ \mathrm{tr}\, \boldsymbol{M}_3\boldsymbol{A}_1^2,\ \mathrm{tr}\, \boldsymbol{M}_1\boldsymbol{A}_2^2,\ \mathrm{tr}\, \boldsymbol{M}_2\boldsymbol{A}_2^2,\ \mathrm{tr}\, \boldsymbol{M}_3\boldsymbol{A}_2^2,\\ &\mathrm{tr}\, \boldsymbol{M}_1\boldsymbol{A}_1\boldsymbol{A}_2,\ \mathrm{tr}\, \boldsymbol{M}_2\boldsymbol{A}_1\boldsymbol{A}_2,\ \mathrm{tr}\, \boldsymbol{M}_3\boldsymbol{A}_1\boldsymbol{A}_2,\\ &\mathrm{tr}\, \boldsymbol{A}_1^2\boldsymbol{A}_2,\ \mathrm{tr}\, \boldsymbol{A}_1\boldsymbol{A}_2^2,\ \mathrm{tr}\, \boldsymbol{A}_1^3,\ \mathrm{tr}\, \boldsymbol{A}_2^3).\end{aligned} \tag{B.11}$$

References

Boehler JP (1987) Representations for isotropic and anisotropic non-polynomial tensor functions. In: Boehler JP (ed) Applications of tensor functions in solid mechanics. Springer, Wien, pp 31–53

Chadwick P (1999) Continuum mechanics: concise theory and problems, 2nd edn. Dover, Mineola, New York

Liu IS (2002) Continuum mechanics. Springer, Berlin

Spencer AJM (1987a) Anisotropic invariants and additional results for invariant and tensor representations. In: Boehler JP (ed) Applications of tensor functions in solid mechanics. Springer, Wien, pp 171–186

Spencer AJM (1987b) Isotropic polynomial invariants and tensor functions. In: Boehler JP (ed) Applications of tensor functions in solid mechanics. Springer, Wien, pp 141–169

Truesdell C, Noll W (2004) The non-linear field theories of mechanics, 3rd edn. Springer, Berlin

Author Index

R. Staroszczyk, *Ice Mechanics for Geophysical and Civil Engineering Applications*, GeoPlanet: Earth and Planetary Sciences, https://doi.org/10.1007/978-3-030-03038-4

T

V

W

X

Y

Z

Subject Index

R. Staroszczyk, *Ice Mechanics for Geophysical and Civil Engineering Applications*,
GeoPlanet: Earth and Planetary Sciences, https://doi.org/10.1007/978-3-030-03038-4

MIX
Papier aus verantwortungsvollen Quellen
Paper from responsible sources
FSC® C105338

If you have any concerns about our products,
you can contact us on
ProductSafety@springernature.com

In case Publisher is established outside the EU,
the EU authorized representative is:
Springer Nature Customer Service Center GmbH
Europaplatz 3, 69115 Heidelberg, Germany

Printed by Libri Plureos GmbH
in Hamburg, Germany